# SOUTHEAST ASIA

While many working in Latin America and Africa have been challenging the entire basis of development as modernization, the growth economies of Southeast Asia have been presented by the World Bank and others as exemplars of development – 'miracle' economies to be emulated.

This book examines the miracle thesis in the light of new post-developmental work and addresses such issues as poverty, rural–urban interactions and export-oriented industrialization. With the former command economies of Vietnam, Laos, Cambodia and Myanmar (Burma) also firmly embarked on the market road, the book incorporates a discussion of their experiences and prospects.

Jonathan Rigg begins by critically assessing the conceptual foundations of development, including an examination of the miracle thesis, the ideas of the post-developmentalists, and indigenous notions of development based on, for example, Buddhism, Islam and the so-called 'Asian Way'. This leads to a detailed assessment of poverty in the region and the various ways in which the poor can be identified and viewed. The central chapters examine the processes and change in the rural and urban 'worlds' and the strengthening interactions which bind the two as 'farmers' make a living in the urban-industrial sector and factories locate in agricultural areas. The final section of the book challenges the notion that development has been a mirage for many and a tragedy for some.

*Southeast Asia: The Human Landscape of Modernization and Development* presents an up-to-date, original analysis which will be of great value to all those interested in Southeast Asia and 'Third World' development more generally.

**Jonathan Rigg** is Reader in Geography, University of Durham.

# SOUTHEAST ASIA

## The human landscape of modernization and development

*Jonathan Rigg*

London and New York

First published 1997
by Routledge
11 New Fetter Lane, London EC4P 4EE

Simultaneously published in the USA and Canada
by Routledge
29 West 35th Street, New York, NY 10001

© 1997 Jonathan Rigg

Typeset in Garamond by Keystroke, Jacaranda Lodge, Wolverhampton
Printed and bound in Great Britain by Biddles Ltd, Guildford and King's Lynn

*British Library Cataloguing in Publication Data*
A catalogue record for this book is available from the British Library

*Library of Congress Cataloguing in Publication Data*
A catalogue record for this book has been requested

ISBN 0–415–13920–1 (hbk)
ISBN 0–415–13921–X (pbk)

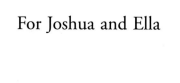

For Joshua and Ella

The sun rises and the sun sets,
and hurries back to where it rises.
The wind blows to the south and turns to the north;
round and round it goes,
ever returning on its course.
All streams flow into the sea,
yet the sea is never full.
To the places the streams come from,
there they return again.
All things are wearisome,
more than one can say.
The eye never has enough of seeing,
nor the ear its fill of hearing.
What has been will be again,
what has been done will be done again;
there is nothing new under the sun.

I have seen all the things that are done under the sun;
all of them are meaningless,
a chasing after the wind.

What is twisted cannot be straightened;
what is lacking cannot be counted.

Ecclesiastes 1: 5–9, 14

Kings conquer whole lands,
reigning over realms that stretch from ocean to ocean,
yet they are not content with simply this shore
– they want the other side as well.
Both Kings and ordinary people must die in the midst of want,
never reaching an end to desire and craving.

When it is time to die,
no one, neither relative nor friend,
can forestall the inevitable.
Possessions are carried off by the heirs
while the deceased fares according to his kamma.
When it is time to die, no one thing can you take with you,
not even children, wife, husband, wealth or land.
Longevity cannot be obtained through wealth,
and old age cannot be bought off with it.
The wise say that life is short, uncertain
and constantly changing.

Buddhist scriptures

# CONTENTS

CONTENTS

## Part III Change and interaction in the rural and urban worlds

## Part IV Chasing the wind: modernization and development in Southeast Asia

CONTENTS

x

# PLATES

All photographs by the author except Plate 5.7, by Jonathan Miller

# BOXES

# FIGURES

# TABLES

# PREFACE

When this book was just an idea scribbled on a few sheets of paper, the intention was to write a volume that would, perhaps rather ambitiously, span a number of divides. The first was the divide between regional scholarship on Southeast Asia – the work of the area studies specialists – and the wider, conceptual literature in the fields of development studies and development geography. The second was between 'indigenous' or 'local' scholarship and views, and what might be termed exogenous or extra-regional scholarship. And the third key divide which I wanted to explore was that which separates the grand view of macroeconomic theory and structural change and the local realities of the *wong cilik* and *khon lek* – the 'little' people that make the world what it is. The book was intended to be a regional synthesis of local-level studies. Or, to put it another way, an attempt to bring a regional perspective to local studies, and a paddy-field and factory-floor view to regional work.

But perhaps more than anything the intention was to address a personal difficulty. For every time I step off a plane from Northeast England and into the heat of Southeast Asia the words of scholarship seem to be left behind in the overhead locker, rendered meaningless by the realities of people and place. Perhaps everyone who has been fortunate enough to have worked in another country has thought while reading a paper or book, or listening to some seminar on a subject or place with which they are familiar, 'it wasn't like that for me'. The great value of first-hand experience is that it gives people the power and confidence to reject the supposed authority of the written word. The words in this book are also destined for the metaphorical overhead locker. They are linked to my personal experiences, greatly informed and amplified by the work of others. Readers will doubtless find themselves exclaiming 'it isn't like that' – or worse. But, with luck, there will also be times when the words might induce a quiet 'yes!'.

Jonathan Rigg
Durham

# ACKNOWLEDGEMENTS

As with most books, this is a reflection as much of other people's work and ideas as it is a reflection of my own. Accepting that no piece of work is produced in a vacuum, there is a long list of people who have helped, sometimes unwittingly. Of those who have offered particular assistance and inspiration are colleagues in the Department of Geography at the University of Durham and former colleagues at the School of Oriental and African Studies in London. At SOAS, I am particularly grateful for the language and regional expertise offered by Anna Allott (Myanmar), Ulrich Kratz (Indonesia and Malaysia) and Rachel Harrison (Thailand). It was their input which changed the section on the 'Semantics of development' into something more substantial than it would otherwise have been. I would also like to acknowledge the help of Philip Stott and Helen Clover, both Thai specialists working in the Department of Geography at SOAS. My fourteen-odd years in the Department at SOAS, first as a student, then a research fellow and finally as a lecturer, made me a regional geographer in the classic and traditional sense, something for which I am very grateful. The Department of Geography at Durham, to which I moved in 1993, is a very different place and it brought to me another – and I would like to think, complementary – vision of geography. At Durham, the members of the Development Studies Group, particularly Janet Townsend, Mike Drury, Peter Atkins, Liz Oughton and Jo Rowlands, have offered their critical comments on parts of the book or the ideas that underpin it. More broadly, the Department at Durham has led me to think critically about Southeast Asia in the context of wider debates within geography and beyond, which formerly I had been happy to ignore as I wallowed in regional scholarship.

Other people who have helped in various ways include: Thomas Enters who commented on a draft of Chapter 5 and William Sunderlin who sent me papers on the Philippines, both working at that time at CIFOR in Bogor, Indonesia; Carrie Turk of ActionAid Vietnam and Juliet Eddington at the ActionAid office in London who generously permitted access to ActionAid's unpublished reports; Becky Elmhirst of Wye College, University of London who allowed me to read her unpublished papers on transmigrant strategies in Lampung province, Sumatra; Mike Parnwell and Pauline Khng at the Centre for South-East Asian

Studies in Hull who helped track down various studies; Philip Courtenay of James Cook University, Queensland, who sent me copies of his papers and reports on rural change in Malaysia; Anne Booth of the School of Oriental and African Studies who sent me one of her recent publications; and Harvey Demaine and Siriluck Sirisup with whom I shared a fascinating foray into the Central Plains of Thailand. Raymond Bryant, a fellow Southeast Asianist in the Department of Geography at King's College, London read the draft manuscript far more carefully than duty required and I am indebted to him for his many perceptive and constructive comments.

At Routledge I am grateful to Tristan Palmer who commissioned the book, Sarah Lloyd who took over after Tristan left, and Matthew Smith and Diane Stafford who helped see it through to the end. The maps and figures were expertly drawn by Arthur Corner, David Hume and Steven Allan in the Cartography Unit of the Department of Geography at Durham, and the illustrations were converted from transparencies to black and white prints by Michele Johnson.

Most book acknowledgements end with a note of thanks from the author to their spouse, partner and/or childen along the lines of 'without whom this book . . . '. In this instance it would be fair to write that the book would have been completed a lot more quickly had it not been for spouse/children. Fortunately, in the grander scheme of things, this book is relatively unimportant and I am pleased that they did 'get in the way', and doubtless will continue to do so.

# A NOTE ON THE TEXT

This book concerns itself with the economic progress of the ten countries that make up the Southeast Asian region. But it is also an attempt to relate the Southeast Asian experience to the wider literature on developmentalism and post-developmentalism, and vice versa. It is therefore intended to be both a regional geography of development, and a regionally-rooted development geography. The three main sections that comprise the volume address themselves to different, but related, issues. The differences they embody are partly of emphasis, partly of scale, and partly of perspective. In a sense, they view the same object – 'modernization and development' – but differently.

The two chapters in Part I, 'Southeast Asian development: the conceptual landscape of dissent', address the various ways in which scholars have viewed development in the region, but emphasizing the conceptual perspectives. Some of these, like the 'miracle thesis', are apparently based on firm empirical information, much of it economic in character. Other perspectives, like Buddhist and Islamic economics, and alternative and endogenous visions of development, are based much more on cultural foundations and are generally less dependent on 'hard' empiricism. None the less, all of these views represent attempts to bring a conceptual understanding to development and to carve out distinctive explanatory niches.

Part II, 'Marginal people and marginal lives: the excluded', brings a grounded perspective to some of these conceptual approaches. Thus the miracle thesis draws on the results of studies of income-poverty in the region, while alternative development and post-developmentalism scrutinize, among other things, the impacts of development on tribal peoples and those at the 'edges' of economy and society. The emphasis here, then, is on the impacts and effects of modernization and development on people and places. A particular concern is on those who have 'missed out' on development – often styled the 'excluded'.

The three chapters in the third part of the book, 'Change and interaction in the rural and urban worlds', take a step further into the mud of the paddy field and the cacophony of the factory floor. Rather than examining the impacts of development on people and places – the emphasis in Part II – the emphasis in Part III is on the varied responses of people to the challenges and opportunities that development has provided. It is a people-centred perspective where macro-perspectives and generalizations are eschewed.

The final part of the volume, 'Chasing the wind: modernization and development in Southeast Asia', is an attempt to link the previous three parts. The manner in which people have responded to the challenge of development is linked back to notions of exclusion, and they in turn are used as a basis on which to discuss conceptual issues like urban bias and post-developmentalism. Thus the discussion comes full circle back to its beginnings: the making and unmaking of a miracle.

# GLOSSARY AND LIST OF ABBREVIATIONS

| | |
|---|---|
| AAV | ActionAid Vietnam |
| *adat* | Custom, tradition or locally accepted code of behaviour |
| ARD | Accelerated Rural Development – a programme of regional development undertaken in Thailand, and especially Northeast Thailand, during the 1970s |
| Asean | Association of Southeast Asian Nations (Indonesia, Malaysia, Philippines, Singapore, Thailand [original signatories of the Bangkok Declaration in 1967], Brunei [joined in 1984] and Vietnam [joined in 1995]). |
| Asean-4 | the four larger Asean economies minus the newest member, Vietnam (i.e. Indonesia, Malaysia, Philippines and Thailand) |
| *asrama* | dormitory, barracks or boarding house (Indonesian) |
| *ban* | village (Thailand, Laos) |
| *barangay* | village (Philippines) |
| *barrio* | village (Philippines) |
| *becak* | tricycle taxi/trishaw (Indonesia) |
| BMR | Bangkok Metropolitan Region |
| *bomoh* | spirit healer (Malaysia) |
| BPS | *Biro Pusat Statistik*, Indonesia's Central Bureau of Statistics |
| *Buddhapanich* | pejorative term meaning 'commercial' or 'commercialized Buddhism' associated with credit card-carrying and amulet-selling monks (Thailand) |
| *bumiputra* | 'sons of the soil', Malaysia's native groups, including the Malays and the indigenous non-Malay, non-Muslim groups of Sarawak, Sabah and the Peninsula. The East Malaysian native peoples are often grouped together under the umbrella term 'Dayak'. |
| *chin thannakaan mai* | 'new thinking', the Lao PDR's economic reform programme, better known in English as the New Economic Mechanism (NEM) |
| CPT | Communist Party of Thailand |
| *cyclo* | tricycle taxi (Cambodia, Vietnam) |

| | |
|---|---|
| Dayak | collective term for the native (or 'tribal') peoples of Borneo |
| *desa* | village (Indonesia) |
| *desakota* | distinct zones of intense rural–urban interaction |
| *doi moi* | 'renovation'; Vietnam's economic reform programme |
| EMR | Extended Metropolitan Region |
| EPZ | Export Processing Zone |
| FDI | foreign direct investment |
| FIES | Family Income and Expenditure Surveys; Philippines' longitudinal consumption surveys, conducted since 1961, on which poverty estimates are based |
| GDP | Gross Domestic Product |
| GNP | Gross National Product |
| GRP | Gross Regional Product |
| GSP | Generalized System of Preferences; US trade law allowing products from developing countries to enter the USA at a reduced tariff |
| HDB | Housing Development Board (Singapore) |
| HES | Household Expenditure Surveys; Indonesia's longitudinal consumption surveys, conducted since 1963, on which poverty estimates are based. Better known in Indonesia as the Susenas surveys |
| HPAEs | High Performing Asian Economies (a World Bank designation including the Southeast Asian countries Indonesia, Malaysia, Singapore and Thailand, along with the East Asian economies Hong Kong, Japan, South Korea and Taiwan) |
| ICMI | Organization of Indonesian Muslim Intellectuals (*Ikatan Cendekiawan Muslim Indonesia*) |
| IDT | *Inpres Desa Tertinggal* or Presidential Instruction Programme for Less Developed Villages (Indonesia) |
| ILO | International Labour Office |
| IMF | International Monetary Fund |
| Indochina | Vietnam, Laos and Cambodia |
| INPRES | *Intruksi Presiden* (Presidential Instruction). Central government programmes in which funds, largely earmarked for infrastructural improvements, are directed to provincial, district and village-level administrations in Indonesia |
| JABOTABEK | *Ja*karta, *Bo*gor, *Ta*ngerang, *Bek*asi. The extended metropolitan area surrounding Jakarta including the city of Bogor and the regencies or *kabupaten* of Bogor, Tangerang and Bekasi. This area is now treated as a planning region |
| *kampung* | village, hamlet or residential area (Indonesia and Malaysia) |
| *Kanpatihup setthakit* | 'reform economy', the Lao PDR's economic reform programme, better known in English as the New Economic Mechanism (NEM) |

| | |
|---|---|
| *kanpattana* | development or modernization (Thailand) |
| *Khor Jor Kor* | Land Redistribution Programme for the Poor Living in Forest Reserves (Thailand) |
| *kota* | town (Indonesia, Malaysia) |
| *kotadesasi* | the process leading to the emergence of distinct regions of agricultural and non-agricultural activity and interaction |
| *krupuk* | deep-fried prawn cracker (Indonesia and Malaysia) |
| *kwai lek* | 'iron buffalo' or rotavator (Thailand) |
| LDC | Less (or Least) Developed Country |
| LECS | Lao Expenditure and Consumption Survey. Laos' first such survey, undertaken in 1992–93 and released in 1995 |
| LLDC | Least Developed Country |
| LPDR | Lao People's Democratic Republic, the country of Laos |
| LPRP | Lao People's Revolutionary Party. The ruling communist party of Laos |
| LSMS | Living Standards Measurement Survey (Vietnam) |
| *mandor* | intermediary or labour sub-contractor (Indonesia) |
| *martabak* | savoury meat pancake (Indonesia and Malaysia) |
| MDU | Mobile Development Unit. Thai military development initiative aimed at areas where the CPT was influential |
| MNC | multi-national company |
| *mukim* | sub-district (Malaysia) |
| MRE | milled rice equivalent |
| MV | modern variety (of seeds or grain) |
| *myo* | village (Myanmar) |
| *nat* | animist spirit (Myanmar) |
| NDP | New Development Policy (Malaysia, 1991–) |
| NEDB | National Economic Development Board (the precursor to Thailand's National Economic and Social Development Board) |
| NEM | New Economic Mechanism, Laos' economic reform programme, better known in Laos as *chin thanakaan mai* or 'new thinking' |
| NEP | New Economic Policy (Malaysia, 1971–90) |
| NESDB | National Economic and Social Development Board (Thailand's national economic planning agency) |
| NEZ | New Economic Zone, Vietnam's resettlement areas established after the victory of the North in 1975 |
| NGO | non-governmental organization |
| NIC | Newly Industrializing Countries (traditionally viewed as consisting of Hong Kong, Korea, Singapore and Taiwan. Some scholars would now wish to include Malaysia and, perhaps, Thailand. It has become a somewhat confused term. The four original NICs are more popularly known as the Four Tigers) |

| | |
|---|---|
| NIE | Newly Industrializing Economies (the World Bank designates Indonesia, Malaysia and Thailand as NIEs) |
| NSC | National Security Council (Thailand) |
| NSO | National Statistical Office (Thailand) |
| NWC | National Wage Council, Singapore's tripartite wage determining body |
| *ojek* | motorcycle taxi (Indonesia) |
| OPM | Organisasi Papua Merdeka (Free Papua Movement, Indonesia) |
| Outer Islands | islands of Indonesia beyond the metropolitan island of Java (and Madura) |
| PA | Public Assistance (Singapore) |
| *Pancasila* | President Sukarno's five guiding principles, effectively the Indonesian state's ideology. They are: belief in one supreme God; the unity of the nation; just and civilized humanity; social justice for all the people of Indonesia; and democracy guided by the inner wisdom of unanimity |
| PAP | People's Action Party, Singapore's ruling party |
| *pegawati* | employees (Indonesia) |
| *pekerja* | workers or labourers (Indonesia) |
| *petani* | peasants (Indonesia) |
| *phi* | animist spirit (Thailand) |
| *phum* | village (Cambodia) |
| RUAS | Rent and Utilities Assistance Scheme (Singapore) |
| *saam lor* | trishaw (Thailand) |
| *sangha* | Buddhist monkhood |
| *sawah* | wet rice land |
| SLORC | State Law and Order Restoration Council, Myanmar's ruling 'council', established in 1988 |
| SPSI | All-Indonesia Workers Union (*Serikat Pekerja Seluruh Indonesia*) |
| *surau* | Muslim prayer hall |
| Susenas | see HES |
| *tentrem* | calm, peaceful, safe and tranquil; also spelt *tenteram* |
| *tentrem batin* | 'peace in one's heart' |
| TFP | total factor productivity |
| *thon* | village (Vietnam) |
| *tuk-tuk* | motorized, three-wheeled tricycle taxi (Thailand) |
| USDA | Union Solidarity and Development Association (Myanmar) |
| VCP | Vietnamese Communist Party |
| *wat* | Buddhist monastery (Thailand, Laos) |
| *wattanatham* | development (Thailand) |
| *zakat* | Muslim wealth tax |

Southeast Asia

# Part I

# SOUTHEAST ASIAN DEVELOPMENT
The conceptual landscape of dissent

There are numerous ways to view Southeast Asian development. It is possible to take an empiricist, largely economic position where the focus is squarely on rates of growth, structural changes in the economy, and returns to investment, for instance, and where the viewpoint is, by and large, macroeconomic. Alternatively, it is possible to take a local perspective and to broaden the scope of analysis to include such issues as rural livelihood strategies, the differential implications of development for men and women, and the effects of modernization on family relations and social norms. Those scholars of a more radical turn of mind have couched their assessments in terms of dependency, vulnerability and super-exploitation, many in recent years also adding a green tinge to their writings as the environment has held centre stage. Equally, there are other scholars who have attempted to eschew the language of developmentalism altogether, analysing development texts as indicative of the hegemony of Northern viewpoints over the cultures and countries of the South. In some cases they have found support in theories of 'alternative' development. While scholars and others in the countries of the South, and in this instance in those of Southeast Asia, can also be slotted into these various approaches, there are also some individuals who have tried, apparently, to forge a distinctively Southeast Asian and indigenous path.

To a significant extent, the means by which development has been conceptualized determines the message that is broadcast. Development can be a noun, a verb or an adjective. It is therefore a state, a process and a characterization all encapsulated in a word which, perhaps more than any other, deserves to be corralled within inverted commas. There are many developments in the real and scholarly Southeast Asian world, and the two chapters in this first section of the book map out this range of views. The discussion attempts to bring together an increasingly diverse and heterogenous literature, one where the certainties of the

1

past have been replaced by schism and dissent. Inevitably, given this diversity of voices, there is a tendency to hear a Babel rather than a debate. None the less the profusion is not chaotic: the authors, commentators and scholars are bound together by their mutual interest in the sources, processes and ramifications of development for people and places.

# 1

# CHASING AFTER THE WIND
## The making (and unmaking) of a 'miracle'[1]

## Introduction

From academic papers to popular journalism, Southeast Asia – usually in tandem with East Asia – is often depicted as a developmental success story. The terminology used to describe the region and individual countries within it resonates with hyperbole: these are the countries of the 'magical' Pacific Rim, 'miracle' economies that defy accepted wisdom, 'tigers' or 'dragons' unleashed upon an unsuspecting world. In a survey of Asia, *The Economist* took this hyperbole to truly cosmic dimensions, writing:

> It is now likelier than not that the most momentous public event in the lifetime of anybody reading this survey will turn out to have been the modernisation of Asia.

> (*Economist* 1993b: 5)

For the World Bank, the International Monetary Fund, and many mainstream economists, the countries of Southeast Asia are paragons of development – exemplars to the rest of the developing world. They are part of the Asian Miracle, and as such should be imitated if they are to be emulated. Their economic success has been founded – so the argument goes – on a combination of sound, market-based, foreign investment-friendly, export-oriented, and generally equitable policies. As Hill suggests with respect to the five original members of the Association of Southeast Asian Nations (Asean)[2] (i.e. the current membership minus Brunei and Vietnam), their economic success demands that scholars examine the countries' development practice and experience for: 'unless the good performance is due wholly to luck or good fortune, there must be something about these economies, their organization and their public policies, that other, poorer performing countries could well emulate' (Hill 1993: 1).[3] Although Hill, like many others, applies his comments to a sub-set of the Southeast Asian region, it seems that the countries of Indochina and Myanmar (Burma), and especially Vietnam, are also being gradually drawn into the characterization. The sobriquet 'tiger' is already being tentatively applied, for example, to Vietnam (reflected in the tendency to refer to it as a tiger 'cub') while Laos is seen as an object lesson in hard-headed structural adjustment and economic reform.

3

Yamazawa has compared the countries of Asia to a flock of geese: industrial Japan in the lead, followed by the Four Tigers of East and Southeast Asia (Taiwan, South Korea, Hong Kong and Singapore, the original NICs), and they in turn by a phalanx of emerging industrial economies (including Indonesia, Malaysia and Thailand [the NIEs], and possibly the Philippines). In Yamazawa's view, not only are the countries of the Asia-Pacific region enjoying a shared experience of development as they fly over the same terrain, but the flock is also becoming increasingly interdependent as industries transfer from 'early starters to late comers' (1992: 1523). He goes on to state that the 'continuing success of the flying geese pattern of industrial transfer in the future is the key to determining the future success or failure of industrialization in Asean and more generally in the Pacific region' (Yamazawa 1992: 1523). As Booth observes, the fear among the Asean countries who are already in flight is that so many other countries are hoping to join the flock in the journey to the Golden Land that competition may become ferocious (Booth, A. 1995a: 30).

Yet, at the same time as the World Bank and many economists have been searching for the elixir of development in the countries of Asia, other scholars have been questioning the miracle thesis. This questioning tends to fall into two camps. First, there are those who challenge the assumption in its own terms. In other words, they question, for example, whether growth has been advantageous to the mass of the population, whether there have not been serious side-effects linked to environmental deterioration, and whether the style of development is sustainable in the long run.[4] Some even question whether there is anything miraculous about the Asian growth experience. Second, there is a camp of scholars who are engaged in exploring the possibility that the development discourse itself is a chimera invented by the West, which has then been imposed on the unsuspecting countries of the South. This critique is more fundamental in the sense that it challenges the very assumptions on which modernization is based. It is a critique which is less about economics and more to do with political and cultural hegemony.

## THE MAKING OF A MIRACLE

### The growth economies

The countries of Southeast Asia, and in particular the growth economies of Asean (Indonesia, Malaysia, Singapore and Thailand), have consistently out-performed the countries of other developing regions over the last 30 years (Table 1.1). It is this simple, yet powerful, observation which has done most to shape the economic miracle thesis. It also does much to explain why those who sub-scribe to the thesis are overwhelmingly concentrated in the field of economics and why the miracle is perceived largely in economic terms. Anthropologists, political scientists, geographers and others may not refute the statement in its own terms, but would wish to add a series of significant caveats. In their turn,

4

supporters of the Southeast Asian growth experience, while accepting these criticisms as largely valid, would say that they do not detract fundamentally from the growth record – or the centrality of this record in the gauging of 'success' (e.g. Hill 1993: 9).

So much has been said and written about the Asian miracle that it is difficult, not just to say anything new, but to prevent a collective sigh of, 'not again'. In the same way that Africa has become stereotyped as the globe's economic laggard, and the analysis of its problems and diversity of experience suffered as a result, so Southeast and East Asia's categorization as a success story has led to a similar tendency towards reductionism and generalization, shorn of debate and difference. There may be, as the World Bank seems to maintain, an important general lesson, but this should not disguise the extent to which the countries of Southeast Asia have become the objects of approbation having followed very different development trajectories.

With the emphasis so clearly on the region's exemplary growth record, it is also all too easy to forget the extent to which Asia was peceived to be beset by problems in the mid- to late-1960s, the period from which the miracle is usually traced. Gunnar Myrdal's influential two-volume work *Asian drama*, published in the middle of this period, is a sober and generally pessimistic study, significantly subtitled 'an inquiry into the poverty of nations' (Myrdal 1968). There is little sense, leafing through its 2,300 pages, that a large proportion of the region it covered would be defined, within a generation, more in terms of success than failure.[5] Indeed, a sense of portentous tragedy permeates the study:

> The lofty aspirations of the leading actors [in South and Southeast Asia's 'drama'] are separated by a wide gap from the abysmal reality. . . . And that gap is widening. The movement of the drama is intensified as, through time, aspirations are inflated further by almost everything that is printed and preached and demonstrated, be it planned or not, while positive achievements lag. Meanwhile, populations are increasing at an even faster pace, making the realization of aspirations still more difficult.
>
> (Myrdal 1968: 34–5)

Myrdal's work, despite its undoubted clarity and breadth of vision was not, in retrospect, prescient. Instead, it represented a skilful exposition of the wisdom of the time. The date of its publication marked the end of stagnation in Southeast Asia and the beginning of growth. As such it also marked the onset of the gradual erosion of the existing stagnation paradigm and its replacement by the new 'growth' or 'miracle' paradigm.

In dissecting the experiences of the eight so-styled High Performing Asian Economies (HPAEs), of which four are Southeast Asian, the World Bank contends that 'they do share some economic characteristics that set them apart from most other developing countries' (World Bank 1993a: 2). In acknowledging the existence of shared characteristics, it was but a small step, despite the cautionary footnotes and clauses, for the authors of *The East Asian Miracle* to

Table 1.1 Growth trajectories: Southeast Asia in global perspective

| | Population (millions, 1994) | GNP per capita average annual growth | | GDP average annual growth | | | Real GDP per capita (PPP$ 1992) | GNP per capita (PPP$ 1994) | Real GNP per capita (US$ 1993) |
|---|---|---|---|---|---|---|---|---|---|
| | | 1965-80 | 1980-93 | 1970-80 | 1980-90 | 1990-94 | | | |
| Brunei | 0.3 | – | – | – | – | – | 20,589 | – | – |
| Cambodia | 9.4 | 0.6 | – | – | – | – | 1,250 | – | – |
| Indonesia | 190.4 | 5.2 | 4.2 | 7.2 | 6.1 | 7.6 | 2,950 | 3,600 | 740 |
| Laos | 4.7 | 0.6 | – | – | – | 6.2 | 1,760 | – | 280 |
| Malaysia | 19.7 | 4.7 | 3.5 | 7.9 | 5.2 | 8.4 | 7,790 | 8,440 | 3,140 |
| Myanmar | 45.6 | 1.6 | – | 4.7 | 0.6 | 5.7 | 750 | – | – |
| Philippines | 67.0 | 3.2 | -0.6 | 6.0 | 1.0 | 1.6 | 2,550 | 2,740 | 850 |
| Singapore | 2.9 | 8.3 | 6.1 | 8.3 | 6.4 | 8.3 | 18,330 | 21,900 | 19,850 |
| Thailand | 58.0 | 4.4 | 6.4 | 7.1 | 7.6 | 8.2 | 5,950 | 6,970 | 2,110 |
| Vietnam | 72.0 | 0.6 | – | – | – | 8.0 | 1,110 | – | 170 |
| All low and middle income countries | | 4.6 | 4.0 | 5.2 | 2.2 | 0.2 | | | |
| South Asia | | – | – | 3.5 | 5.7 | 3.9 | | | |
| Sub-Saharan Africa | | 1.4 | -1.8 | 3.8 | 1.7 | 0.9 | | | |
| Latin America and Caribbean | | – | – | 5.4 | 1.7 | 3.6 | | | |

Table 1.1 continued

| | Agriculture 1970 | Agriculture 1980 | Agriculture 1994 | Industry 1970 | Industry 1980 | Industry 1994 | Manufacturing 1970 | Manufacturing 1980 | Manufacturing 1994 | Services 1970 | Services 1980 | Services 1994 | Exports 1970-80 | Exports 1980-90 | Exports 1990-94 | Net FDI 1980 | Net FDI 1993 |
|---|---|---|---|---|---|---|---|---|---|---|---|---|---|---|---|---|---|
| | Distribution of GDP (%) | | | | | | | | | | | | Average annual growth rate of exports (%) | | | Net FDI (million US$) | |
| Brunei | – | – | – | – | – | – | – | – | – | – | – | – | – | – | – | – | – |
| Cambodia | – | – | – | – | – | – | – | – | – | – | – | – | – | – | – | – | – |
| Indonesia | 45 | 24 | 17 | 19 | 42 | 41 | 10 | 13 | 24 | 36 | 34 | 42 | 6.5 | 5.3 | 21.3 | 180 | 2,004 |
| Laos | – | – | 51 | – | – | 18 | – | – | 13 | – | – | 31 | – | – | – | 0 | 48 |
| Malaysia | 29 | 22 | 14 | 25 | 38 | 43 | 12 | 21 | 32 | 46 | 40 | 42 | 3.3 | 11.5 | 17.8 | 934 | 4,351 |
| Myanmar | 38 | 47 | 63 | 14 | 13 | 9 | 10 | 10 | 7 | 48 | 41 | 28 | 0.2 | –7.0 | 27.2 | 0 | 4 |
| Philippines | 30 | 25 | 22 | 32 | 39 | 33 | 25 | 26 | 33 | 39 | 36 | 45 | 7.2 | 2.9 | 10.2 | –106 | 763 |
| Singapore | 2 | 1 | 0 | 30 | 38 | 36 | 20 | 29 | 27 | 68 | 61 | 64 | – | 12.1 | 16.1 | – | – |
| Thailand | 26 | 23 | 10 | 25 | 29 | 39 | 16 | 22 | 29 | 49 | 48 | 50 | 8.9 | 14.3 | 21.6 | 190 | 2,400 |
| Vietnam | – | – | 28 | – | – | 30 | – | – | 22 | – | – | 43 | – | – | – | 0 | 300 |

Sources: World Bank and United Nations Development Programme annual reports

*Plate 1.1* Kuala Lumpur, the capital of Malaysia, one of the world's fastest-growing economies and a member of the World Bank's exclusive club of High Performing Asian Economies (HPAEs), more popularly known as the 'Miracle economies'

arrive at a loose recipe for economic success, or the 'essence of the miracle' as they termed it (Table 1.2). The 'essence' seems to comprise the admonition to 'get the basic right' and in this regard the authors mirror Hal Hill's advice to 'get the fundamentals right' (Hill 1993: 37). Indeed, the lessons that Hill draws from the Southeast Asian experience are almost identical to those listed by the World Bank for the East Asian experience more widely (see Table 1.2). Anne Booth, while accepting the critical role of investment and exports in generating growth, takes issue with the assumption that the Asean experience is merely a continuation of the East Asian experience (i.e. that of Taiwan, Korea and Japan) (Booth A. 1995a). She highlights, particularly, the lower levels of capital formation and deeper inequalities in the Asean countries. The technological 'deepening' which is central to the flying geese metaphor is being hampered, in her view, by a very real lack of skilled and educated workers and she would clearly like the miracle

*Table 1.2* The 'essence of the miracle'

### Characteristics of the miracle

Rapid economic growth
Equitable distribution of income
Dramatic improvement in human welfare, including:

- increased access to basic services
- declining poverty
- better nutrition
- rising life expectancy
- improving education

### Roots of the miracle

| World Bank (1993a) | Hill (1993) |
| --- | --- |
| 'Getting the basics right' | 'Getting the fundamentals right' |
| High levels of domestic saving and investment | High savings and investment rate |
| Declining population growth rates | – |
| Efficient public administration | A stable and predictable policy environment |
| Fundamentally sound macro-economic policy | Conservative macro-economic management |
| Disciplined government intervention | Pragmatic and effective role for the state |
| Investment in building human capital | Sustained investment in social services, especially education |
| Investment in physical capital | Sustained investment in physical infrastructure |
| Openness to foreign technology and investment | Outward orientation |
| Export-orientation | – |
| | A generally benign and accommodating international environment |

*Sources*: adapted from World Bank 1993a and Hill 1993.

thesis to be more nuanced to take account of differences between countries. Even so, although the proponents of the miracle may not speak with one voice, they all seem to be singing the same tune.

---

*Box 1.1* Criticisms of the World Bank's *The East Asian Miracle*

No single book has been more influential in promoting the Asian miracle thesis than the World Bank's *The East Asian Miracle* (1993a). Given its high profile and potentially long-lasting and insidious influence on national policymaking and world opinion, it was quickly attacked by economists, and others, on a number of fronts. Although the economists' criticisms were focused more on the book itself than on the experience of development in the countries concerned (which is the focus of this chapter), the points are relevant to the wider debate and are summarized here.

- The study is as much a political as an economic document, and was written to reflect the beliefs and predilections of the World Bank's upper management.
- The book omits reference to work which refutes the general interpretation of East Asia's miracle as being market-based – possibly because a conclusion that stressed the effectiveness of interventionist industrial policy would have undermined the whole market-friendly ethos of the Bank.
- It dismisses work and omits examples that stress the role of government intervention over the market by saying that the relative balance between the two is almost impossible to ascertain. None the less, the authors find sufficient evidence to decide that government intervention was largely ineffective, thereby allowing them to emphasize the importance of the market in generating growth.
- Some of the calculations, and in particular the calculation of total factor productivity, are open to serious question (see note 15 on TFP).
- It does not suggest how the East Asian miracle might be transferred to other developing countries.
- It emphasizes economic measures of success and virtually ignores social and environmental measures of failure.
- It plays down the heterogeneity within the HPAEs. There is not one miracle, but several.

*Collated from*: Amsden 1994; Kwon 1994; Lall 1994; Perkins 1994; Yanagihara 1994.

---

## Reform in Indochina (and Myanmar)

Until recently, discussions of Southeast Asia's development have tended to focus, almost entirely, on the experiences of the market economies. This concern for a sub-set of the region has been driven by a lack of information on the countries of

Indochina and Myanmar, and an assumption that the experiences of the two Southeast Asias are so different that they require separate analysis. The first of these constraints is easing as studies become more widely available while the second can increasingly be contested. The former command economies have embraced economic strategies which, to a considerable degree, ape those of the market economies. What marks the countries of Indochina and Myanmar out from the other nations of the region, today, is not that they are engaged in a fundamentally different experience of development, but their sheer under-development (Table 1.1). Revealingly, Nayan Chanda reports that another catch-phrase used in Vietnam to describe the reform initiative is *tut hau* – meaning 'catch up'. 'The whole nation', he writes, 'resembles a beehive, with Vietnamese of all political backgrounds frenetically engaged in making money and trying to gain ground on their [more] prosperous regional neighbours' (Chanda 1995: 21). Further, and as will become clear, the problems and issues with which the leaderships of the countries of Indochina (this is less true of Myanmar) are having to contend are increasingly those with which the governments of the market economies have also had to grapple.

For the countries of Indochina – Vietnam, Laos and Cambodia – the years since the mid-1980s have been ones of seemingly deep economic reform. Myanmar, too, has embraced elements of the market, albeit to a lesser degree and with apparently less conviction. In Vietnam this process has been termed *doi moi* ('renovation', 'new way' or 'new change'), in Laos *chin thanakaan mai* (new thinking) or *kanpatihup setthakit* (the 'reform economy').[6] In each case the terminology implies a change of kind – of thinking – not just a fine-tuning of the existing system.

## The roots of reform

In broad terms, reform in all the former command economies of Southeast Asia has taken the same, familiar, route: a shift from a centrally-controlled economy to a market-oriented system. Tables 1.3 and 1.4 detail the progress of the reform process which can be dated from the late 1970s and early 1980s in the cases of Vietnam and Laos, and rather later in Myanmar (and Cambodia). With the exception of Cambodia, economic reform has not been accompanied by significant political reform. The Vietnamese Communist Party (VCP), the Lao People's Revolutionary Party (LPRP), and Myanmar's State Law and Order Restoration Council (the unattractively acronymed SLORC) are all still firmly in control of each country's political system.

The rationale for economic reform was remarkably similar in all four cases, although the details inevitably varied. First and foremost, it rested on a broad acceptance that socialism was facing severe difficulties globally. This stood in contrast to the apparent economic success of those countries that had adopted a market-led approach to development, especially in Southeast and East Asia (Kerkvliet and Porter 1995: 3; Vylder 1995: 35). A second reason driving the

*Plate* 1.2 Vietnam is sometimes styled a 'tiger cub' – a miracle economy in the making. Here private enterprise flourishes in the town of Hoian

*Table 1.3* Landmarks of economic reform, Laos, Vietnam and Myanmar

### LAOS 1975–94

**1975**
*December:* Full and final victory of the communist Pathet Lao

**1982**
Reforms first touted

**1985**
Pilot studies of financial autonomy in selected state-run industries

**1986**
Decentralization of decision-making to the provinces including provincial tax administration
Freeing-up the market in rice and other staples
*November:* NEM endorsed by the Party Congress

**1987**
Restrictions on the cross-provincial movement of agricultural produce abolished; barriers to external trade reduced; provincial authorities charged with the responsibility of providing health and education services.
*June:* prices of most essentials market-determined

**1988**
Forced procurement of strategic goods at below market price abolished; reduction in public sector employment; tax reforms introduced; private sector involvement in sectors previously reserved as state monopolies permitted; introduction of new investment law.
*March:* prices of fuel, cement, machinery and vehicles freed; tax reforms enacted; state and commercial banking sectors separated; state enterprises made self-reliant and autonomous; explicit recognition of the rights of households and the private sector to use land and private property.
*June:* nationwide elections held for 2,410 positions at the district level
*July:* multiple exchange rates abolished; liberal foreign investment code introduced; payment of wages in kind abolished.

**1989**
*June:* second tax reform enacted
*October:* first joint venture bank with a foreign bank begins operation, the Joint Development Bank

**1990**
*March:* privatization ('disengagement') law introduced
*June:* key economic laws covering contracts, property, banking and inheritance discussed by National Assembly
*July:* State Bank (Central Bank) of the Lao PDR established and fiscal management of the economy formally handed over to the new bank

**1992**
Thai Military Bank begins operating a full branch in Vientiane
*January:* Commercial Bank and Financial Institutions Act introduced

**1993**
Accelerated privatization programme announced
*December:* removal of last quantitative restrictions and licensing requirements for imports

*Table 1.3* continued

## 1994

*March:* new investment and labour laws passed in March by the National Assembly, to be enforced within 60 days. As an incentive to foreign investors, the investment law lowers some import taxes and the tax on net profit, streamlines the approval process, and ends the foreign investment period limit of 15 years.

*Sources:* various

## VIETNAM 1975–95
### 1975
*April:* end of the Vietnam War

### 1978
*December:* Vietnamese forces invade Cambodia

### 1979
*September:* Resolution No. 6 issued by the Sixth Plenum of the VCP calls for reforms in industry and agriculture, including a loosening of state control, a shift from large-scale industrialization to an emphasis on smaller enterprises, and policies to boost efficiency. Microeconomic reforms are introduced within the context of a command economy.

### 1981
*April:* Directive 100/CT, popularly known as the Contract 100 system, introduces the first stage of decollectivization. Production groups, usually individual households, are allocated plots to farm by contract. The cooperative maintains control over supply of inputs and marketing.

### 1986
*December: Doi moi* – 'new change' or 'renovation' – officially endorsed at the Sixth Party Congress.

### 1987
*December:* liberal foreign investment law promulgated to attract foreign investment

### 1988
*April:* Resolution No. 10, known more usually as the Contract 10 system, makes individual households the basis of agricultural production in exchange for the payment of an agricultural tax. Households are awarded usufruct land rights for 15 years. Resolution 10 still restricts the sale, renting and exchange of land except in certain, limited situations. Real grain prices double and the distribution of inputs is removed from the control of cooperatives.

Banking reforms introduced separating the roles of the Central Bank and commercial banks. Exchange rate unified.

### 1989
*January:* fiscal policy tightened and positive real interest rates introduced. Subsidies to state-operated industries sharply reduced.

*March:* price reforms introduced; price subsidies are gradually abolished for most goods.

### 1990
Record harvest in the 1989-90 crop year allows Vietnam to become the world's third largest rice exporter after Thailand and the USA.

### 1992
'Equitization programme' introduced in which a handful of small state-owned enterprises are sold off to employees and others

14

*Table 1.3* continued

*June and August*: vague reference in revised Land Law to mortgaging and renting of land.

**1993**
*June*: granting of long-term use of agricultural land to peasant families.
*July*: new Land Law passed. Households and individuals are allocated land according to household size (yet recognizing the existing distribution of land between households) and given the right to exchange, transfer, lease, inherit and mortgage land use rights. The role and power of the cooperative in people's lives is reduced but the State has the power to categorize land according to use.

**1994**
New bankruptcy law passed making it possible for banks to seize the assets of state-owned enterprises.
*February*: normalization of trade relations with the USA
*May*: new labour legislation enacted giving workers the right to strike

**1995**
*April*: National Assembly considers new law partially or fully ending the state monopoly in selected sectors
*July*: full normalization with the USA

*Sources*: various

## MYANMAR 1987-95
**1987**
*September*: government implements reforms to remove restrictions on trade in agri-culture, allowing farmers to sell their surplus on the open market. The reforms also decentralize cropping decisions to the farm-level.

**1988**
*November*: Foreign Investment Law enacted which allows 100 per cent foreign equity or joint venture participation with a minimum of 35 per cent foreign equity, a three-year tax break, repatriation of profits, and a guarantee of no nationalization. The law was initially aimed at the development of import substituting and agro-industries. In practice, foreign investors directed their attention to investments in the oil and gas, timber, fisheries and tourism sectors. The Foreign Investment Commission assists potential investors.

**1989**
*September*: State Owned Enterprise Law allows private companies to engage in activities previously restricted to state sector enterprises.

**1990**
*November*: Private Industrial Enterprise Law (or Private Investment Law)

**1992**
Private banks permitted to operate in Myanmar

**1995**
*January*: the SLORC announces that it will set up a new 'privatization' commission.

*Sources*: various

*Note*: In the case of Myanmar, far more so than in Vietnam and Laos (although the issue is also pertinent in those two cases), there is reason to doubt the efficacy of the reforms in changing the structure and functioning of the economic system in line with their stated intentions.

*Table 1.4* Elements of economic reform in Laos, Myanmar and Vietnam (1981–95)

| | Laos | Vietnam | Myanmar |
|---|---|---|---|
| Move to a market determination of prices and resource allocation | ✓ (1986, 1987, 1988) | ✓ (1988, 1989) | |
| End of subsidization of state-run industries | ✓ (1988) | ✓ (1988) | |
| End/partial end of state monopolies | ✓ (1988) | ✓ (1995) | ✓ (1989) |
| Restrictions on trade removed or substantially reduced | ✓ (1988) | ✓ (1989) | ✓ (1987, 1988) |
| Decollectivization of agricultural production | ✓ (1986) | ✓ (1981, 1988) | n.a. |
| Privatization of state-owned industries | ✓ (1990, 1993) | ✓ (1992, 1995)* | ✓ (1995) |
| Reform of banking, investment and property laws | ✓ (1988, 1990, 1994) | ✓ (1988, 1992, 1993) | ✓ (1988, 1990) |
| Tax reforms | ✓ (1988, 1989, 1990) | ✓ (1988, 1989) | ✓ (1988) |
| Exchange rate reforms | ✓ (1988) | ✓ (1988, 1989) | |
| Decentralization of control to industries and lower levels of government | ✓ (1986) | ✓ (1989) | |
| The encouragement of foreign investment | ✓ (1988) | ✓ (1987, 1990) | ✓ (1988) |

*In 1992 Hanoi introduced a 'equitization programme' involving the sale of a handful of small state-owned enterprises to employees and others. In 1995 a law was debated in the National Assembly on how and when to privatize state assets. However, the leadership did not commit itself to a privatization programme as such.

*Note*: the years in parentheses give the date for major reforms in the area specified. Minor reforms may have preceded and postdated the years listed.

reform process was a sheer lack of financial resources – national bankruptcy – coupled with declining production. In Vietnam, for example, the mounting costs of the military occupation of Cambodia (1979–92), the failure of the collectivization programme in the south, the freezing of aid from the West, and the costs of the border war with China in 1979 all contributed, by the end of the 1970s, to an economic crisis. Social instability threatened to undermine the regime's very legitimacy (Beresford 1993: 33). In a similar vein, Cook believes that deteriorating economic condition in Myanmar during the 1980s were instrumental in creating the political climate where reform became indispensable. 'In effect', he writes, '[the reforms] amounted to a survival strategy for the ruling military [in Myanmar]' (1994: 117; see also Cook and Minogue 1993: 1156).

A third factor, which was particularly relevant in Vietnam and Laos, was the increasing pressure that peasants were exerting on their leaderships to loosen or abandon centralized control over agricultural production. In Laos, cooperativization simply did not materialize in many areas such was the degree of resistance, and the central authorities proved powerless to impose the system on a recalcitrant peasantry. In Vietnam, where the commune system was apparently well established in the north, villagers' resistance was 'conveyed not through large organizations and overt political protest but through [the] cumulative effect of unobtrusive foot dragging . . . ' (Kerkvliet and Porter 1995: 26). In the south of Vietnam, the cooperative system was only ever partially implemented such was the degree of obstruction and resistance. Dang Phong goes so far as to suggest that the growth of Hanoi's population from 300,000 in 1954 to 2 million in 1990 was a direct result of the failure of the cooperative system. In his view villagers simply had no other choice but to abandon the countryside for the city (Dang Phong 1995: 176).[7] Thus the pressure for reform was coming both from the top down, as the leadership battled to restore vitality in a stagnant and bankrupt economy, and from the bottom up, as farm households and agricultural cooperatives engaged in what became known as 'fence-breaking', bypassing the state planning system (see Vu Tuan Anh 1995).

To begin with, the reforms in Vietnam and Laos were intended not so much to deconstruct the existing system, but rather to refine and reconstruct it in order that the transition to socialism might be continued (Beresford 1993: 34). The March 1989 Plenum of the Vietnamese Communist Party, for example, concluded that 'the private, individual, small owner, and private capitalist economic forms are still necessary in the long run for the economy . . . in the structure of the commodity-based economy for the advance toward socialism' (quoted in Kolko 1995: 21).[8] The General Secretary of the Lao People's Revolutionary Party, Kaysone Phomvihane, similarly stated at the critical Fourth Party Congress in 1986, that:

> In all economic activities, we must know how to apply objective laws and take into account socio-economic efficiency. At the present time, our country is still at the first stage of the transition period [to socialism]. Hence the system of economic laws now being applied to our country is very complicated. It includes not only the specific laws of socialism but also the laws of commodity production. Reality indicates that if we only apply

the specific economic laws of socialism alone and defy the general laws pertaining to commodity production, or vice versa, we will make serious mistakes in our economic undertaking during this transition period.

(Lao PDR 1989: 9)

---

*Box 1.2* The history of collectivization and reform in Phuc Loi Village, Son La Province, Vietnam (1960–90)

| | |
|---|---|
| 1960 | Establishment of village cooperative |
| late 1970s | Further centralization with creation of cooperative unions and, finally, one large community-wide 'commune'. Land was allocated to work teams of as many as 50 households and each team was given a quota to meet. Teams could keep, and distribute to its members, 80 per cent of any production over this quota |
| 1981 | Beginning of dismantling of cooperative structure |
| 1981–83 | Implementation of Contract 10 (known locally as Contract 3) allowing the contracting out of land to households or groups of households. Households were allowed to keep 65 per cent of production, the remaining 35 per cent going to the cooperative to pay the agricultural tax and to meet the costs of maintaining cooperative functions (e.g. raising buffalo) |
| 1988–90 | Allocation of land as stipulated under the 1988 Land Law to individual households for 5 years |
| 1990 | Cooperatives effectively extinguished |

*Source*: ActionAid Vietnam 1995

---

This notion that the reforms were some sort of stop-gap remains pertinent to a limited degree in the case of Myanmar and also applied to Vietnam in the late 1980s when some commentators saw the clamp-down on the free market in Ho Chi Minh City as indicating that the reforms were merely a tactical manoeuvre to get over the immediate economic crisis (Vylder 1995). What subsequently became clear in Vietnam and Laos though, was that at the local level the reforms were being used to legitimize existing market activities. They also then became a means to push forward still further the frontiers of market activity – and in some instances to push back the limits of the State. Although party cadres may occasionally talk of such hybrid forms as 'market socialism', it is hard to find many who sincerely believe that the reforms are a means – a stepping stone – to some undefined socialist end.

## How 'commanding' were the command economies?

Even at the height of state economic planning and control, there were social, economic, and political activities in Vietnam that the state did not authorize. . . . there were numerous 'black markets' . . . many cooperatives in the 1970s seemed to conform to the state sanctioned model but in fact worked in different ways, which local people themselves had devised. 'Pluralism' . . . has been around in Vietnam for some time.

(Kerkvliet 1995a: 86)

since 1986 – under the Doi Moi slogan of the Sixth Party Congress – Vietnam has carried out the most comprehensive and profound renovation in its modern history.

(Le Dang Doanh 1992: 1)

There is the important question of how far the countries of Indochina and Myanmar were ever command economies akin to those of Eastern Europe and the former Soviet Union. There seem to be two key issues here: first, what have the reforms changed and, second, where are they headed. As Thayer writes with reference to Vietnam: 'Quite clearly Vietnam is in a transitional period. But, "from what, to what"? Did the transition period begin in the mid 1980s? Or, were initial assumptions about the nature of the state–society relations and the Vietnamese political system mistaken?' (1995: 59, see also Irvin 1995). There are several areas of doubt concerning the question of how truly commanding were these so-called 'command' economies.

To begin with, the countries concerned were, and still are, largely agrarian economies with a large portion of production being retained within what might be termed the subsistence system. In Laos, therefore, the New Economic Mechanism (NEM) did not involve the dismantling of an existing socialist system, but rather the reorientation of a subsistence system to market demands (Rigg 1995c). At the same time, there is reason to question whether the market was ever completely eradicated (Thayer 1995: 46). Dao The Tuan, in collating the scanty material on peasants' living standards in north Vietnam from 1930 to 1990, argues that peasant differentiation was continuing even during the peak period of collectivization in the 1960s. He draws the conclusion from this that 'families had the opportunity to produce their own household income even within the cooperative structure' (1995: 156). Over large areas of Laos and south Vietnam cooperativization never really materialized. In Myanmar, though the civilian government did introduce a Land Nationalization Act in 1948, with the ultimate aim of collectivizing production, it was never implemented and a second similar act promulgated in 1954 proceeded only slowly until its suspension in 1957 (Van Schendel 1991: 212). The military government under Ne Win from 1962 took some new measures, and in particular began to procure from farmers a set quota of production at a fixed (and low) price (Steinberg 1982). None the less, the farm household has remained the basic unit of production throughout the period since Independence.[9]

19

For many, if not most farmers in Vietnam, Laos and Myanmar, the methods and structures of production remained largely unchanged despite the ideological rhetoric and the claims of some official reports. Even in North Vietnam, the successful period of cooperativization, up to the mid 1960s, coincided with the years when production units were based on family, clan or neighbourhood associations.[10] When the Party tried to impose higher-order cooperatives they faced considerable obstruction from the peasantry. Peasants in Vietnam devoted enormous efforts in time and energy to the cultivation of their small private plots, responding to the disincentives inherent in collective farming through go-slows and foot-dragging (Table 1.5). In effect, by undermining the logic of the collective system farmers were ultimately eroding their own livelihoods but, as one villager in north Vietnam explained to Benedict Kerkvliet 'no one mourns for the father of everyone' (*cha chung không ai khóc*) (1995b: 403). By the 1970s many villagers were deriving as much as three-quarters of their income from their tiny private plots (Kerkvliet 1995b: 405). Kerkvliet reports that some families even attempted to expand private production by encroaching on collective land. 'Disinterest and disgust', he suggests, 'were so serious in some areas that tens of thousands of hectares went unplanted' (Kerkvliet 1995a: 69, see also Kervliet 1995b). By the late 1960s, cooperative members in the North were attempting surreptitiously to bypass the collective system through what became known as *khóan chui* or 'sneaky contracts' (Kerkvliet 1995a: 69). It seems that farmers often found allies in local cadres, who were not so much representatives of the Party at the grass roots, but people with strong local affinities who happened to be party members. They occupied the 'vast in-between terrain of "everyday politics"', fighting as much for the interests of farmers and those of the Party and supporting the introduction of modifications that effectively undermined the collective system (Kerkvliet 1995b: 400).

Needless to say, this view of cooperatives in Vietnam as being neither what they seemed (or what rhetoric and ideology would lead one to believe), nor as efficient as they appeared, is contested. Kolko, for example, states that the system had 'succeeded fairly well in Annam and Tonkin [central and north Vietnam]' and uses this as a baseline from which to construct the argument that the agricultural reforms have undermined what was previously a relatively productive and equitable system (1995: 20).

*Table 1.5* Farmers' objections to cooperatives in North Vietnam

- no incentive to work diligently
- manipulation of the system by production team leaders
- deterioration in the condition of collective property (land, machinery and animals) due to lack of incentive
- stagnation or deterioration in living standards
- need to support non-productive officials and cadres
- erosion of traditional extended labour groups

*Source*: adapted from Kerkvliet 1995b: 402-7

While farmers in North Vietnam were apparently attempting to bypass or cheat the cooperative system through foot-dragging and obstruction, the cooperative system was failing to provide anything akin to a comprehensive security net for the poor (see page 140). Whether it was farmers' actions that led to cooperative failure, or cooperative inefficiencies which led to farmers responding in this way is not clear, and most likely varied from place to place. None the less, Vietnam's equivalent of Mao Zedong's 'iron rice bowl' was, it seems, more for rhetorical consumption than for subsistence.

It is common to read that the economic reforms have decreased the role of the state in people's lives. This may be true in areas where collectivization was pronounced, but should not be automatically accepted. In Sam Ta, a Hmong village in northern Vietnam, the reforms resulted in significantly greater intrusion: the state banned farmers' traditional practice of shifting residence, sharply curtailed shifting cultivation, restricted their access to the forest and forest products, and imposed a rigorous taxation system. 'When asked what, in his opinion, was the greatest difference between now and "the old days", the Village Elder said that it was the greater restriction [now] on villagers' freedom' (ActionAid Vietnam 1995: 47). Although Sam Ta, a highland Hmong (minority) community, can be viewed as something of an exception to the broad sweep of rural settlements in north Vietnam (but is probably indicative of many more villages in Laos), it none the less highlights an important point: if collectivization was weak in many areas, and if villagers were able to maintain considerable autonomy in their lives, then although the reforms may have increased the incentive to produce it does not automatically follow that this has been achieved because the State has withdrawn from people's lives. In some senses, and particularly in some areas, the reverse has occurred. It is notable, for example, that one of the key aims of Vietnam's important 1993 Land Law, which is seen as a central plank in the 'freeing-up' of the agricultural economy, has been to increase state control over land. It permits the State to categorize land according to use – whether that be for agricultural, residential, urban, forest, or some other special use (Smith and Tran Thanh Binh 1994: 5). In highland areas of the country, the Land Law is being used as a way of introducing a more sustainable agriculture by defining acceptable land use for mountain tops (banning the cultivation of annual crops, for example), encouraging a shift from shifting cultivation to settled agriculture, and allocating land for reafforestation (Smith 1995: 39).

Broadening the argument somewhat, it is questionable whether even in North Vietnam, the centre (Hanoi) was able to control the provinces. The exigencies of the war and the need to decentralize economic and other activity meant that even after the end of the conflict, the Party in Hanoi had difficulty imposing its will countrywide. Fforde hazards, on the basis of the fact that farmers did not toe the Party line, that the Democratic Republic of Vietnam was a 'weak state' (Fforde 1994: 4). He argues that there occurred a 'systematic adaptation of the neo-Stalinist system in which that system's institutional forms were endogenised, with far greater local independence and capacity to allocate resources in accordance with local interests' (Fforde 1994: 5). Though he does not seem to wish to categorize Vietnam as a weak state as such, Kerkvliet similarly believes that there have always been sources of political power independent of the State (1995b). Thus the economic independence outlined above was mirrored – and was to some extent a consequence of – the inability of the State to dictate terms.

So when the economic reforms were introduced in Vietnam from the early 1980s, the VCP was not introducing radical reforms – at least in agriculture – but merely following a road that the peasantry had already marked out. Directive 100 of January 1981 represented 'proposals for reform [based] upon illegal reality'; the reforms were therefore chiefly within the constitutional order, not within the order of agriculture (Fforde 1994: 7). Some commentators suggest that a similar argument could be applied to the industrial sector. Fforde, for one, argues that 'as early as 1964 [in Vietnam] . . . it was clear that various state-owned enterprises were "running to the market"' (1994: 10). Spontaneous 'fence-breaking' by farmers, industrialists and others became recognized by the State which gave its consent to the process by changing the constitutional land-scape. *Doi moi*, at least in agriculture, was merely official recognition of what was already a reality and to talk of the Party 'introducing' economic reforms is to miss the point. The Party was, more accurately, 'accepting' or 'endorsing' reform.

Just as there are doubts as to how far the period of command economic management was truly commanding, so the obverse is also true, albeit probably to a lesser extent: namely, that many elements of state intervention continue to remain in place, particularly within the state industrial sector, even following the reforms of the 1980s and 1990s (Beresford 1993). It is in Myanmar where the recent reforms have been most piecemeal and least reformist. In 1988, with the accession of the SLORC to power, the State declared that it was embracing an 'open door, market-oriented policy'. Two years later, however, such terms had been dropped from the government's vocabulary (Cook 1994: 134, and see Cook and Minogue 1993). Prices and other financial responsibilities are still centrally controlled by the State, and enterprise managers are said to have very little autonomy (Cook 1994: 127–30). Even the private sector, which has apparently been given a boost with the lifting of some restrictions, operates in a state of considerable uncertainty. In addition, the Private Investment Law of 1990 has not done away with the raft of registration fees and licences within which private enterprise is still constrained. In the light of this, whether the reforms in Myanmar have really led to a sharp increase in private enterprise, or whether the legalizing of some activities has merely brought the parallel economy more fully into the open, is a moot point.

# THE UNMAKING OF THE MIRACLE

## Critiques from within

There is reason . . . to question the use of the term 'development' at all when applied to this pattern of economic growth [in Thailand] as it violates important values of equity, economic democracy, ecological balance, and human decency.

(Bell 1992: 61)

The irony of Vietnam today is that those who gave and suffered the most, and were promised the greatest benefits, have gained the least. The communists are abandoning them to the inherently precarious future of a market economy.

(Gabriel Kolko quoted in *FEER* 1995: 28)

There have been many critiques of the Southeast Asian growth experience and the intention here is not to trawl through that literature in detail.[11] However it is necessary to highlight the main elements that make up these cries of dissent as they represent a useful counterpoint to the more laudatory perspectives. Some authors, like Bell (1996), build an alternative case to challenge the growth thesis. In other words, to posit the view that the HPAEs of Southeast Asia are not success stories, not exemplars to the rest of the developing world, but indicative – and therefore a warning – of the dangers of over-rapid, export-led, capitalist growth. Bell uses the term 'maldevelopment' to describe Thailand's experience, writing that this 'refers to a pattern of development with strongly negative socio-economic consequences in terms of inequality, unevenness, cultural frag-mentation, and a negative impact on women and the environment' (1996: 49). There are many more scholars though who take a less polemic view, and merely call for greater balance in the assessment of the Southeast Asian growth record. Thus Parnwell and Arghiros, after having listed a litany of Thailand's problems, remark that 'whilst not wishing to deny or decry the Kingdom's undoubted economic, social and welfare achievements, the notion of "success" and the accomplishment of "development" have both to be qualified' (1996: 2). With the countries of Indochina embarking on the market road, studies expressing similar sentiments regarding their experiences are also emerging in the literature (e.g. Kolko 1995). Though these critiques are numerous, the essence of the case can be encapsulated in just three words: 'Modernization without devel-opment'. The principal components of this case are presented in brief below, although many of the issues are returned to and discussed in greater detail in later chapters.

## *Development that is vulnerable, dependent and subordinate*

A central area of debate concerns whether Southeast Asian development, based as it is on foreign investment-led, export-driven industrialization, is not creating dependency and vulnerability. Writing of Singapore's development strategy, Rodan, for example, contends that 'there is an externally-imposed precariousness about participation in this [global economic] structure over which even the most astute policy-makers . . . have very little control' (1989: 113). Chant and McIlwaine, with reference to the Mactan Export Processing Zone (EPZ) in the Philippines, concur suggesting that its 'uncertain' future is linked to 'the inherent vulnerability of multinational export manufacturing.' (1995b: 167). Kolko expresses the same fears with regard to Vietnam's recent industrialization (1995, and see the quote at the beginning of this section). There are a number of fears subsumed within this general observation though. One focuses on per-turbations in the global economy, and their effects at the national level. Singapore's recession of 1985, for example, was partly due to a simultaneous decline in key export sectors brought on by events in the international market-place (see Rigg 1988b). A second area of concern is linked to the perceived rise in global protectionism and the assumption that the countries of Southeast Asia would find themselves poorly placed should entry to their main export markets in the developed world be barred or restricted. A third issue is linked to the commitment of foreign companies to the countries they invest in, especially when the industries are footloose.[12] It has been noted, for example, that garment and footwear industries have continually shifted their operations to those countries with the lowest wage rates – from Taiwan, South Korea and Hong Kong to Malaysia; from Malaysia to Thailand; from Thailand to Indonesia; and from Indonesia to Vietnam and China. In Thailand, the textile and clothing industry suffered thousands of lay-offs and a rise in labour disputes in the early 1990s as buyers and investors shifted their attention to other, lower wage locations (*Economist* 1993c). After the rapid growth of the 1980s, the garment sector has stagnated and the share of garments in total exports has declined from a peak of 12.1 per cent in 1987 to 7.3 per cent in 1995 (Table 1.6). Even in Vietnam, the government is concerned that they are pricing themselves out of the market (Schwarz 1996). It is for these reasons that Forbes states that '[t]he longer-term viability of . . . dynamic Pacific Asian growth must be critically questioned' (1993: 60).

Krugman, like those scholars quoted above, also argues that the high eco-nomic growth rates experienced by some Asian economies will not be sustained far into the future, but takes a rather different line in making this assertion (Krugman 1994). He argues that East and Southeast Asia's growth – like that of the former Soviet Union which he offers as a parallel – has been based largely on a remarkable mobilization of resources. In Singapore, for example, the economically active proportion of the population rose from 27 per cent in 1966 to 51 per cent in 1990 (1994: 70). There was also a massive investment

*Table 1.6* The rise (and demise?) of Thailand's garment industry

| | *Value of garment exports (bn baht)* | *Total value of exports (bn baht)* | *Share of garments in total exports (%)* | |
|------|------|---------|------|------|
| 1980 | | 133.2 | | |
| 1981 | | 153.0 | | |
| 1982 | | 159.7 | | |
| 1983 | 8.8 | 146.5 | 6.0 | ⎫ |
| 1984 | 12.2 | 175.2 | 7.0 | ⎪ |
| 1985 | 14.7 | 193.4 | 7.6 | ⎬ GROWTH |
| 1986 | 20.2 | 233.4 | 8.7 | ⎪ |
| 1987 | 36.3 | 299.9 | 12.1 | ⎭ |
| 1988 | 45.6 | 403.6 | 11.3 | ⎫ |
| 1989 | 57.9 | 516.4 | 11.2 | ⎪ |
| 1990 | 65.8 | 589.8 | 11.1 | ⎬ STAGNATION |
| 1991 | 86.7 | 725.5 | 11.9 | ⎭ |
| 1992 | 86.8 | 824.6 | 10.5 | ⎫ |
| 1993 | 89.6 | 940.9 | 9.5 | ⎪ |
| 1994 | 100.7 | 1,137.6 | 8.9 | ⎬ DECLINE |
| 1995 | 102.0 | 1,404.5 | 7.3 | ⎪ |
| 1996* | 39.4 | 696.6 | 5.7 | ⎭ |

*Sources*: National Statistical Office data, Bangkok Post Economic Review (various), Tasker and Handley 1993
*Notes*
* = Jan–Jun
For much of the above period US$1 = 25 baht

in physical infrastructure and an upgrading of human resources as increasing numbers of young people received secondary education. 'These numbers', Krugman writes, 'should make it obvious that Singapore's growth has been largely based on one-time changes in behaviour that cannot be repeated' (1994: 71).[13] He asserts that there is little evidence that the NICs' growth has included any great increases in efficiency; the primary impetus behind growth has been the mobilization of resources on a grand scale.[14] This has two main implications. First, it indicates that the growth economies of Southeast (and East) Asia, like the former Soviet Union, will experience a sharp slowdown in growth as further resource mobilization becomes impossible. And second, it undermines the notion that the experience of these countries – which he terms 'Paper Tigers' – is in any sense 'miraculous'.

The belief that growth will slow down in the growth economies of Asia is generally not disputed. However Krugman's suggestion that there has been little or no increase in efficiency is contested, as the figures on which he makes the assertion are called into question. While Singapore, Malaysia and Indonesia show only a modest increase in total factor productivitiy (TFP), Thailand, as well as Hong Kong, Taiwan and South Korea show a very large increase (Page 1994: 617, World Bank 1993a: 54–8).[15] It has also been pointed out that the

Asian growth economies have made massive investments in education which should feed through into technological upgrading and efficiency improvements. The Thai, Malaysian and Indonesian governments are all too aware that sustaining high growth is dependent on a continual upgrading of the labour force through investments in education and training. Furthermore, it is unfair to compare the economic growth of Asia with that of the Soviet Union in the 1950s and 1960s and then use the subsequent failure of the Soviet Union to sustain such growth to suggest that the same might happen in Asia. There were evident and fundamental political and economic differences between the two which would intuitively lead one to reject such a parallel. And moreover, a high proportion of investment in the Soviet Union went into defence, while in Asia military expenditure has been modest by comparison.[16] In a reply to Krugman, Rostow writes that Asian growth rates 'will decelerate as they approach the technological frontier, but the industrialization of Asia will shape the next century, miracle or not' (Rostow 1995: 184).

### Development that perpetuates poverty and poor working conditions

In order to maintain competitiveness within the context of a global economy, employers are forced to pare labour costs and overheads to the bone. This, it is argued, has tended to perpetuate low wages and poverty, and created the economic exigencies in which poor and dangerous working conditions can flourish. The process has been termed 'super-exploitation' or, playing on the World Bank's 'market-friendly' approach, a 'not-so-friendly-to-labour' approach (Jenkins 1994: 75; Amsden 1994: 632; Bell 1996: 54–5; see also Hewison and Brown 1994). In Vietnam, the minimum wage rate at foreign-invested companies was set at US\$30–35 per month in 1992. With rising labour unrest during 1995 there was considerable pressure to raise this minimum wage. The Minister of Labour, Tran Dinh Hoan, though sympathetic to workers' difficulties felt unable to act, however: 'Vietnam cannot set its minimum wage higher than other regional countries. . . . Otherwise foreign investment will not come to Vietnam, but will go elsewhere' (Schwarz 1996). Vietnam, like other countries whose industrialization strategies are dependent on foreign investment, is trapped by the apparently inescapable logic of the global economy which channels investment to the lowest wage locations (see Figure 1.1).

There is also a gender angle to the process, when it is observed that in many export-oriented enterprises the great majority of shop floor workers are women. Peter Bell has gone so far as to state that 'development [in Thailand] has been literally built in large measure on the backs of Thai women. Women are at the cutting edge of "export-oriented growth"' (1992: 61). It is on the basis of such views that scholars have referred to a 'feminization of poverty' in the region. These issues are dealt with in greater detail in Chapter 6, but it is notable that the series of human catastrophes to have affected Thailand in recent years are linked, in the eyes of many Thais, to the emphasis on profit and economic

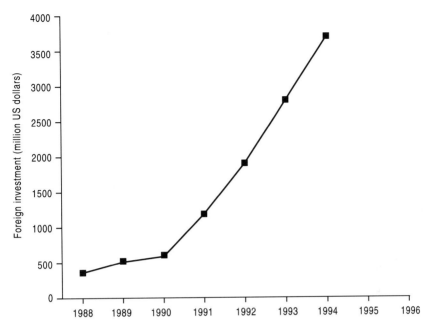

*Figure 1.1* Annual investment of foreign capital in Vietnam (1988–94)
*Sources*: Vu Tuan Anh 1995, Eliot 1995b, Gates 1995: 388

growth at the expense of health, safety and equity (Table 1.7). Critics have also made the observation that during a period of unprecedented economic expansion, the incidence of poverty in Thailand has, seemingly, barely declined (see page 79 for a fuller discussion).

### Development that infringes human rights and undermines human decency

The need to maintain an investment climate that is attractive to multinational firms not only encourages the perpetuation of low wages, but also creates a political and industrial climate where trades unions are banned or institutionalized, workers' rights are infringed, and where labour activists are persecuted (see Chapter 6). At a rather wider level, some scholars would deploy the same arguments to explain the persistence of authoritarian government in Asia (see the collection of chapters in Hewison *et al.* 1993).[17]

### Development that is destructive of the natural environment and environmentally non-sustainable

There is a widely held belief that Southeast Asia's environmental 'crisis' is a direct result of the economic policies that have been pursued. Bryant and Parnwell, for example, in writing of the:

27

explosive mix of rapid but uneven economic growth and pervasive environmental degradation [contend that] [h]ere [in Southeast Asia], perhaps more than anywhere else in the developing world, the contradictions between environment and development, economic growth and environmental conservation, are visible.

(Bryant and Parnwell 1996: 2)

The implication seems to be that if development had taken another course then the environment would have remained, if not pristine then at least more so. Bell takes the same line of argument, contending that Thailand's growth is environmentally non-sustainable and that the root cause of the Kingdom's environmental crisis, and the struggles that this has precipitated, can be found embodied in the direction that development has taken (1996: 59).

*Table 1.7* Human tragedies in Thailand

---

Thailand's hyper-growth of the late 1980s and 1990s has been accompanied by a series of human tragedies. Critics of the kingdom's economic development contend that the cause of these tragedies lies in the systematic evasion of health and safety standards in the interest of profit, and that the government has been complicit in these evasions.

*1988*: floods in the South, near Chumphon, and hillside deforestation cause landslips which engulf a village.
Loss of life: 300

*1990, September*: A *tuk tuk* (three-wheeled motorized taxi) and a LNG tanker collide in Bangkok.
Loss of life: 91

*1991, February*: A lorry transporting dynamite near Phang-nga, south Thailand, explodes.
Loss of life: 171

*1992, March*: A ferry carrying Buddhist pilgrims to Si Racha, off the Eastern Seaboard, sinks after being involved in a collision with an oil tanker.
Loss of life: 119

*1993, May*: A fire breaks out at the Buddha Monthon (Kader) toy factory on the outskirts of Bangkok. Many corridors are blocked and doors locked.
Loss of life: 188

*1993, August*: A hotel in the Northeastern town of Korat collapses after the owner adds additional floors and the contractors cut corners.
Loss of life: 120

*1994, February*: A ferry carrying Burmese workers back to Myanmar from Thailand capsizes in the Andaman Sea.
Loss of life: about 200

*1995, June*: A crowded jetty on the Chao Phraya River in Bangkok collapses after being struck by a ferry.
Loss of life: 28.

---

*Sources*: various

*Development that undermines local cultures and creates an ethic of consumerism*

Part and parcel of rapid capitalist growth is a culture of consumerism, individualism, greed and acquisitiveness. These traits are seen to be replacing local cultural traits which stress community action, consensus, and moderation. Needless to say, this process of cultural change is perceived as wholly retrograde. Mulder, in writing of social and cultural change in Java, states that 'The new order is a working order, not a moral order . . . but an order in which one strives for survival rather than fulfilment'. He continues: 'The experience of social life has become less wholesome, the compulsion of power and the exigencies of survival setting some people apart, potential feelings of class solidarity thwarted by official policy at the same time that [traditional] patronage is out of reach of most' (1989: 147). Sulak Sivaraksa expresses similar sentiments in writing that the 'present trend of development [in Thailand] is wrong because the people at the top equate modernization with Westernization and the gross materialistic values of a consumer culture' (1990: 171). This has led to the belief, widely held in local circles, that indigenous cultures are inferior and that so long as local peoples adhere to their cultures of birth then they will remain 'backward' and 'primitive'. The hegemony of the development discourse ensures that there will occur a gradual homogenization of Southeast Asian culture in which the region's cultural form will become a mirror of the West's (see Escobar 1995a: 52–4).

*Development that widens inequalities between people, between rural and urban areas, and between regions*

Critics of the Southeast Asian growth experience have observed that the benefits of growth have accrued to particular groups in society and to particular regions. This has been expressed in class terms, in terms of rural–urban bias, and in terms of core and periphery, but the general implications are much the same: that inequalities have widened. This extends from the growth economies of Asean to the reforming economies of Indochina. 'Thailand', Parnwell and Arghiros write, 'is as striking an example of uneven development as it is of economic achievement' (1996: 2). In Vietnam, the economic reform programme or *doi moi* has bestowed wealth on a few people in a few areas, but left most people in poverty. Premier Vo Van Kiet was quoted in the Vietnamese Communist Party daily *Nhan Dan* warning that the 'confrontation between luxury and misery, between cities and country' could cause problems and talked of the need to 'establish a new order of sharing' (quoted in Williams 1992). These issues will be addressed in more detail in Chapters 3 and 4, but it is notable how far growing inequality and rapid economic growth are perceived to be two, inescapable sides of the same capitalist coin.

## The 'impasse' in development studies

There has been a great deal written about the so-called 'impasse', 'gridlock' or 'crisis' in development studies. In using terms such as these though, authors often allude to rather different sets of issues. Some scholars use them to denote a failure of development to achieve its objectives; others to highlight a perceived crisis in Marxist and post-Marxist development theory; still more to a need to make development sensitive to local peoples and cultures; and yet others to the unhealthy hegemony of the development discourse. The first of these links with the discussion in the previous section while the last is addressed in the following section.

For Booth, whose paper to some extent marks the beginning of the debate, the impasse refers to the shortcomings of much Marxist development sociology (Booth D. 1985, see also Schuurman 1993, Booth D. 1993 and Kiely 1995). Importantly, his thesis is generic; it is not linked to any single theory, but rather to the entire edifice. He sees the impasse as lying in three key shortcomings. First Marxist development sociology's preoccupation with seeing and explaining the world as it *should* be.[18] Secondly, the failure to adapt and to internalize criticism (Booth sees this as particularly true of dependency theory). And thirdly, a tendency to divorce theorizing from real-world problems. Vandergeest and Buttel, in a reply to Booth's paper, further diagnose the problem and in the process highlight the stagnation within Marxist development sociology. While non-Marxist theorizing has moved on from the rigidities of modernization theory, Marxist practitioners, they argue, remain strangely trapped within an intellectual tradition that viewed the North and South as two largely homogeneous blocks which could be subjected to monocausal explanation (Vandergeest and Buttel 1988: 685). The heterogeneity and dynamism within and between the countries of the developing world – and which is becoming more apparent with every passing year, and perhaps nowhere more so than in Southeast Asia – was eschewed in the interests of theoretical symmetry.[19] In a third sortie, Corbridge widens the debate to argue that the crisis in Marxist development sociology is just one part of a much wider, and more significant, crisis of development studies and developmentalism (Corbridge 1990).

Arising chronologically rather than directly from Corbridge's paper, have been a spate of books and articles that have challenged the bases of developmentalism. Many have drawn upon post-modernist critiques, have focused on the power of language and the need to look at development 'texts', and have highlighted the importance of power relations in understanding the discourse of development. These issues are discussed briefly in the next section. However, running alongside this more conceptual critique has been one with which it shares parallels in terms of the identified cure(s), but which is firmly rooted in, and informed by, the real world. Michael Edwards, in his provocatively titled paper 'The irrelevance of development studies' (1989) proffers the thought that development is not about growth but about 'processes of enrichment, empowerment and participation, which the technocratic, project-oriented view of the world simply cannot accommodate' (1989: 120). To some extent Edwards' views mirror those of the radical critiques noted above in the section entitled 'Critiques from within'. But in an important sense he is moving the debate on from such tangible issues as inequality, health, education and environment, to a landscape where the debate is about, for instance, power, knowledge and popular participation. In this schema, the context and means of development are more important than the ends (see Chapter 2 for a fuller discussion). None the less, these alternative developmentalists are still very much concerned with what is happening to real people in the real world; to individuals, households and communities within distinct social and physical environments. Edwards' concerns are also reflected in the outpouring of literature on new social movements (see page 145).

The 'crisis of developmentalism' challenges the assumption of uni-linear progress that underpins modernization theory. To begin with, the vicarious development (or maldevelopment) experiences of non-Western and Western societies implies that there is no single ahistoric path to riches (Nederveen Pieterse 1991). Second, the evidence from the South is that modernization is marginalizing and excluding at least as many people as it is including and developing. The hegemony of modernization theory has not just allowed the benefits of development to be accepted as a self-evident truth, but also to assume that the means and path to development are both ahistorical and transculturally applicable. This debate has clear parallels with that of the post-developmentalists discussed below. It also, though, has parallels with those scholars who are still committed to 'development' – the notion that broad-based development should be promoted and welcomed – but would wish the means and ends to be made more inclusive, more indigenous, and more humane (see Chapters 3 and 4).

### The unmaking of development

> Something seems to be amiss. . . . Granted the vast sums invested in trying to find a solution to . . . the problem of underdevelopment . . . matters should be getting better rather than worse. Instead it would seem that development projects often contribute to the deterioration.
>
> (Hobart 1993: 1)

The 'critiques from within' are, to a large degree, endogenous. They fight the scholarly battle on the same ground, often from within the sphere of economics, and usually with the same weapons, occasionally hardened with different ideological alloys. But, and to take the metaphor of conflict further, it could be argued that the main scholarly battle is being fought on a different field entirely, and to some extent the combatants barely come to grips with one another. Scholars within anthropology, sociology and geography, for example, are frantically deconstructing developmental texts, building theories of Orientalism and the 'other', analysing vocabularies of development, and rather more generally focusing on the 'discourse' of development. This group might be rather uncomfortably grouped together under the heading 'post-developmentalists'. Economists – or 'developmentalists' – meanwhile, although they may be sharply at odds in their interpretations of the evidence, continue to focus on aggregate growth rates, levels of foreign direct investment, incidences of poverty, and the fiscal rectitude of governments: in other words, the panoply of development-speak. The concerns are different, the tools of analysis are different, and the fields of study are different. Even their purpose is sharply at odds. While Crush, for example, in the introduction to his edited volume *Power of development*, writes that '[t]his book's primary concern is to try to make the self-evident problematical' (1995a: 3), many economists and others at what might be termed the cutting edge of development would see such statements as word games with little to commend them beyond the thick walls of academia.[20] Post-developmentalists might retort that economics' tendency to present itself as rational and realist is a deceit. Economics is a cultural force which has colonized the developing world, surpressing alternative ways of understanding (see Escobar 1995a: 58–63). As such it is imperative, if the hegemony of economics is seriously to be challenged, that the debate be moved onto other fields of conflict where economics is not in the ascendancy. Sachs, for example, argues that development 'is much more than just a socio-economic endeavour; it is a perception which models reality, a myth which comforts societies, and a fantasy which unleashes passions' (1992: 1).

Although to stereotype such a wealth of material invites criticism, while the post-developmentalists are intent on dissecting 'development' as an idea and as a political, economic and cultural force, developmentalists use the term merely as short-hand for modernization, and are concerned with how it can be promoted, managed, controlled and measured. The gulf between the two can be seen played out in seminar rooms across the Western hemisphere, as academics from different disciplines fill the same space but occupy different mental worlds.

Mark Hobart, an anthropologist specializing in Southeast Asia, in the quote at the beginning of this section, appears to suggest that things are actually getting worse in the 'developing' world. How can this be, it is fair to ask, when many Southeast Asian economies at least are expanding rapidly and the incidence of poverty is declining? There seem to be two possible answers to this question. First, many post-developmentalists would suggest that 'under-development' has been wrongly diagnosed as a problem in search of a cure. If under-development, or an absence of modernity, is seen as a positive attribute, or at least one from which no negative connotations of backwardness can be drawn, then how can 'development' lead to a superior state? Some scholars would take this further and say that development is the Trojan horse for commercialization, a process which makes people dependent on the market, and countries subservient to the interests of the North and of multinational capital. What is advanced and good, therefore, is interpreted by the North and imposed on the people and states of the South. That the South should become more like the North is taken as a self-evident truth. Development, in short, has created problems which were never previously thought to exist. These so-styled problems – 'small farmers with low productivity', 'illiteracy', 'poverty', 'under-production' – then demand interventions by governments and aid agencies, creating a vast industry of experts and technologies whose existence is predicated on a set of problems that they have invented and defined and which often have no locally-rooted reality. Escobar writes that the end result of the 'seemingly endless specification of problems' was the 'creation of a space of thought and action the expansion of which was dictated in advance' (1995a: 42). The élites of the developing world, and most people in the developed world, embraced the fact of under-development and were, in the process, colonized through the power of the development discourse. As this discourse had its roots in the West, so Western notions of progress, culture and society also gained hegemony over indigenous systems.[21] Sulak Sivaraksa, a well-known critic of Thailand's modernization ethos, argues that in the late 1950s, at the time when the newly-formed National Economic Development Board was formulating the first of the kingdom's development plans (1961–66), American advisers managed to persuade the Thai government to encourage the Supreme Sangha Council to advise monks not to preach on the virtue of contentedness (Sulak Sivaraksa 1996). If people were content with nothing (in material terms), these advisers reasoned, then how could Thailand enter the race to development?

The second, and more orthodox answer to the question 'how can things be getting worse when countries are getting richer?' is that development is erroneously measured and assessed in economic terms. Economic growth and the incidence of poverty – which is usually based on income/consumption data – do not measure development (see Chapters 3 and 4). Many economists use development to mean modernization, while post-developmentalists would suggest that modernization has very little to do with development. This links into the abundant literature, dating from the 1960s, which challenges modernization theory – the belief that the goal of development is to create a modern society built on a modern economy. The assumption that countries would pass through a series of evolutionary stages on the way to the Golden Land of high mass consumption, the notion that trade would bring benefits, and the contention that developing countries should (and would) become mirror images of the developed countries, were all challenged (see Toye 1987; Arndt 1987).

## To speak in tongues: the local and endogenous

Reading the literature of post-developmentalists there is an almost constant refrain that the objects of development must be viewed from within, in terms of their cultural and historical contexts (see for example Crush 1995a: 8–18, Pottier 1992, Escobar 1995a). A central criticism of development is that it strips people and countries of their human and historical places. It makes them almost generic – homogenized and stereotyped. And yet, perversely, in much of the post-developmental literature there seems to be little understanding of local culture and history. Terms like endogenous, indigenous and local pervade the material, and yet local histories and cultures are barely dissected.

Arturo Escobar's *Encountering development: the making and unmaking of the Third World* (1995a) is a good example of this tendency to build towering generalizations on shallow foundations. The general tenor of the book is clear from the start:

> This book tells the story of this [development] dream and how it progressively turned into a nightmare. For instead of the kingdom of abundance promised by theorists and politicians in the 1950s, the discourse and strategy of development produced its opposite: massive underdevelopment and impoverishment, untold exploitation and oppression.
>
> (Escobar 1995a: 4)

There can be little doubt that Escobar wrote his book with the intention that it should be a general statement on the plight of the developing world. It represents, in many respects, a compelling and disturbing argument. And yet there is remarkably little here on East and Southeast Asia, or on the Pacific Rim more widely. The index has no entries for Singapore, Thailand, the Philippines, Laos, Cambodia or Vietnam, makes just a single reference to Indonesia, and three to Malaysia. Even more startling, there are no entries for South Korea, Taiwan, or Hong Kong.[22] Given that it is the economies of Asia which might refute Escobar's sweeping assertion quoted above, this seems a significant omission. It is all the more surprising when one of the book's constant refrains is the criticism of the development discourse's tendency to paint all people in the developing world in a single shade of 'poor'. Escobar does much the same when he paints all the countries of the developing world in a single shade of (largely Colombian) 'under-developed'.[23] Of course, should he have included a detailed discussion of the Asian economies the main thrust of the book might have had to have been modified. And as the volume appears to be a polemic of the paradigm-shift variety, this would – so to speak – have spoilt the party.

How far such apparently more nuanced understandings of people and places contained in the post-developmental literature truly reflect local histories and experiences is hard to say. What is to be made of farmer 'experts' in the Northeastern region of Thailand who use the term 'globalization' (*lokanuwat*) to shed light on their plight, and are then patted on the back for 'learning' so well from the NGO facilitators who have worked with them? (Sophie Wigzell, personal communication 1995). The construction of the past to fit an image of our own imagination is not restricted to colonial historians and latter-day developmentalists. It is as much a feature of the post-developmentalists and their agenda. So, when scholars ask for interpretations of development, history and

*Box 1.3* Indigenous imaginings: the local creation of local worlds

The earliest piece of Thai literature, and perhaps the best-known, is the so-called Inscription no. 1 of King Ramkhamhaeng (Rama the Brave) of Sukhothai, dated to 1292. Its lines, inscribed on a stone pillar, tell of a fecund country, a flourishing state, and a benevolent monarch:

In the time of King Ramkhamhaeng, this land of Sukhothai is thriving. In the water there is fish, in the fields there is rice. The Lord of the realm does not levy toll on his subjects for travelling the roads. . . . When any commoner or man of rank dies, his estate – his elephants, wives, children, granaries, rice, retainers and groves of areca and betel – is left in its entirety to his son. . . . So the people of this *müang* [state] of Sukhothai praise him.

Not only has the authenticity of the inscription been challenged by some scholars who maintain that its discovery in 1833 served the purposes of the future King Mongkut in his attempts to prevent the colonization of his country by the encroaching British and French, but even if authentic it was clearly a piece of self-aggrandizing propaganda (see Gosling 1991, Chamberlain 1991)

Arguably, it is on the basis of the Sukhothai myth that the Thai (and more particularly, T'ai) sense of themselves has been forged. Chatthip Nartsupha has written at length on the pre-capitalist 'village-community'. It was, in his view, an 'ancient institution', 'naturally' set up by the people, based on 'subsistence production', 'self-sufficient', 'communal', 'self-sustaining', and 'relatively autonomous' (1986: 157-9; see also Chatthip Nartsupha 1991). Much alternative development in Thailand aims to re-create this history, and to re-build lives and traditional institutions supposedly undermined and eroded by commercialization and capitalism.

But work by scholars such as Bowie and Koizumi indicates that even in traditional, pre-modern Thailand rural people were differentiated, dynamic, and engaged in market exchange. 'To suggest', Bowie writes, 'that the traditional nineteenth-century economy [of Northern Siam] was a subsistence economy is to miss the significance of wealth in the context of poverty' (1992: 815 and Koizumi 1992). Terweil admits in his study that 'one of the rather unexpected findings has been the discovery of the size and efficacy of the nineteenth-century government apparatus' in Siam (1989: 251). Even those people living some distance from Bangkok could not escape the long reach of the government, and were expected, if so registered, to take up their corvée responsibilities. Even the existence of the 'village', with its sonorous associations of timelessness and community, has been challenged by scholars who deem it to have been a construct of the central authorities, and then imagined into life by academics:

The term *muu baan*, variously translated as 'village', 'administrative village' or 'administrative hamlet', refers to a formal administrative

continued . . .

division which in at least some areas was arbitrarily imposed on the landscape in a way which either combined or divided existing communities however defined, or provided a wider social framework where none had hitherto existed.

(Kemp 1988: 8)

We are . . . at the 'roots', not of the village community as a historical phenomenon but of its theoretical invention arising out of early colonial writings on India and their incorporation into both Marxist and non-Marxist social theory.

(Kemp 1991: 325)

The experience of Siam/Thailand shows the degree to which 'indigenous' may be created and re-created at the local level to serve the interests of élites and other groups. To assume that there is one 'natural' indigenous vision which can somehow be accessed to shed light on the reality of local places, and which at the same time truly represents the interests of the whole, is highly unlikely. Even local visions are unrepresentative, are subject to manipulation, have been created and moulded by powerful interest groups, and are subject to constant change.

culture to be rooted in, and based on local/indigenous visions and experiences, it is fair to ask 'which local?' (see Box 1.3). Hobart wonders whether 'there is a risk of these [academic] writings becoming part of the processes of hegemony, which they ostensibly set out to criticize' (1993: 13), noting rather later that there is a worrying tendency towards the programmatic, polemical, and grand-scale perspective. The focus on 'textual, literary and linguistic concerns' leaves such work largely Eurocentric in its origins and articulation so 'replicating in subtler form the presuppositions they set out to criticize' (Hobart 1993: 17). At least, then, some post-developmentalists accept that there is a certain deceit in criticizing Eurocentric visions of development in terms which are, themselves, Eurocentric (both Hobart 1993 and Crush 1995a can be counted among these navel-gazers). As Crush observes, if development – as an idea – should be wished into oblivion, then to talk of '*alternatives* to development is non-sensical' (1995a: 19, emphasis in original). Escobar, in presenting the case for an 'alternative' to development, also seems self-consciously aware that he is articulating his ideas within an intellectual milieu – a discourse of development – which he has previously set out to denigrate as a malign influence (1995a: 151–2).[24]

With even so-styled 'peasants' in Northeastern Thailand committing themselves to the language of 'globalization', 'ecology' and 'community forests' (see Chapter 2), it would seem clear – and notwithstanding the differences that divide households and individuals in terms of livelihood, wealth and power – that most people in Southeast Asia have climbed aboard the modernization bandwagon, whether they be for or agin it. Even the language of the indigenous is framed in terms which are part and parcel of the development discourse. It seems from the paddy fields and swidden plots to the shanty towns and factory floors of Southeast Asia that people are increasingly comfortable with notions of 'development' as modernization. Alison Murray in her analysis of *kampung* life in Jakarta challenges the assumption that the prostitutes of her study are victims of development. They may be 'treated as commodities and belittled as immoral or pathological deviants, but . . . they are actually making a rational choice in response to the economic prospects of the city, and in selling their bodies as commodities are exploiting the capitalist system for their own purposes . . . to satisfy [their] consumerist aspirations'. While admitting the prostitutes' precarious position, she favourably contrasts their position with that of petty traders: 'they are selling themselves to buy into the system (however temporarily), while the traders and their *kampung* are being sold in the name of development' (Murray 1991: 125). Although there are important differences in the semantics (see Chapter 2), to a large extent the ethic of, and allegiance to modernity, consumerism, progress and development have been internalized. Development and alternative development, like development and underdevelopment, are two sides of the same coin. To grasp one, is to grasp them both (see Watts 1995).

## Notes

1  This title draws on Escobar's book (1995a) *Encountering development: the making and unmaking of the Third World.*
2  In 1996 the membership of Asean consisted of: Indonesia, Malaysia, the Philippines, Singapore, and Thailand (the original five signatories of the Bangkok Declaration in 1967); Brunei, which joined the Association in 1984; and Vietnam which became a member in 1995. Laos and Cambodia applied for membership in early 1996 and are due to become full members in 1997. Myanmar (Burma) is likely to become Asean's tenth member at the same time.
3  Warr, also an economist, writes much the same of the Thai experience: 'The rest of the world needs to learn about Thailand and its economy, and learn it quickly, if it is to do business successfully with this booming neo-NIC' (1993: 2).
4  Henderson writes: 'Under the hegemony of the neoclassical paradigm, economics has evolved into the least introspective of the social sciences' (Henderson 1993: 200).
5  Myrdal's study covers the countries of 'South' Asia. These include, though, all the countries of South (India, Bangladesh, Pakistan, Sri Lanka) and Southeast Asia.
6  In most English-language reports and publications, *chin thanakaan mai* is referred to, rather more prosaically, as the New Economic Mechanism or NEM.
7  This, though, does not take account of the decline in Hanoi's population between 1954 and 1967 in response to the US bombing campaign. See page 101.
8  At the end of 1990, the VCP released a document significantly entitled the *Draft platform for the building of socialism in the transition period.* It is, as Forbes observes, contradictory, highly generalized, and with no clear programme of implementation (1995: 802).
9  The government in Myanmar has controlled the trade in paddy ever since Independence.
10  Among many highland minority peoples collectivization never occurred. The

household always remained the basic unit of production (ActionAid Vietnam 1995: 40).

11 Chapters 5 and 6 provide a more detailed critique of agricultural modernization and foreign-investment driven industrialization in the region. However, it is worth noting that, from a regional perspective, the critical literature is much richer for Africa, South Asia and Latin America than it is for Southeast Asia. It is tempting to see this as a reflection of Southeast Asia's economic vitality; the observation, for example, that during the 1980s poor people in Africa and Latin America suffered a real decline in their standards of living is hard to extend to Southeast Asia (see Schuurman 1993: 9–11).

12 James Clad, formerly a journalist with the *Far Eastern Economic Review* writes of Singapore, for example, that '[f]oreign banks and companies tend to think of Singapore as a parking-place rather than a nation' (1989: 126).

13 *The Economist* framed the same argument rather more prosaically in terms of the 'Myth of the Sausage-makers': 'If you invest in more sausage machines and employ more sausage-makers, of course you will make more sausages. Where's the miracle? Growth will slow down when you run out of extra sauage-makers' (1995b: 71).

14 Krugman's thesis has links with the earlier work of Yoshihara Kunio (1988) who suggests that Southeast Asia's development is based on 'technologyless' industrialization. He uses the technical weakness of industrial capitalism in the region, and the absence of indigenous innovators, to suggest that growth will not be sustained. Like Krugman, he also highlights the inefficiencies in many industrial activities.

15 TFP is usually calculated by deducting from output growth the contributions to growth made by increases in the labour force, capital accumulation, and human capital accumulation. Per worker, it is the residual of growth after subtracting capital and human capital accumulation. However TFP calculated in this manner does take a neoclassical stance and assumes that markets are competitive, that there are constant returns to scale, and that there is a long-run equilibrium between factors of production. Some economists regard all three assumptions as being highly questionable in the context of the fast-growing Asian economies.

16 Though the region has experienced an arms race in recent years.

17 The emergence of democractic systems in Taiwan and South Korea, not to mention Thailand, during periods when development was firmly based on export-led growth would lead one to challenge this view.

18 Booth writes: 'Curiosity about why the world is the way it is, and how it may be changed, must be freed not from Marxism but from Marxism's ulterior interest in proving that within given limits the world *has* to be the way it is' (Booth D. 1985: 777).

19 Corbridge's book *Capitalist world development* mirrors much that is contained in Booth's paper and in the reply of Vandergeest and Buttel. He writes that: ' . . . if radical development geography is to maintain its credibility it must give up four particular failings which now scar it . . . a tendency to oppositionism, a tendency to determinism, a tendency to spatial over-aggregation, and a tendency to epistemological confrontation' (1986: 8).

20 Indeed, Crush anticipates such a response when he writes that his concerns might be dismissed as 'another form of faddish intellectualism destined, like all others, to bloom and fade. . . . The developer will say that there is no time for such esoterica' (1995a: 4). Anticipating a response does not, of course, render that response null.

21 The best-known study of cultural hegemony is Edward Said's *Orientalism* (1979).

22 There is a cursory mention to the NICs on page 93. China also does not get any mention in the index (though there is a reference to China on page 73 and another on page 160 which are not indexed). This omission might be justified on the

grounds that China excluded itself from mainstream development from the Revolution in 1949 through to the beginning of the economic reforms in 1978 and as such is not pertinent to the general discussion. This would not seem to be true for the period since 1978 however.

23 *Encountering development* is a volume which draws almost all its detailed discussion from the Latin American, and especially the Colombian, experience. Lesotho also gets disproportionate treatment, drawing largely on James Ferguson's *The anti-politics machine: 'development', depoliticization, and bureacratic power in Lesotho* (Cambridge University Press, 1990). It should, perhaps, have been sub-titled *The making and unmaking of (just a small part) of the Third World.*

24 Chapman, in his study of autobiographies of islanders in the Pacific, asks rhetorically: 'Why does the western intellectual tradition so prefer the scholarly strategy of the outsider looking in and generally avoid the converse of the insider looking out?' (1995: 251).

# 2

# THINKING ALTERNATIVELY
# ABOUT DEVELOPMENT IN
# SOUTHEAST ASIA

## Introduction

With the growth economies of Southeast Asia often paraded as exemplars of orthodoxy, it is perhaps surprising the degree to which 'alternative' development strategies are debated. Alternative here is used to encapsulate all those theories, concepts and methods which lie outside the mainstream of economics and development planning and practice – ranging from community development strategies, to the Asian Way, to Buddhist and Islamic economics. The discussion which follows is not intended to suggest that there has been a paradigm shift in the region away from orthodoxy and towards some vicarious mix of radical and reactionary. Rather the intention is to highlight the multiple debates that are occurring within the region, both at the national and more local levels, regarding the meaning(s) and practice(s) of development.

There are a range of possible reasons why alternative visions of development have seen such growth during the 1980s and 1990s. First, some commentators have argued that the Asian Miracle has not been a case of the mere transplantation of Western approaches to the East. There has been a critical Asian ingredient in the recipe, one which has more to do with culture and history than with economics (Mehmet 1995 takes this line, for example). Thus, they argue, it is necessary to view Asian – and in this case Southeast Asian – success as different and possibly unique within the global context. Such a view would have significant implications so far as the 'transferability' of the Asian experience to other countries is concerned. Second, there are scholars who maintain that since the early 1980s, rapid modernization has created severe tensions and dislocations in society, economy and environment. Modernization, in their view, has undermined traditional mores, broken down existing structures of authority, corroded communities, and destroyed lives and livelihoods (see Chapter 1 and Bell 1992 and 1996 for examples of such a view). In most cases, they would maintain that a preoccupation with economic growth has led to the neglect of the quality of development. Third, there is also a body of scholars and policymakers whose interest in alternative development does not, in most instances, reflect a perceived failure of orthodoxy, but rather a search for indigenous concepts of development, like the Asian Way, rooted in the Southeast Asian experience. They are engaged in an attempt to bring a distinctly Asian perspective to modernization. Fourth, and finally, there are those who would refute the miracle thesis, and instead propose that (Western) mainstream economics, largely because of its Eurocentric basis, has simply failed to deliver (see Mehmet 1995).

There is clearly a good deal of overlap between the four lines of argument, and with some of the issues raised in the last chapter. It is also true that there are multiple sources of debate. Non-governmental organizations (NGOs), village development groups, religious leaders, radical and reactionary scholars, journalists, and opposition politicians form a heterogeneous and fractious coalition challenging the accepted wisdom that mainstream development is good for you. Many economists would argue that this babel of comment hardly constitutes a coherent alternative to existing development theory and practice. But perhaps that is missing the point: whatever the validity of the arguments, there are many groups and individuals, for an assortment of reasons, who wish to stamp their mark on the development debate. The existence of such a body of dissent is significant *in itself*; in some cases, it is what they stand for, rather than what they say, which is the more important.

The previous chapter dwelt on the extent to which development studies is being reassessed among (Western) scholars. A criticism that could reasonably be levelled at this process is that it once again places the emphasis on Western reappraisals of development, using them to illuminate the debate as it is emerging in Southeast Asia, and elsewhere. This chapter will try to approach the issue from the other end, by starting with the regional/local context and then relating this to the ongoing debates in the West. By taking such an approach, the intention is not to deny the importance of the recent literature emanating from the pens and wordprocessors of Western scholars. Indeed, to a significant extent, the debates in Southeast Asia are adapted reflections of debates in the West, and the claim for uniqueness and local origin is often hard to sustain (see below). None the less, many Southeast Asians would find little in the Western literature that resonates with their own ideas and concerns. There is a convincing argument that despite the claims of revisionist Western academics that they are embedding development studies more firmly in the reality of the developing world, that the debate remains closed and incestuous – conducted by, among, and for, a limited audience of Western, Western-trained, Western-inclined, and Western-based scholars. The same criticisms have also been levelled at indigenous scholars trained in the West and could, of course, be applied to this very volume.

## THE LANGUAGES OF ALTERNATIVE DEVELOPMENT

### 'Alternative' development: what's in a word?

Given the range and sheer quantity of literature that can be grouped under the heading 'alternative development studies', it is clear that there are many alternative developments. In some instances they have more in common with mainstream, orthodox approaches, than they do with one another. In addition, the 'alternative' has a trying tendency of becoming mainstream, and vice versa, as intellectual fashions and shades of politics wax and wane. Thus the ideas of the New Right, sometimes called Neoliberalism or (rather confusingly) Neoconservatism, which began life as a 'counterrevolution' in development studies in the late 1970s had, by the 1980s, been embraced by governments and by agencies such as the IMF and World Bank (see Brohman 1995, Friedmann 1992, Toye 1987). Notably, it was the perceived success of countries in Asia, including Singapore, Malaysia, Thailand, and Indonesia, to achieve development through embracing such policies that led to them being hailed as the answer to the woes of the developing world. Even Laos and Vietnam, two of the poorest countries in the world, have been applauded by the World Bank and IMF as exemplars of structural adjustment. But almost coincident with the time when the New Right was infiltrating the corridors of the establishment so, stage left, notions of community development, empowerment, sustainability, and participatory development were also leaving the radical ghetto where they had been nurtured and were being incorporated into mainstream thinking. Those on the left and right may cry that the essence of their ideas have, in the process, been distorted or diluted but it is none the less important to grasp the dynamism, both intellectual and practical, that exists as approaches and ethoses shift their ground:

> NGOs now find themselves in the novel and rather uncomfortable position of representing the New Orthodoxy. It is difficult to find voices arguing the case against participatory, sustainable, decentralising, and empowering strategies of development.
>
> (Rigg 1994b: 24)

The following discussion focuses on 'left-inclined' alternative development. Although there is an enormous degree of diversity even with this subset (see Farrington *et al.* 1993 for a good summary and range of studies), there are common threads which bind the various approaches and visions, and give the following discussion some common ground (Table 2.1). Non-governmental organizations have played a critical role in promoting alternative development, although it should be noted that not all NGOs necessarily espouse alternative development and not all alternative development is done by NGOs. They therefore play a central, but not an exclusive or definitive, role. For Friedmann, alternative development encapsulates three 'doctrinal beliefs' (1992: 6):

- The belief that the state is part of the problem and that alternative development should occur outside, and perhaps even against, the state
- The belief that the people can do no wrong
- The belief that community action is sufficient for the achievement of alternative development

*Table 2.1* Mainstream versus alternative visions of development

| Mainstream development | Alternative development |
|---|---|
| 1 Treats people as 'objects' of development | Treats people as 'subjects' of development |
| 2 Applies exogenous theories and methods | Applies endogenous theories and methods |
| 3 Top-down | Participatory |
| 4 Interested in ends of development | Interested in the means and ends of development |
| 5 Concerned with practicalities | Concerned with ethical and moral issues as well as practicalities |
| 6 Applies modern technology | Applies 'appropriate', sometimes 'intermediate', technologies |
| 7 Undertaken with full support of the State | Bypasses the State, and is sometimes anti-State |
| 8 Increases the role of the market in people's lives | Sometimes aims to decrease the role of the market and promote self-reliance |
| 9 Centralizing | De-centralizing |
| 10 Stresses the empirical | Stresses the cultural |

NGOs are also part of a wider debate on the role of new social movements in challenging and resisting the established discourse of development (see page 145 for a fuller discussion).

## Eurocentrism in development

Perhaps the single most important criticism of development theory and practice is that it is Eurocentric.[1] In using the word in a critical fashion, rather than merely as a descriptive noun, authors are usually alluding to a number of inter-locking problems ranging from universalism, to a lack of sensitivity to cultural variation, to the tendency towards reductionism and racial superiority (Table 2.2). These criticisms can be usefully grouped into three: first, there are those that highlight the reprehensible deprecation that underlies such Eurocentric visions; second, those that focus on the ineffectiveness of policies and programmes which are formulated on such visions; and third, those which see a Eurocentric world-view perpetuating the hegemony of the West, ideologically, economically and culturally.

It is the second of these areas of criticism which most concerns the discussion here. The neglect of culture and regional distinctiveness has been a major short-coming in the development debate. For Brohman, 'development is contextually defined', and the tendency to overlook the local context, he argues, is 'not merely an oversight . . . but should be seen as a paradigmatic blind spot' (1995: 124). This does not merely encompass the particularities of culture and history, but also such factors as the lack and poor quality of much data in the developing world and, often, an absence of sufficiently-trained personnel to implement

*Table 2.2* Criticisms of Eurocentricity

- Denigration of other people and places
- Ideological biases
- Lack of sensitivity to cultural variation
- Setting of ethical norms
- Stereotyping of other people and places
- Tendency towards deterministic formulations
- Tendency towards empiricism in analysis
- Tendency towards male-orientation (sexism)
- Tendency towards reductionism
- Tendency towards the building of grand theories
- Underlying tones of racial superiority
- Unilinearity
- Universalism

development of this variety. So, it is not just that Eurocentric approaches are inappropriate to developing countries, but also that developing countries are not suited to Eurocentric approaches. McGee suggests such a sequence of constraints when he discusses the inapplicability of Eurocentric views on Asian urbanization, and suggests that his own vision of Asian extended metropolitan regions (EMRs – see pages 264–6) is both more appropriate and more Asia-centric (1991b).

Eurocentrism is also sometimes applied to the rarefied air of academe: that Western scholars have intellectually colonized the rest of the world (see McGee 1991b: 337). Researchers have mined Southeast Asia for material and data for papers and books rarely written in local languages and which, if they are available locally (and often they are not), are priced beyond the reach of most local scholars. Local people and places are used as guinea pigs on which Western academics can test their Eurocentric theories, and then write up their results in a form and style which makes them inaccessible to those people the works purport to describe and explain. It is true, admittedly, that there has been a concerted effort from some quarters to change this state of affairs. Many funding bodies insist that local researchers be involved, that results are circulated and workshops convened locally, and that reports and executive summaries be published in local languages. Even so, the great bulk of published material remains in English and other European languages and most journals and books are not easily available to local scholars and development workers.

This debate over Eurocentrism does tend towards depicting Southeast Asia as the hapless victim of all-conquering Western concepts. Yet (and this is a theme which will crop up more than once in the book), in many respects this is far too condescending a viewpoint. Southeast Asians, as they have arguably done throughout history – for example in the adoption and adaptation of Hindu, Muslim, Buddhist and Chinese systems of thought and governance – have selectively used European ideas for their own ends. In doing this, they are not embracing foreign ideas wholesale, but with discretion according to their perceived needs and conditions (see Box 2.1). Perhaps the most obvious example is that while the countries of the region, and especially Indonesia, Malaysia and Singapore, have enthusiastically subscribed to the modernization ethos, they have either studiously avoided, and in some cases actively challenged, the assumption that development should also embody Western notions of democracy and human rights (see the section below, The Asian way).

---

*Box 2.1* Internalizing the external

King Mongkut of Siam (r. 1851–68) was instrumental in introducing Western scientific knowledge to Thailand. He went out of his way to promote modern astrology, and his death from malaria was a direct result of an expedition which he led to Wako in a forested area of southern Thailand in 1868 to witness a full solar eclipse. The King mounted the expedition after he had predicted the event using modern techniques. The date of the eclipse, 18 August, is celebrated as National Science Day and Mongkut is regarded as the Father of Thai Science. What is particularly striking about Mongkut's fascination with Western science is the manner in which Western thought was assimilated into the Thai system of thought and explanation. For Mongkut is not just regarded as the Father of [modern] Thai Science but also as the Father of [traditional] Thai Astrology. Mongkut's science was allied with, and assimilated into, indigenous systems of thought. In this way, Western science did not replace Thai 'folk' science, but reinforced and extended it. Intellectually incompatible to the outsider, to the insider they were functionally compatible. Thus, the modern study of *phumisat* or geography in Thailand built upon the geography of the *Traiphum Phra Ruang*, or *The Three Worlds according to King Ruang*. The terminology of modern geography borrowed from the cosmology of the *Traiphum*. The 'indigenous taxonomy', Thongchai Winichakul argues, 'provided the means by which modern geography could be understood' and 'despite different conceptual systems, the indigenous taxonomy also became the vocabulary of modern geography' (1994: 59). The same is true of Buddhism more generally, which King Mongkut as well as his counterpart in Burma, King Mindon (r. 1852–78), undertook to modernize and rationalize in the light of theories of physics and evolution (Schober 1995).

---

'Eurocentrism' also suffers from its increasing use as a catch-all. Its uncritical and universal application has left it vacuous, shorn of explanatory power. In a sense, by explaining everything, it explains nothing. Fred Halliday writes with reference to the Middle East, but which is relevant here:

> the fact that a particular discovery or idea was produced by a particular interest group, or context-bound individual, tells us nothing about its validity. Medicine, aeronautics or good food may be produced in such contexts of time, place, culture: they are not therefore to be rejected. . . .

The terms 'eurocentric' and 'ethnocentric', at present far too easily bandied about, confuse a statement of historical origin with a covert assessment that needs justification in its own terms.

(Halliday 1996: 212)

Not only does the use of the term 'Eurocentric' in this universalist and all-embracing manner detract from the analysis of the utility of ideas of Western origin, it also tends to discourage and hinder trenchant and critical analysis of Eastern/Southeast Asian ideas. The very reason why concepts with indigenous roots are so often accepted without critical debate is because they are indigenous and therefore escape the label Eurocentric.

### The importance of semantics

It is common to read papers and books arguing that mainstream development theory and practice is inappropriate to the developing world. It is rather less common to find texts which focus on the issues of semantics and language. Are we, in short, talking the same language? Terms likes 'development', 'sustainability', and 'empowerment' are often coined in the West, translated into local languages, and then transposed onto the local developmental landscape (see Table 2.3). Western researchers (probably not conversant in the local language) may then find such terms being translated back to them as if they have one and the same meaning in each language. They can then return to their desks to write reports or papers describing how participatory development is being achieved, or otherwise, without being aware that 'participation' in the local context may mean something rather different from its meaning in the English language lexicon of development.

One set of terms, much used, but which have different nuances of meaning in the region are those for 'environment', 'forest', 'nature' and 'park'. The commonly used Thai word for both 'nature' and 'environment' is *thammachaat*. But this should not be viewed as meaning environment in the Western sense. It encompasses both the human and natural worlds, and has an almost poetic connotation which divorces it from the physicality normally attached to 'environment' in the West. The Burmese word for nature, *thaba-wá* (from the Pali) likewise is applied to both the physical world and to human nature.[2] Stott, when alluding to untamed nature in Thailand, prefers instead to use *pa thuan* – 'forest, wild, uncivilised' (Stott 1991: 144).[3] In Myanmar, the word *tàw* also has such a dual meaning for while it ostensibly means 'forest, jungle or wood' at the same time has connotations of 'wild, rustic and untamed', and even an association with the evil spirits of the jungle (Anna Allott, personal communication). *Thammachaat*, by contrast, might be viewed as nature tamed and manipulated in the interest of humans; 'savages', people beyond the civilized world (in the Thai context sometimes applied to hill tribes), are known as *khon pa* (people of the wilds/forest); they would not be referred to as *khon thammachaat*. A similar set of differences

Table 2.3 The semantics of development

| | Thai | Malaysian/Indonesian | Myanmar |
|---|---|---|---|
| **Development** | *kanpattana*<br>This stresses the economic; social/human development are more exactly referred to by the word *wattanatham*, but this is rarely used in the mainstream development studies context – although it is popularly employed by grassroots activits | *pembangunan*<br>From *bangun* meaning 'to rise up' in a planned and coordinated manner. It has become increasingly associated with the State and particularly with the New Order. There is little sense of autonomous development. | *bpún-bpyò*<br>From two words, *bpún* (be/grow fat) and *bpyò* (plentiful, abundant). It does not mean 'modern', however. The ruling SLORC has co-opted the word and, in the process, made it ingreasingly closely linked with State actions, and associated negative connotations. |
| **Participation/participatory development** | *kaan khao maa mii suan ruam*<br>('to come in and take part'). Hirsch (1989) notes how this Thai term has little of the spontaneity and individuality that is inherent in the English word. Vandergeest (1991) sees development implying obligations, in particular the obligation to provide labour for the 'common good', *suan ruam*. | | *ko-htu-ko-htá*<br>This word is sometimes translated as self-help. It consists of a number of words: 'self-erect, put up+self+stand up'. The State employs the term to indicate that villagers must pay for an amenity (like a school) themselves, and erect it themselves. |
| **Environment** | *thammachaat*<br>This is a rather loose and poetic word, embracing both natural and human phenomena. It applies, though, more to the human-influenced environment than to the 'wilds'. *Pa thuan* would probably be used to denote wild/uncivilized space. | *Hutan* is wild and untamed 'jungle' white *taman* is 'park' tamed and managed in the interests of humans. | *tàw*<br>This means 'forest, jungle or wood' but also, 'wild, rustic and untamed'. |

*Sources*: Hobart 1993, Demaine 1986, Anna Allott and Ulrich Kratz, personal communication

can be identified in the words for National Park in Indonesia and Malaysia, *Taman Nasional* and *Taman Negara* respectively. *Taman* is normally directly translated as meaning park, and yet it is more akin to the park in Regent's Park or Central Park. It is park or garden in an artificial and controlled sense. The word for wild and untamed forest is *hutan*. This is a place to be entered only with care. Though, in the past, it was important as a source of wood and wild products like honey, resins and meat (less so today), it was also a hazardous space where a lack of care could place humans in jeopardy, whether from spirits or wild animals.

Even more striking are the differences in meaning between such English words as 'development' and 'participation' and their equivalents in Southeast Asian languages. The Thai word for development, or at least the one most commonly used, is *kanpattana*. As Demaine points out, *kanpattana* was coined relatively recently and only seems to have been embraced by the Thai state in 1957 when General Sarit Thanarat became prime minister. Sarit created the National Economic Development Board (NEDB), giving it the task of drawing up Thailand's first five-year development plan (1961–66).[4] *Kanpattana* is probably better translated as 'modernization' rather than 'development', or it should at least be made clear that it refers to economic growth. This is evident in its common pairing in Thailand with another term, *khwaam charoen*, which means prosperity in the consumerist sense (Hirsch 1989: 50). Interestingly, before *kanpattana* came into widespread use in the early 1960s, *wattanatham* was the word usually employed to denote 'development'. This word, though, is more in keeping with the multidimensional gloss given to 'development' in most contemporary Western texts as it encompasses all the various elements that might come under the heading 'human advancement'. It is also notable that many NGOs prefer to use *wattanatham* to *kanpattana*, and one influential Thai grassroots movement, the Community Development Perspective, is known as *Wattanatham Chumchon* (see Hewison 1993, Rigg 1991a and 1993).

The official term for development in Indonesia is *pembangunan*. *Bangun* means to develop or to rise, while the affixations *pe-* and *-an* imply that this is a process or activity (Ulrich Kratz, personal communication). However there are also associations of planning, control and direction implicit in the word. Hobart maintains that the term has connotations of guidance and support, the implication being that development cannot be achieved without the helping hand of the State. It emphasizes 'the need for guidance by those with power and knowledge, in this case the government officials who elaborated the notion in the first place' (Hobart 1993: 7). This, though, is not what the word originally meant; rather, it is what the word has come to mean. As in the case of the verb *membangun*, meaning to rise up according to a plan (as in the building of a house), the identity of the architect is not clear. But over time the New Order has, as it were, appropriated *pembangunan* so that the identity of the archiect – in this instance the State – has become intimately associated with the word (the same has occurred in the case of the Burmese word for development, see below).

In selecting *pembangunan*, the Indonesian state has rejected other terms like *perkembangan* which suggests spontaneous growth, and *kemajuan* or 'progress'. *Kemajuan* also implies political and economic advancement as well as modernity, secularization and urban sophistication (Ulrich Kratz, personal communication). It is interesting that while Sukarno, Indonesia's first president (1950–65) is known as the 'Father of the Country' his successor, President Suharto, is usually given the appellation the 'Father of Development'. It has been widely suggested that his legitimacy is based on the country achieving healthy economic growth. This creates an explicit link between state legitimacy and modernization.

There are also some quite interesting semantic implications behind the introduction in 1994 in Indonesia of the three-year Presidential Instruction Programme for Less Developed Villages, or *Inpres Desa Tertinggal* (see Box 4.1 page 116 for more background). Although *desa tertinggal* has been translated into English as meaning 'less developed villages' it more accurately means villages that have been 'left out', 'left behind' or 'forgotten' (Ulrich Kratz, personal communication). Here the fact of their poverty is linked to their lack of development, the implication being that poverty is something that exists when development is lacking and only through their incorporation into the development process can it be tackled. Furthermore, there is the hidden implication that these villages have missed out through accident. This view – that poverty is about an absence of development – links back to the discussion in Chapter 1 about poverty being a creation of the development discourse (see page 33).

In Myanmar, the word for development is *hpún-hpyò* (as in developing nations). This is derived from two words, *hpún* (be/grow fat) and *hpyò* (plentiful, abundant). Anna Allott notes that although it implies prosperity it does not mean modernization, which is more closely associated with an alternative word meaning 'progress' (personal communication). She also suggests that *hpún-hpyò* was well established by the 1960s. More interesting still is the increasingly state-centric association of development in people's minds. In 1993 the State Law and Order Restoration Council (SLORC) established the Union Solidarity and Development Association (USDA) to replace the discredited Socialist Programme Party. In Burmese this was called *Kyán-hkaing-yày Hpún-hpyò-yay Athìn*, using the word for development (*hpún-hpyò*) noted above. As many people resent the heavy-handed activities of the USDA, Anna Allott considers that 'development' may well have begun to acquire not just an intimate link with the State but also (and not coincidentally) unpleasant associations for many.[5]

Moving from broad concepts such as 'development' to more specific terms like 'participation', it becomes clear that the differences in meaning remain significant. Hirsch notes that in official development discourse the Thai term for participation is *kaan khao maa mii suan ruam* – 'to come in and take part'. This 'connotes participation as a willingness of villagers to conform with projects initiated by government development agencies . . . [and] . . . is devoid of the principles of initiative, variety or spontaneity' (Hirsch 1989: 51). Vandergeest (1991) sheds some light on this view of participation when he suggests that development, from very early on, was framed as the government's 'gift' to the people. Importantly though, this gift also made obligations of those who benefited from it: to receive the development 'gift', people would also have to donate labour for the common good (*suan ruam*). Thus 'participation' in development became a variant form of forced labour (see also Tapp 1988). For Myanmar, the language of development is, as it were, rather less developed, but none the less there appear to be important differences which mirror those found in the Thai language. Anna Allott reports that she has seen the word *ko-htu-ko-htá* used to mean 'self-help' in the context of village schools. These schools, though, are built and paid for by villagers after the government has claimed that it does not have sufficient funds. The word itself reveals the extent to which self-help in Myanmar really means 'self-pay' and 'self-do': 'self+erect, put up+self+stand up' (personal communication).

## Local and global: languages of action

But the issue is not just one of semantics – of the meaning of words. It is wider than this and also includes different ways of understanding, of resistance, and of action. The example of the Buddhist monk and environmental activist Phra Prajak Khuttajitto described in Box 2.2 brings into focus the multiple sources and bases of conflict over resources. Structuralists may claim that this local example of resistance against the interests of the State and of big business is both symptomatic and a product of the corrosive effects of dependent modernization. The demand for pulp to feed Thailand's modernization resulted, in this instance, in an unholy alliance of state and business attempting to evict poor people from their land, under the hypocritical guise that it was aimed at preserving the forest. However, from the local level, the resistance mounted was clearly driven by, and intimately linked to, the presence of an informed activist monk, respected by local people, and willing and able to mount a local compaign that had national ramifications. The *Khor Jor Kor* débâcle also illustrated the contrasting ways that groups interpreted and explained the programme and the confrontation that it caused. At the end of 1992 in the wake of demonstrations in the Northeastern city of Korat, the Deputy Governor of the province of Nakhon Ratchasima, Poj Chaimaan, was quoted as saying: 'I don't know if the people are really going to help preserve the forest or destroy it. We don't have any way out and the society at large is confused. But if proper zoning is not done soon, more people will move into the forest, which could mean the end of the forest. And villagers don't understand the importance of the forest like academics do'. By contrast, Tinnakorn Arjharn, the village head of Ban Daan Lakor, one of the communities threatened with eviction, placed his argument clearly in the realms of emotion, not science, when he responded: 'It's time to decide which side is right and what rights people have' (Pravit Rojanaphruk 1992).

Larry Lohmann's work on the languages of resistance and engagement in Thailand offers a far more nuanced appreciation of the diversity and complexity that lie behind what may, at first glance, appear to be fairly simple (though often intractable) conflicts of interest (Lohmann 1995). It was suggested earlier that there are difficulties connected with the differing meanings of words in English and in Southeast Asian languages. But added to this should be the differences that also exist between élites and non-élites in the way that language is used. Lohmann emphasizes that while élites tend towards literacy, non-élites favour orality; while the former tends to be impersonal, the latter is firmly bedded in personal relations; while élites are usually not space-bound and emphasize the universality of ideas, non-élites are locally bedded and stress morality and (local) community relevance. Lohmann suggests that it is when interest groups make strategic alliances, but accept that each will speak in, and on its own terms, that alternative voices are loudest and most powerful.

> Thus, villagers can speak in their 'own' voice at meetings and demon-
> strations . . . while newspapers expose abuses, dissident academics speak
> credibly in scientific or economic language against corporate consultants,
> students take the political offensive, bureaucrats fight turf wars within
> ministries, *phuu yai* [big/powerful people] approach *phuu yai* at the top
> levels and non-governmental organizations arrange forums at which the
> diverse members of alliances learn how to co-ordinate with and use one
> another better, look at themselves from the points of view of the other
> groups present, and maintain mutual respect across systems of thought.
>
> (Lohmann 1995: 226)

None the less, just as some academics have been quick to espouse local ideas, so local people have embraced scientific terminology to promote their causes. Thus, villagers in Thailand now talk of *niweet witthaya* (ecology), *paa thammachaat* (natural forest) and *paa chumchon* (community forest), all terms which have been imported from the élitist lexicon (Lohmann 1995: 222–3).

## RELIGION, CULTURE, ECONOMICS AND DEVELOPMENT

### Buddhist and Islamic economics

The last twenty years have seen a burgeoning of interest in the contribution that religion can make to economics and development. Islamic and Buddhist economics, particularly, are becoming increasingly influential in determining both alternative means of achieving development, and in identifying desireable ends. To generalize about what is becoming an increasingly diverse literature, the aim of religious economics is to bring a consideration of issues of morality and ethics to a subject which, to date, has studiously ignored such abstractions. The difficulty which students of these religious economics face, however, is that they are often forced to use language and concepts rooted in Western economic theory. The Buddha, for example, never taught about the subject of economics;

thus Buddhist economics attempts to extract his teachings and place them within an economic context.

Buddhist economists argue that Western economics is artificial. It constructs rational solutions to the human condition which is, in most part, irrational. In this sense, economics is unreal. Nor does it sufficiently take account, they would argue, of the extent to which all living things are interrelated and interdependent. Thus the Buddhist scholar the Venerable Prayudh Payutto proposes that 'economics is grossly out of touch with the whole stream of causes and conditions that constitute reality', going on to suggest that '[w]hile economists scrutinize one isolated segment of the cause and effect process, the universe manifests itself in an inconceivably vast array of causes and conditions, actions and reactions' (Prayudh Payutto 1994: 18, see also Schumacher 1973: 44–51). To simplify, Buddhist economics attempts to do three things: first to inject moral and ethical considerations into the study of economics; second, to accept that all actions have an outcome, and to strive to minimize the effect of such actions; and third, to bring a holistic vision to economics, reflecting the holistic nature of existence.

Buddhism attempts to integrate a philosophy of life into economics. Only by giving due importance to Right Livelihood, it is argued, can the vacuity of market economics be minimized. Thus, while mainstream economists might argue that the key benefit of development and economic growth is that it increases human choice (see Arndt 1987: 177), a Buddhist would say that if choice means *tanha* (blind craving), then it needs to be controlled. Choice should not be endorsed at all times and may, indeed, be decried. Moderation – the Middle Way – is seen to be the key to Right Livelihood. Contrary to popular opinion, poverty is not commended nor is wealth condemned (Mendis 1994: 200). Rather the onus is on the way in which wealth is acquired and used.

The flowering of Islamic economics, both theoretical and practical, has been initiated for many of the same underlying reasons as has Buddhist economics: a concern for inequality and intractable poverty, a disgust with rampant materialism and consumerism, and a fear that mores are being undermined and ignored. Like Buddhism, Islam enjoins all believers to avoid exploiting others and to embrace moderation in consumption; the Qur'an has numerous injunctions against the concentration of wealth while it advocates social justice: *al-'adl wa-ihsan* – 'the oppressed of the earth [shall] become the inheritors of Allah's bounties' (Engineer 1992, Naqvi *et al.* 1992). More specifically, the Qur'an commands that the poor, handicapped, the unemployed and orphans be aided through the payment of a wealth tax, *zakat*, while glaring inequality is censured.[6] There has been considerable debate over the Qur'anic prohibition regarding the taking or giving of 'interest' (*riba*). Some scholars have interepreted this to mean 'usury' (the practice of doubling or trebling a debt when repayments are missed) rather than interest in the modern commercial sense. However rather more have taken it to imply real, predetermined interest on capital. The alternative to interest generally proposed by Islamic economists is some form of profit-sharing (*mudaraba*), where those extending the loan would also share the risk and would not receive any predetermined return.

One of the most notable recent attempts in Southeast Asia to bring an Islamic tenor to bear on issues of development was the creation in 1990 of the Organization of Indonesian Muslim Intellectuals (ICMI). ICMI's power and influence belies its comparatively small membership of just 40,000. However it would be wrong to see ICMI as subversive. Far from it; the power of the organization stems from the backing and support that it has received from the State and the fact that it has a membership – albeit small – that cuts through many strategic organizations. In addition, though ICMI is expressly 'Muslim', it enthusiastically endorses a modernist Muslim perspective (see McBeth 1996a). It is not a body which rejects notions of modernization, but rather one which is engaged in an attempt to embed modernization within a Muslim milieu. What is interesting in the Indonesian context is that the very success and influence of ICMI has led to the flowering of other groups established partly to challenge ICMI's agenda and perspective: the National Brotherhood Foundation, the Indonesian Muslim Congregation or Masyumi, the Association of Intellectuals for Pancasila Development (a secular group), and Indonesian National Unity, for instance (McBeth 1996a).

Critics of Buddhist and Islamic economics tend to address their concerns in four ways. First, they would argue that much that is presented as 'Islamic' or 'Buddhist' is not distinctive and frequently ambiguous when it comes to implementation. The consideration, for example, given to ethically sound policies, to assuring that the outcome of development does not harm the environment, society and individuals, and that all people should have a minimum quality of life (the Four Requisites in Buddhist teaching: food, clothing, shelter and medicine) is not intrinsically Buddhist or Islamic. The Four Requisites displays more than a passing resemblance to mainstream ideas on basic human needs. Western economic theory and some of mainstream development practice have, these critics would argue, already taken account of and incorporated such issues.[7]

*Homo islamicus, Homo christianicus* and *Homo buddhisticus* are all enjoined, in a generalized sense, to be altruistic, just, moderate in action, and socially responsible. The growth of Buddhist approaches to development has occurred in large part due to the perceived failure of economic development. Thus the highly respected Thai Buddhist commentator Sulak Sivaraksa writes:

> in the name of progress and development the state told [monks] that economic development was good. The monks have realized that this is all a lie. They realize that their lives have become much worse: the environment is spoilt, the animals have gone, there are more roads and dams and more electricity. Yet only the rich reap the benefit. To counter the wrong trends of development, these monks who are working with the people, bring Buddhism back to life for the masses. Meditation has become collective

meditation, meditation with social analysis. They want self-reliance, not growing for sale, and they are reviving their old traditions. Instead of using tractors they work together with buffaloes.

(1990: 263, see also Parnwell 1996: 288–90)

The identification of the problems facing Thailand in the above quote are not distinctively Buddhist in any sense. Nor for that matter are the solutions – although they are articulated within the Buddhist context of moderation and meditation. But take away the Buddhist façade and what is left is a vision of alternative development which finds echoes across the world: an emphasis on local knowledge and wisdom; a desire for self-reliance and a de-linking from the market; a concern for morality; and a wish to recreate the past.

The second area of doubt concerns the extent to which religious economics presents issues like poverty and redistribution as zero-sum-games. Thus the Venerable Prayudh Payutto, perhaps Thailand's foremost scholar-monk, suggests that over-consumption and greed 'apart from doing . . . no good . . . deprives others of food' (Prayudh Payutto 1994: 44–5). Arndt, rather disparagingly, writes that 'it simply does not occur to them that poverty can be alleviated more effectively by development than by redistribution, by increasing the size of the cake than by sharing it out more equally' (1987: 169).

A third objection concerns the motivation behind these initiatives. Commentators have argued that xenophobia, nationalism and anti-imperialism are often more important than religious zeal *per se*. In Burma/Myanmar some anti-colonial politicians espoused the cause of 'Buddhist socialism'. Foremost among them was U Nu, prime minister of Burma between 1947 and 1958 and again between 1960 and 1962. U Nu linked the emergence of private property, class conflict and greed with the end of the age of abundance symbolized by the Buddhist story of the Padeytha or 'wishing' tree from which all desires were granted. Pridi Phanomyong, a civilian member of the 114-strong clique that ended the absolute monarchy in Siam in 1932 and later went on to become prime minister (albeit for only a few months between March and August 1946), also envisaged socialism as a means to return to the 'pure' socialist past which he saw reflected in the teachings of the Buddha (Cohen 1984). In his socialist economic plan of 1933, Pridi extolled the virtues of a socialist world where 'at last [the people] will be able to feast on the fruits of happiness and prosperity in fulfilment of the Buddhist prophecy to be found in the story of Ariya Mettaya' (quoted in Cohen 1984: 201).[8] In releasing his ideological tract *The Burmese Way to Socialism* in 1962, General Ne Win too was seemingly engaged in an attempt to fuse socialism and Buddhism into a unique Burmese amalgam (Yong Mun Cheong 1992: 446–50, Steinberg *et al.* 1985: 400–1). Schumacher, in *Small is beautiful: a study of economics as if people mattered,* uses Burma as an example of a country that has remained faithful to its heritage (1973: 44). However in many respects the state that Ne Win created and the ideals which underpinned it were more secular than those of the previous democratic government. Ne Win was careful not to alienate the largely non-Buddhist minorities by declaring Buddhism the state religion, for example, as U Nu had threatened to do. The state ideology as it emerged under Ne Win was of mixed parentage, and only 'Buddhist' in the loosest of terms. As Taylor writes:

> The ideology of the state as developed by [Burma's] Revolutionary Council and promulgated through the organizations and publications of the state and Party is sufficiently vague and general to appeal to a large proportion of the population without tying the state to specific policies other than general goals such as socialism and affluence. The ideology is grounded in Buddhist epistemology but gains its analytical language from Marxism and, to a certain extent, its political ideas from Leninism. The ideology combines classical Buddhist notions of the origin and purpose of the state with twentieth-century notions of political organization and state practice.
>
> (Taylor 1987: 360)

While Burma's 'way to socialism' masqueraded as being sensitve to the country's religious and cultural inheritance, in reality it ended up being little more than a mix of extreme nationalism and repressive authoritarianism – and has brought Burma/Myanmar great human suffering.

The fourth objection focuses on relevance and applicability in a modern economy. For example, *zakat* was designed to work in the context of a primitive agrarian society. *Zakat* rates apply to animals like goats and camels,[9] but not to wage earners. Thus, in Malaysia, while *zakat* is extracted from rice-producing households, many of whom live close to the poverty line, office workers and property-owners are exempt. The tax in this instance is regressive, and does not achieve the objective of redistribution (Kuran 1992: 23). Likewise, the desire to replace interest-bearing bank accounts with profit-sharing accounts might cause retired people to put their savings at excessive risk. It is because of these tensions between the Islamic ideal and 'the realities of the market place [that Islamic banks are] finding it prudent to abandon profit and loss sharing for interest – concealed, of course, in Islamic garb' (Kuran 1992: 38–9). More generally, scholars like Parnwell wonder whether the idealized visions can be converted into reality. He argues that the mere espousing of Buddhist approaches will fall largely on deaf ears in Thailand. Or, even more worryingly, he fears that they will be 'hypocritically commandeered . . . as a public relations camouflage for [key actors'] very un-Buddhist actions' (1996: 290). In Thailand, a pejorative term has been coined to describe the commercialized Buddhism associated with credit card-carrying and amulet-selling monks: *Buddhapanich* (Rachel Harrison, personal communication).

## Moral and indigenous ecologies

A sense of the environment has a long history in Southeast Asia. In Thailand, for example, 'forest monks and the teachings of the Buddha, traditional human–land systems and modes of thought, and the communal management of village forests and wildlands all pre-date the rise of "modern" environmental management and modern environmentalism' (Rigg 1995a: 8, see also Rigg and Stott forthcoming). Latterday environmentalists in the region have drawn on these indigenous conceptions to provide local meaning and relevance to concerns which may have been initiated in the West.

Buddhism, especially, is felt to have a particular contribution to make to the study of the environment, and the environmental impacts of modernization. Yenchai Laohavanich writes, for example, that 'Buddhism is so close to nature that the religion deserves to be called a "religion of nature"' (1989: 259). Through the integration of Buddhist thought into the study of the environment it is possible to create, its proponents hazard, a 'moral ecology'. Key to this is the notion of the Middle Way which encapsulates both moderation and equilibrium – central tenets of sustainable development. It is also argued that Buddhism allows an ecocentric perspective to be adopted, where human beings are just one element in a highly complex web of interactions.[10] The Buddhist principle of dependent origination highlights the interrelationships that exist between all things. Phra Prayudh Payutto summarizes this principle as follows (1995: 89):

> All things have a relationship dependent on common factors;
> all things exist in an interrelated fashion;
> all things are impermanent, existing only temporarily;
> all things do not exist unto themselves, that is, there is no real self;
> all things do not have a 'first cause' that brought them into existence.

Sponsel and Poranee Natadecha-Sponsel write that '[o]ur basic thesis is that environmental ethics are inherent in Buddhism . . . but that modernization has undermined adherence to Buddhism in nations like Thailand, and precipitated an unprecedented environmental crisis . . . ' (1994: 80 and see Sponsel and Poranee Natadecha-Sponsel 1995). Not only is Buddhism viewed as having an intimate association with nature, but it is also seen to be relevant at a practical level for solving Thailand's environmental problems.[11] This extends from offering a new approach to development, one which favours balance and equilibrium, to providing tools in the fight against the forces of environmental degradation. Across Thailand, for example, trees have been ordained (*buat*) to protect them from the chainsaw by encircling them in saffron cloth, while sacred groves (*phai aphaithaan*) have been created by enclosing them in sacred thread (*saai sin*) (see Taylor 1993: 11; and Box 2.2). But, the ecofriendly claims made of Buddhism have not gone unchallenged, and they are disputed on many of the same grounds that Buddhist and Islamic economics have been contested. In particular, while the concern of individual monks for conservation is not disputed, whether Buddhism has anything truly distinctive and coherent to offer, beyond banalities, is questioned. Thus Pederson, writing of attempts to co-opt religion in the interests of conservation in general, maintains that in trying to project contemporary problems back in time, academics 'substitute a modern crisis-ridden knowledge about nature for whatever knowledge about nature . . . was contained in the scriptures'. In the process 'they make up a thing that never was' (Pederson 1992: 156). It is difficult to argue that Buddhism, in word and deed, has always been environmentally-friendly, but that only now has the globe's environmental crisis brought such facts into the light of day. Historical records show that Buddhism has been used and manipulated by powerful groups for centuries creating a religion that is as much a product of historical perturbations as it is a reflection of the scriptures (see O'Connor 1993 and Taylor 1993). Thus the outpouring of works extolling the environmental virtues of the religion says more, arguably, about pressures and concerns in civil society than it does about Buddhism's environmental credentials. A parallel can be discerned in the management of Thailand's highly successful family planning programme. The programme's director, the charismatic Mechai Viravaidya, used the *sangha* (the monkhood) and the teachings of the Buddha as tools (see Krannich and Krannich 1980).[12] Monks were requested, for example, to bless condoms and rural people enjoined to acknowledge that 'many births cause suffering' – a slogan taken from the Buddhist scriptures. But these actions should not be taken to mean that the Buddha was a population planner ahead of his time.

In interrogating Buddhist ecologies (and Buddhist and Islamic economics) in this manner may, though, be pushing the debate too far into the realms of academic discourse and away from that of popular discourse and action. The environment–development debate in Thailand:

> has become a conflict between interests groups played out in the media and on the Kingdom's political stage. Theravada Buddhism has been co-opted to argue the case for a more environmentally-friendly approach to development. It is attractive because the arguments are apparently sourced locally, not 'imported' from the West. Therefore, this is Thailand's (or Asia's) environmental revolution. Whether or not the claims stand up to academic scrutiny is neither here nor there; if such claims can help in the maintenance of Thailand's environment, then the ends will have been achieved.
>
> (Rigg 1995a: 12)

Notwithstanding the question of whether Buddhism can be convincingly presented as an inherently ecocentric religion, it is certainly true that Buddhist monks find themselves on the front lines of the environmental conflict. In Thailand, forest monasteries, originally established to create a religious haven away from the diversions of the secular world, now protect the few remaining wild areas in the country. As Box 2.2 describes, monks have found themselves actively engaged in guarding the forests against encroachment. Taylor concludes in his case study of Wat Dong Sii Chomphuu ('Pink Forest Monastery') in Sakon Nakhon province: 'With an awareness of the ecosystem and the importance of sustainable cultural practices, forest monks like Ajaan Thui have in recent times become a vital means of protecting the remaining peripheral forests' (Taylor

*Box 2.2* The monk, the poor and the forest: Phra Prajak Khuttajitto and the *Khor Jor Kor* land settlement programme

> The only monk in the history of Thailand
> ever sent to prison
> Imprisoned for trying to save the virgin forest of Buriram.
> A monk who has come to save the forest he lives in,
> Imprisoned by the very officials responsible for saving the forest.
> This is our story, our story,
> The story of Phra Prajak,
> Buriram, Thailand
> Phra Prajak, Buriram

> A popular Thai song, *Luang Por Prajak*, written and performed by
> Yeunyong Ophaakhun (alias At Carabao), translated in Taylor 1993

There have been numerous cases in recent years of Buddhist monks in Thailand confronting established business, military and civil service groups to defend the rights of the poor and the interests of the environment. Perhaps the single most controversial scheme has been the Land Redistribution Programme for the Poor Living in Forest Reserves, known in Thai as *Khor Jor Kor*.[a] The programme was proposed by army chief General Suchinda Kraprayoon in 1990, and initiated through the Internal Security Operations Command (ISOC), an agency that was created to coordinate the fight against the Communist Party of Thailand. The programme's remit was to resettle 250,000 families illegally living on 2.25 million hectares of forest reserve land on 800,000 hectares of degraded forest land. Well over half the abandoned forest reserve land was earmarked to then be turned over to private plantation companies for replanting to fast-growing trees, such as *Eucalyptus camaldulensis*. It quickly became clear that the programme had severe weaknesses: the land allocated for resettlement, for example, was of poorer quality and smaller in area than the forest reserve land from which farmers were to be moved. The land was also often already occupied while some farmers, due to a lack of formal registration papers, would not be allocated land at all. Critics, delving beneath the programme's environmentalist cloak, quickly identified what they took to be its true *raison d'étre*: the *Khor Jor Kor* programme was an attempt by the army to make money by cooperating with private tree plantation companies, while at the same time garnering political kudos by disguising it under the banner of environmentalism (e.g. Handley 1992). As Rigg and Stott have written, '[t]he irony of the army and its business associates protecting the forest by taking it out of the hands of those [the farmers] who had been protecting it in the first place, was palpable' (forthcoming 1997).

  Faced with this programme, Phra Prajak Khuttajitto, a monk living in Dongyai forest in Buriram province in the Northeastern region, near the

continued . . .

border with Cambodia, decided to stay and help the poor farmers who were threatened with eviction by the Royal Forestry Department and a private plantation company:

> In the first instance Prajak had to convince the villagers of the importance of maintaining the ecosystem and protecting the forest for the future well-being of the local community. In addition, although he was constrained to a great extent from direct action because of his monastic disciplinary rules, he outlined participatory strategies for self-help among the local villagers, including the important concept promoted by national non-governmental organizations of sustainable community forestry.
>
> (Taylor 1993: 7)

In this way, Prajak brought to bear in Dongyai a philosophy of Right Livelihood based on Buddhist principles juxtaposed with a strategy of action founded on notions of alternative development. Needless to say, Prajak's role did not endear him to the authorities. He was arrested for encroaching on Forest Reserve land in April 1991, and then again in September of the same year. The army-controlled television stations portrayed him as a renegade monk and local villagers as environmental villains. The largely independent print media, by constrast, depicted Prajak and his followers as heroic figures standing up to the might of the army and big business. Following wider demonstrations in Bangkok and Nakhon Ratchasima (Korat), Prime Minister Anand Panyarachun abandoned the *Khor Jor Kor* programme in July 1992.

Note

[a] For background to the Khor Jor Kor programme and the subsequent debate over its implementation see TDSC 1992; Apichai Puntasen *et al.* 1992; Taylor 1993; Rigg and Stott [forthcoming 1997]; Lohmann 1991 and 1992; Tasker 1994b.

1991: 121). A growing interest in Forest Buddhism has also led to a veritable cascade of religious literature of a conservationist bent (e.g. Seri Phongphit 1988; Shari 1988; Chatsumarn Kabilsingh 1987). It may be true that this is oriented primarily to Thailand's educated élite and growing middle class, but it represents an important trend – an 'indigenization' of environmental thought (Stott 1991: 150–2).

## The 'Asian Way'

While it has tended to be alternative groups who have embraced and developed ideas like Buddhist and Islamic economics, it has been governments in the region who have been at the forefront of espousing what has become popularly known as the Asian Way. In part, interest in a distinctly Asian approach to development has come about because of the apparent ability of the countries of Asia to achieve high rates of economic growth while apparently avoiding many of the perceived social costs of progress such as rising crime and rampant individualism. But, while those in positions of authority have tended to present the Asian Way as an alternative – and better – path to development than that forged by the West, many of its opponents have seen it as nothing less than a justification for repressive government.

In confronting these views, the first difficulty lies in identifying what exactly constitutes the Asian Way. Many politicians in the region would say that Asian values are a key element. In turn, a common shorthand for Asian values has been 'Confucianist'. Confucian values are seen to include respect for elders (filial piety) and the law, hard work, and the recognition that the interests of society may transcend those of the individual. In Singapore, the National Ideology is based on a number of Asian – as opposed to Western – core values and in 1991 the *Straits Times* listed these as (quoted in Tremewan 1994: 146):

1 Nation before community and society above self
2 Family as the basic unit of society
3 Community support and respect for the individual
4 Consensus not conflict
5 Racial and religious harmony

Former Singapore prime minister Lee Kuan Yew observed that '[a] Confucianist view of order between subject and ruler – this helps in the rapid transformation of society . . . in other words you fit yourself into society – the exact opposite of the American rights of the individual' (quoted in Eliot 1995a: 24). In explaining Singapore's economic success, Lee has remarked: 'we were fortunate we had this cultural backdrop, the belief in thrift, hard work, filial piety and loyalty in the extended family, and, most of all, the respect for scholarship and learning' (Zakaria 1994: 114). Kim lists a similar, though broader, range of characteristics that he notes have been highlighted by scholars as 'Confucian' in orientation (1994: 99):

1 This-worldly orientation
2 Importance of self-cultivation
3 Lifestyle of discipline including social disciplines of hard work and frugality
4 Duty consciousness and reciprocity
5 Respect for authority
6 Public accountability of authority
7 Centrality of family in social harmony and stability
8 Primacy of education
9 Political order as a moral community
10 Necessity of state leadership
11 Group orientation and aversion to individuality

Two central problems with the Asian Way thesis are, first, that it ascribes to Asia a common set of values and second, that it assumes that Asian equals Confucian.[13] The diversity of historical and cultural experiences within Southeast Asia would seem to suggest that a single explanatory model would be inadequate. As Ian Buruma, a journalist and long-standing critic of the Singapore government, writes:

> It is in new, insecure, racially mixed states, such as Malaysia and Singapore, that you most often hear officials talk about Asia, or Asian values, or the Asian Way. Indeed, the phrase 'Asian values' only really makes sense in English. In Chinese, Malay, or Hindi, it would sound odd. Chinese think of themselves as Chinese, and Indians as Indians (or Tamils, or Punjabis). Asia, as a cultural concept, is an official invention to bridge vastly different ethnic populations.
>
> (Buruma 1995: 67)[14]

In addition, the Confucian model is drawn from the lessons of East Asia, in particular Japan, Korea and Taiwan. Here, a historical, cultural and philosophical Confucian tradition has, perhaps, been influential. In Southeast Asia, despite the large Chinese population, the Confucian tradition is far less pervasive and although the Chinese minority may enjoy a disproportionately powerful role in the economies of the region, the cultural landscape is far more complex and differentiated (see page 123 for a discussion of the Chinese diaspora in the region).[15] Nor, for that matter, is Confucianism a clearly defined philosophy and practical guide to life. In the process of its dissemination across Asia, and its embracing by various groups, Confucianism has undergone significant change. The Confucianism of Japan is substantively different from that of Korea, and within Japan the 'high' Confucianism of the élite was substantively different from the 'popular' or 'bourgeois' philosophy of the lowlier groups (Kim 1994: 96–8). There are, in other words, many faces to Confucianism. Even in Japan it is hard to argue that Confucianism was a driving force behind modernization. Indeed, modernization undermined orthodox Confucianism during the Meiji period.

It is in the light of these objections that Kim Kyong-Dong writes that while authors have done much to identify 'attitudinal–behavioural tendencies that they claim to be Confucian in origin', it is unclear whether 'these elements can really be located in *the* Confucian portion of the culture [emphasis in original]' (Kim 1994: 88). In addition, there is the secondary question of what to do about all those elements of culture in Southeast Asia which have nothing to do with a Confucian tradition but which must, it is reasonable to assume, also impact on the work ethic. To reject them as unimportant would be dishonest; yet to incorporate them would undermine the Asian values thesis. As Kim explains:

> It is true that East Asians are very much Confucian in many respects; there are very much something else as well. If one attributes everything to the

Confucian origin because Confucianism was once so pervasive in these societies, one does not really explain anything.

(Kim 1994: 103)[16]

Critics would also maintain that not only are the justifications of the proponents of Asian values disingenuous, but that their very success is open to dispute. With reference to Singapore, Tremewan argues that the core values that comprise the National Ideology are merely a legitimation of the PAP's policies, 'carefully chosen to accord with the PAP's past and present political strategies to reproduce the division of labour amenable to its alliance with foreign capital' (1994: 146). Culturalist justifications for authoritarian government also risk becoming tautologous: Asian governments are authoritarian because their core cultural values are authoritarian. At the same time, it is hard to sustain the argument that Thailand, for example, is safer than the West because of its core Asian values: Thailand has a murder rate of 9.7/100,000, higher than the USA and over eight times greater than England's.

Mahbubani, Permanent Secretary of Singapore's Ministry of Foreign Affairs, has emerged as a key defender of the Asian values thesis. He maintains that many Western scholars and commentators are simply incapable of stepping into the Asian mind. 'Their minds', he writes, 'have never been wrapped in colonialism' (1995: 103). He challenges the view that democracy is an unmitigated good and rejects the assumption, so entrenched in Western thought, that Asia, as it develops and modernizes, must become more like the West. There is, he contends, no 'natural progression of history that will lead all societies becoming liberal, democratic, and capitalist' thus implying that the Asian way (or Pacific way) is distinctive and different (1995: 105; see also Svensson 1996). He asks that Westerners who visit Asia show more humility and not arrive merely to learn of Asia, but to learn from Asia. In this way, rather than Asia becoming more like the West, Asia and the West will become more like each other.

In a number of important respects, the arguments of Mahbubani and others dovetail with those of the post-developmentalists. They are both arguing that the Asian experience not be viewed through a Western cultural lens; that Asia be assessed in, and on, its own terms. They also both explicity call for a more culturally-rooted assessment of development and accept that progress need not be unilinear. Only by making the perspective more endogenous, both seem to be saying, albeit with very different political agendas, can the self-serving hegemony of the development discourse be evaded.

Perhaps the most important single theme that underlies the Asian model debate is the role of culture in economic development. In discussing the World Bank's *The East Asian Miracle* (1993a), Lee Kuan Yew stated that its main short-coming was a failure to address the issue of culture.[17] It is the unique cultural context of East Asia which makes, he argues, the transferability of the miracle to other developing countries so difficult (Zakaria 1994: 116–17). In saying this, though, Lee illustrates the shifting cultural sands which underpin the debate.

Three of the so-styled miracle economies are Southeast Asian and patently not culturally Confucian yet the miracle has, apparently, been successfully transferred there. Opponents of the culturalist explanation also find themselves wading through this quagmire of cultural determinism, though apparently against their will. Thus Kiely, while rejecting the Confucianist explanation for East Asia's economic success does argue that 'culture is central to the process of capitalist development'. However in his terms 'culture' is a much broader beast and refers to the 'specific social history of the [East Asian] region, and the social relations that emerged out of this history' (1995: 117).

To end this section, it is worth pointing out the irony of Confucian culture being used as an explanation for Asia's economic success, when Max Weber in the 1960s was using the same cultural construct to account for China's economic backwardness and failure to industrialize (McVey 1992: 9). In other words, it seems that 'culture' is sufficiently malleable and opaque to be used to explain just about anything.

## RADICAL, MAINSTREAM, REACTIONARY: WHAT'S ALTERNATIVE?

### A conceptual vacuum? Scholarly endeavour in Southeast Asia

What is perhaps most striking about Southeast Asian studies is the general lack of a truly indigenous, radical vision of development and society.[18] In 1995, the *Journal of Southeast Asian Studies* published a twenty-fifth anniversary special issue celebrating a quarter-century of regional scholarship (*JSEAS* 1995). Two themes come through time and again in the disparate papers it contains: first, that despite twenty-five years of work, the essence of Southeast Asian studies has remained remarkably unchanging. Modernization continues to be a largely uncontested goal and attempts to 'delineate indigenous ways to nationhood and modernity . . . seldom amounted to more than providing the paradigmatic structure with a slightly different decor' (McVey 1995: 3). A second theme is that there is a persistent sense that the role and task of scholarship is not to challenge the status quo and to be subversive, but to support it: 'to fill in the blanks rather than to test the framework' (McVey 1995: 3).[19] Kanishka Jayasuriya has argued much the same in accusing supposedly non-state actors of being little more than '"cheer leaders" for government policy' (cited in Huxley 1996: 220). It is striking the degree to which scholars in the region, in economics, geography, political science, international relations and anthropology, are primarily engaged in applied work, much of it with a policy-related slant. There is compartively little theoretical scholarship and even less that could be described as ground breaking.

What, then, of the 'indigenous' visions with which this chapter is concerned? First, it is possible to hazard that, and in the main, Western concepts have merely been provided with a Southeast Asian façade. Perhaps the single most innovative and challenging area of debate in the region has been that stimulated by Thai scholars regarding the nature of Thai history and society (see Reynolds 1987; Thongchai Winichakul 1995; Chatthip Nartsupha 1991). Scholars such as Jit Poumisak, Chatthip Nartsupha and Nithi Aeusrivongse, and the other academics who took up their call to arms, not only appeared to redefine Thai history, political economy and society in Thai terms, but in doing so they also challenged existing royalist interpretations. Perhaps the most important of these radical histories is Jit Poumisak's *The real face of Thai feudalism today*, published in 1957 and regarded, 'as a break (*waek naew*) in Thai historical studies, because of the way it departed from conventional historical practice' (Reynolds 1987: 10). Seemingly, their ideas were at once innovative, indigenous and subversive – and for many royalists, also seditious.[20] However not only is their revisionist work exceptional in regional terms, but questions have been raised about how far they have really rejected Marxist and Weberian concepts and created an indigenous vision of Thai history and society (see Thongchai Winichakul 1995: 104/footnote 19 and 106). Hong Lysa touches on a similar series of debates in her assessment of *Warasan Setthasat Kanmu'ang* (*Journal of Political Economy*), a Thai language journal which concentrated on Marxist and leftist work and explicity tried to promote local views suited to Thailand's unique socio-economic and political configurations (Hong Lysa 1991). Much of the work published in the journal involved a reworking of radical, largely Western, viewpoints in the light of the Thai reality. But these radical counterpoints were, from the start, accepted as largely unproblematic. As a result, Hong suggests, there is little sense of intellectual development over time – merely a continual re-elaboration of what had gone before (1991: 102–3).[21]

It seems that scholars of the region find themselves in a rather uncomfortable dilemma. On the one hand they would wish to maintain that truly indigenous visions and scholarship exist not just as fragments, but as a coherent whole, with both direction and unity. If the impetus for rethinking Southeast Asian studies came from beyond the region, then, as McVey writes 'we might rightly question whether the search for new visions is not at best the aping of international academic fashion, and at worst an example of foreign intellectual domination aimed at diverting Southeast Asian scholars from completing their nation-building and modernizing task' (McVey 1995: 6). Yet, taking a dispassionate overview of scholarly work, it is hard to see more than nuggets of riches. Indeed, just two pages on from the previous quote, McVey writes that 'the study of Southeast Asia was conceived in a powerful but narrow ideological framework . . . of foreign origin [which though internalized] . . . is still largely unquestioned and holds much scholarship in its service' (1995: 8). When Michael Aung-Thwin writes of the Classical period in Southeast Asian history, for example, he asks whether even the work of indigenous scholars trained in the West can escape the label of 'Orientalist'. Is his history of Pagan, he muses, 'not a representation of Pagan itself, but a paper model that reflects the thinking of Karl Polanyi, Edmund Leach, Clifford Geertz, Melford Spiro, and Marc Bloch . . . ?' (Aung-Thwin 1995: 85). The attempt to 'authenticate' Southeast

Asian studies is, as Reynolds suggests, 'very much a Western, post-colonial project' (1995: 437). Authenticity, localization, indigenization, vernacularization and so on are not part of Southeast Asian scholars' repertoires. The rewriting of Southeast Asian history is primarily being pursued by Western scholars.

Expecting a truly indigenous intellectual tradition to evolve is, perhaps, asking too much. Alatas, for example, deplores 'nativism' – the rejection, wholesale, of Western scholarship – and instead makes a plea for 'indigenisation'. 'Indigenisation means filling these voids [that Western culture has left] by looking at the various non-Western philosophies, cultures and historical experiences as sources of inspiration, insights, concepts and theories for the social sciences' (Alatas 1995: 135). This is really an argument for greater pluralism and interdependence – rather than the current situation of overdependence of Southern academics on Northern scholarship. However this pluralistic vision seems unlikely to emerge in such an unequal scholarly setting – or at least not for some decades.

## Where is, and what is, 'alternative'?

Taking a rather Eurocentric vision of what is 'alternative', and focusing on the work of establishment scholars and researchers, it seems that Southeast Asia – unlike Latin America – is notably lacking in truly indigenous, radical visions of development and modernization. However, there are three ways in which this conclusion can be challenged.

First, that in concentrating on the work of academics in mainstream journals we are missing the most vital, imaginative, challenging and subversive work. It is among NGOs, within religious organizations, and by people outside the mainstream of scholarly endeavour that the regional frontier in development studies is being advanced. Although many such thinkers maintain university links, it is their separation from establishment which is critical. It is notable, though not suprising given their perceived destructive agenda, that even those academics like Chatthip and Nithi mentioned above were, for a while, cold-shouldered by the scholarly establishment. In writing of environmentalism in Thailand, Rigg and Stott argue that the search for an intrinsically Thai philosophical and ethical basis has focused on animism, Buddhism, 'myth-making' about the Thai past, traditional human–land relationships, and local, often village-level, environmental issues. The characters who make up this disparate alternative movement 'range from key figures in the former hunting community to students, short-story writers and novelists, intellectuals, Buddhist thinkers, newspapers like *The Nation*, members of the wider royal family, local villagers, and foreigners, living and working in Thailand, or visiting and acting from outside the country' (Rigg and Stott, forthcoming 1997). Overt reference to academics is notably absent from the list.

A second possibility is that in looking for the radical and indigenous, scholars are searching for the wrong thing. For what is truly distinctive about scholarship in the region is its very tendency to support the establishment and the status quo. This links to the discussion above about the Asian Way. For while universities in the West have tended to nurture – or at least to accept – radical thought, and many would see the role of scholars as embodying intellectual confrontation and subversion, in Southeast Asia this is not the case. Southeast Asia's institutes of higher education and research, and their staffs, are there to support the modernization efforts of the State. In the most part this is seen in very practical terms: to undertake surveys, to process data, and to interpret results. Not, in other words, fundamentally to challenge the basis of modernization.[22] When Western scholars call for an 'indigenization' of development theory in Southeast Asia, this is often seen in terms of producing comparative work that builds upon existing – and often radical – conceptual foundations (this appears to be the argument in Schmidt forthcoming). Why such studies should be any less derivative than those that take an orthodox modernist perspective is not clear.

Lastly, and to return to a point made towards the beginning of the chapter, we should not belittle the extent to which Western concepts have been manipulated and adapted into new forms in the region. Reading, especially, the historical, anthropological and archaeological literature, it is striking how far the 'genius' of the region has been in making initially foreign ideas locally relevant. Religions, for example, are notably syncretic: imported cults like that of the holy mountain, Mount Meru, were merged with local cults of mountain worship and moulded to tradition, animist spirits (*nat* in Myanmar, *phi* in Thailand) remain central to popular religion as do amulets, talismans, soothsayers and healers. At one level it is possible to say that Buddhism, Hinduism, Brahmanism, Islam and the various Chinese 'religions' are extrinsic; at another, that they have been made intrinsic and therefore indigenous. In the nineteenth century, Buddhism in both Burma (Myanmar) and Siam (Thailand) was modernized and rationalized by kings Mindon (r. 1852–78) and Mongkut (r. 1851–68) respectively as they sought to come to terms with their countries' encounter with the European 'other'. The *sangha* (monkhood) was centralized, court control over monastic life was extended, and traditional Buddhist cosmologies were rationalized in the light of Western science (see Box 2.1 and Schober 1995). This same process of adaptive internalization is true of modernization, an 'idea' which has its origins in the West, but which has been variously adopted and adapted by non-Western societies like those of Southeast Asia as they have had to respond to the modernization challenge (see Kim 1994: 89). It is in the synthesis of the imported with the local, the overlaying of the indigenous with the exogenous, that creativity – and originality – has occurred. This conclusion leads one to pronounce that Southeast Asia – and the same could be argued to be true of any region, country or people – can *only* be understood in the light thrown by the analysis of *other* cultures. In other words, although Southeast Asia has unique qualities, these need to be viewed in comparative, not in isolated, perspective.[23]

## Notes

1 Most authors use the term Eurocentric to mean West- or North-centred, including North America as well as Europe.

2 The Burmese word for environment is *pat-wùn-kyin* which can be translated as 'surrounding conditions' (Anna Allott, personal communication).

3 *Pa thuan* is usually linked, in opposition, with *müang*. *Müang* can be loosely translated as 'town' or 'state', but in essence it means civilized space. *Pa thuan*, meanwhile is uncivilized space that lies beyond the bounds of (civilized) human control.

4 Significantly, the NEDB only became the National Economic *and Social* Development Board (NESDB) rather later.

5 As in: 'Oh God, we've got to go on another "development" parade' (Anna Allott, personal communication).

6 In 1996 President Suharto of Indonesia sent a booklet on the Prosperous Self-reliant Fund to 11,000 wealthy individuals and companies (defined as those with an after-tax income of more than 100 million rupiahs [US$2 million]) asking these 'affluent taxpayers' to contribute – voluntarily – to the poverty alleviation scheme, and suggesting an amount of 2 per cent of after-tax income. In the booklet, the President explained: 'you should consider yourselves fortunate to have been able to seize the opportunities presented during the over-all development acitivities in this Land of Pancasila . . . [now] you have the opportunity to carry out the noble task of poverty alleviation together with the government' (Cohen 1996).

7 Of course there are many critics of Western approaches who would say that just lip service is paid to such concerns.

8 In Indonesia, Sukarno advocated an alliance between Islam and Marxism in the war against capitalism, explaining that 'profit is the same as surplus value which is the very breath of capitalism' (quoted in Jomo 1992: 133).

9 A person who owns up to twenty-four camels, for example, pays *zakat* of one goat for every five camels. Those fortunate to herd elephants or mules, however, are not expected to pay *zakat* on these animals because they do not breed (Kuran 1992: 20).

10 This links back to the earlier discussion of Buddhist economics.

11 Thailand has a long history of forest monks; the Buddha himself gained enlightenment under a tree – the *bodhi* tree (*Ficus religiosa*); and a large number of Buddhist texts concern themselves with nature in the broadest sense (see Taylor 1993: 11).

12 Mechai was later appointed as head of Thailand's AIDS programme, another job to which he brought a unique combination of plain speaking, clever public relations, and tight administration.

13 This characterization of the Asian model is not shared by all its proponents. Lee Kuan Yew, for example, links it specifically with East Asian (i.e. Confucian) societies, and would wish to exclude Southeast Asian societies (with Singapore, presumably, representing one of the former) from the formulation. He also expresses doubts whether there is an Asian model as such. None the less, in the same interview he later included Thailand, along with Korea, Hong Kong and Singapore in responding to a question on the Asian model (Zakaria 1994: 113 and 118). There therefore seems to be some confusion over which countries are 'in' and which are 'out' and both proponents and opponents resort to a degree of appropriate selection depending on the point that is being argued.

14 Buruma quotes a bookshop owner in Singapore expressing the view that 'the only people talking about Asian values are PAP [People's Action Party, the ruling party] politicians', adding that 'After breaking down communities, languages, and cultures, they now want to recreate Asian culture artificially' (1995: 68). David Marshall, former chief minister, put it more succcinctly to Buruma, calling Asian values: 'phony baloney' (Buruma 1995: 70).

15 Anne Booth points out that as the countries progress and educational opportunities become more widespread, it would be reasonable to assume that Chinese economic dominance will be reduced (Booth A. 1995a: 33–4).

16 This argument is very similar to Pederson's criticism of Buddhist economics (see pages 56–7).

17 Many academic critics of *The East Asian Miracle* argue just the reverse (see Chapter 1).

18 By Southeast Asian studies it is meant the work of resident scholars of the region. Henderson argues much the same with regard to East Asian scholars, when he laments the dearth of indigenous scholars who are 'theoretically engaged with the analysis of development issues' (Henderson 1993: 202). More broadly, Alatas writes that 'we have yet to see [in the human sciences] the emergence of indigenous alternative theoretical traditions outside of the West' (Alatas 1995: 124). He asserts that intellectual progress has been been characterized by 'under- and unachievement' and a distinct absence of creativity and originality. Fred Halliday sympathetically refers to Said's critique of intellectual life in the Middle East and his observation that the rulers of Middle Eastern countries, while they have built numerous international airports, have not funded a single decent library (1996: 213–4).

19 This resonates with the notion of the individual's place within society esposed by supporters of the Asian Way (see above).

20 Jit was shot and killed in the Northeastern province of Sakhon Nakon in May 1966, around seven months after he had fled into the jungle to join the Communist Party of Thailand and take up the struggle against Prime Minister Sarit Thanarat. It seems that he was shot by a village headman, although the circumstances of his death are disputed. Some reports state that one of his hands was cut off and sent to Bangkok so that his finger prints could be identified (Reynolds 1987).

21 Hong Lysa maintains that this acceptance of orthodox Marxist theory as unproblematic is also a characteristic of the Chatthip School (1991: 103).

22 It is notable that modernization has never been rejected as a worthwhile objective by many Southeast Asian states. While scholars in the West from the mid-1960s were challenging modernization theory, and by association the modernization ethic, governments and many scholars in the growth economies of Southeast Asia were continuing to build a vision of development that, with one or two local overlays, was modernist in tone and approach.

23 This mirrors Fred Halliday's personal viewpoint on the Middle East outlined in *Islam and the myth of confrontation* (1996).

# Part II

# MARGINAL PEOPLES AND MARGINAL LIVES
## The 'excluded'

The first two chapters of the book allude to what has become, in all its profusion of guises, probably the central concern in Southeast Asia's economic development, indeed in economic development in general. This is the relatively simple question: who has missed out? Scholars have concerned themselves with the widening gap between rich and poor, rural and urban, and core and periphery. Politicians have become vexed that such inequalities may lead to instability and have increasingly voiced their fears that modernization is creating social tensions with serious political ramifications. And many ordinary people have become convinced that development creates 'losers' as well as 'winners' and that social justice is the victim of rapid growth.

The two chapters in this section of the book aim to dismantle the question 'who has missed out?' and to show the numerous ways in which we can define, understand, and view the 'excluded'. The so-styled 'view from above' takes the perspective that has traditionally been reflected in government reports and academic texts: the challenge of accurately gauging income-poverty, the distribution of the poor in society and across space, and the reasons for the failure (and success) of poverty alleviation programmes, for instance. The following chapter, 'The experience of exclusion', challenges some of the assumptions on which 'the view from above' is based. Is income-poverty, for example, an adequate measure of exclusion and how do different groups in society experience exclusion? As the case studies in the chapter show, groups and individuals in society are often rendered 'excluded' by their very inclusion in the modernization process. Nor is exclusion–inclusion a simple binary relationship. Prostitutes often find that by climbing aboard the modernization bandwagon and becoming commodities – selling themselves in the name of development – they become social outcasts, 'deviants' that mainstream society may on occasion use but, at the same time, would wish to repudiate and reject. It becomes clear as one scratches the experiential surface that conceptual structures reflect reality only dimly.

# 3

# THE GEOGRAPHY OF EXCLUSION
## The view from above

## Introduction

A hallmark of many of the critiques of Southeast Asia's economic development is the extent to which the benefits of growth have been unequally distributed. This is a theme that is so worn that it is almost venerable. Dudley Seers, in his famous address to the Eleventh World Congress of the Society for International Development in New Delhi in November 1969 admitted that 'it was very slip-shod of us [development economists] to confuse development with economic development and economic development with economic growth'. 'The questions to be asked about a country's development', he continued, are 'what has been happening to poverty? What has been happening to unemployment? What has been happening to inequality?' (quoted in Arndt 1987: 91).

Despite the venerability of the 'uneven' growth thesis, it still represents the lodestar of many critical studies. This was made clear in the opening two chapters of this book and is also reflected in the titles of many recent publications.[1] There seem to be two reasons why 'uneven' growth remains a central topic of debate. First, because it continues to be such a recalcitrant presence – the *alter ego*, so to speak, of development and growth. Respected Thai economist Medhi Krongkaew, for example, talks of poverty and income inequality as the 'flip side' of economic development, implying that there is some unholy union between the two (Medhi Krongkaew 1995: 63). And second, because 'uneveness' can be conceived of in so many overlapping ways, from the culture and politics of social exclusion through to the economics of lagging regions. Scholars from all the social science disciplines can find their own haven of academic endeavour in this rich seam of enquiry. This chapter adopts the more traditional, economics-centred perspective on uneven growth – which is termed here 'the view from above'. The next chapter will address the more human-centred and local-level perspectives.

# WHO ARE THE EXCLUDED? STATISTICAL VIEWS

## The view from above

The incidence of poverty in Southeast Asia, though still significant, has declined markedly in recent years in all the original members of Asean, with the exception of the Association's economic 'laggard', the Philippines (Figure 3.1). The incidence of poverty also remains notably high in Vietnam (which joined Asean in 1995), Laos, Cambodia and Myanmar. However data – and in particular longitudinal data – for these four countries are notably lacking.

The figures presented in Figure 3.1 and Table 3.1 highlight two important issues in the study of Southeast Asian poverty. First, the degree to which the growth economies of Southeast Asia have each achieved significant reductions in the incidence of poverty during a period of unprecedented market-led growth. If the data were available, they might also show the discrepancy between levels of poverty in the growth economies of Southeast Asia (Indonesia, Malaysia, Singapore and Thailand) and in the former command economies of Indochina (Cambodia, Laos and Vietnam) and Myanmar. As Warr writes with reference to Thailand and its neighbours, 'Compared with the Thais, the vast majority of their Burmese, Lao and Kampuchean neighbours live in poverty. Clothing is shabby, food is poor, transport is inadequate and housing rudimentary at best. Life is a constant struggle for survival' (Warr 1993: 46). And second, the data underline the problems of poverty that continue to remain acute, even in the growth economies: the geographical concentration of poverty in particular 'lagging' regions and, more generally, in rural as opposed to urban areas.

The figures in Table 3.1 are superficially reassuring in their apparent precision. However there are numerous and well-known difficulties associated with their calculation and application. What is perhaps surprising is that even given all the caveats, cautionary footnotes and exhortations for care, poverty statistics continue to be widely used to identify problems, gauge success, and to design development programmes. Their defenders emphasize the extent to which there is no alternative to using official poverty statistics, particularly when trends over time and space need to be ascertained. With no better tool in their kit bag with which to assess progress, development economists are forced, by necessity, to use a method with accepted and undoubted shortcomings.

The problems most often highlighted apply to the difficulties of arriving at – conceptually – an appropriate measure of poverty; the difficulty, then, of gathering the necessary data to calculate poverty on the basis of the identified measure; and, finally, the difficulty of applying that measure between regions. At each stage in the process, numerous, often arbitrary, assumptions have to be made in the interests of statistical symmetry. The degree to which different assumptions can lead to markedly different measures of poverty is clear in the case of Indonesia. In 1980, the World Bank estimated the incidence of rural poverty in Indonesia to be 44.6 per cent; the *Biro Pusat Statistik* (the Indonesian

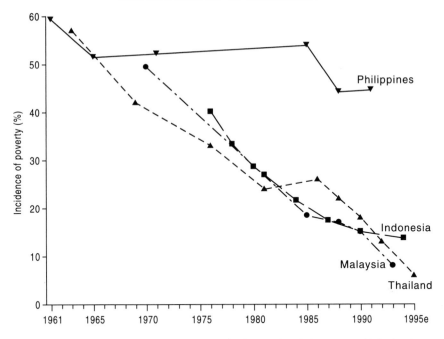

*Figure 3.1* Incidence of poverty in Indonesia, Malaysia, the Philippines and Thailand

*Sources:* Thailand: figures calculated by the National Economic and Social Development Board, quoted in Warr 1993: 46; Amara Ponsapich *et al.* 1993; Medhi Krongkaew 1995. (The 1995 figures are estimates calculated on the basis of the 1988–92 trend. The National Statistical Office only began its two-yearly Socio-economic Survey in 1975/76; the figures quoted here for 1963 and 1969 are based on alternative data.) Indonesia: Biro Pusat Statistik (BPS) and World Bank figures quoted in Booth A. 1993: 65–6; updated from Firdausy 1994; latest BPS figure from Cohen 1995. Malaysia: Rigg 1991b; Booth A. 1992a: 640. Philippines (figures based on the Family Income and Expenditure Surveys): Balisacan 1994: 122.

*Note:* the Malaysia figures apply to West Malaysia only – East Malaysia has significantly higher levels of poverty (see the discussion later in this chapter.)

Central Bureau of Statistics, BPS) arrived at a figure of 28.4 per cent; while Esmara calculated that 43.2 per cent of Indonesians living in rural areas could be classified as poor. The figures for the incidence of urban poverty diverged to an even greater extent: 19.7 per cent, 29.0 per cent and 37.3 per cent respectively (see Box 3.1, 'Identifying the poor in Indonesia'). All these three measures show poverty in Indonesia declining over time. If it were a case merely of different poverty lines resulting in different measures of the incidence of poverty, with the temporal pattern of decline and the spatial distribution agreed upon, then these differences might be accepted without too much difficulty. But not only is the incidence of poverty markedly different for each; so too is the distribution of the poor between regions and between rural and urban areas (see Table 3.1). This means that whichever measure is accepted as the more accurate, is likely to lead to rather different poverty eradication initiatives. The World Bank figures would

*Table 3.1* The poor in Thailand, Indonesia, Malaysia and the Philippines

## THAILAND
### Percentage of the population defined as poor

|      | Total | North | Northeast | South | Central | Bangkok |
|------|-------|-------|-----------|-------|---------|---------|
| 1963 | 57 | 65 | 74 | 44 | 40 | 28 |
| 1969 | 42 | 38 | 68 | 40 | 18 | 11 |
| 1976 | 33 | 35 | 46 | 33 | 16 | 12 |
| 1981 | 24 | 23 | 36 | 21 | 16 | 4 |
| 1986 | 26 | 22 | 41 | 23 | 17 | 5 |
| 1988 | 22 | 21 | 35 | 22 | 16 | 3 |
| 1990 | 18 | 17 | 28 | 18 | 13 | 2 |
| 1992 | 13 | 14 | 22 | 12 | 6 | 1 |
| 1995e | 6 | 8 | 13 | 5 | 0 | 0 |

*Sources*: National Economic and Social Development Board (NESDB) figures quoted in Warr 1993: 46, Amara Ponsapich *et al.* 1993: 9, and Medhi Krongkaew 1995.
*Notes*
1 The 1995 figures are estimates calculated on the basis of the 1988–92 trend.
2 The National Statistical Office (NSO) only began its two-yearly Socio-economic Survey (SES) in 1975/76; the figures quoted here for 1963 and 1969 are based on alternative data.

## INDONESIA
### Percentage of the population defined as poor

|      | | BPS | | | World Bank | |
|------|-------|-------|-------|-------|-------|-------|
|      | Total | Urban | Rural | | Urban | Rural |
| 1970 | – | – | – | | 50.7 | 58.5 |
| 1976 | 40.1 | 38.8 | 40.4 | | 31.5 | 54.5 |
| 1978 | 33.3 | 30.8 | 33.9 | | 25.7 | 54.0 |
| 1980 | 28.6 | 29.0 | 28.4 | | 19.7 | 44.6 |
| 1981 | 26.9 | 28.1 | 26.5 | | – | – |
| 1984 | 21.6 | 23.1 | 21.2 | | 14.0 | 32.6 |
| 1987 | 17.4 | 20.1 | 16.4 | | 8.3 | 18.5 |
| 1990 | 15.1 | 16.8 | 14.3 | | – | – |
| 1994 | 13.7 | – | – | | – | – |

*Sources*: Biro Pusat Statistik (BPS) and World Bank figures quoted in A. Booth 1993: 65–6; updated from Firdausy 1994. Latest BPS figure from Cohen 1995.

## MALAYSIA
### Percentage of the population defined as poor

|      | Peninsular Malaysia | East Malaysia Sabah | Sarawak |
|------|---------------------|---------------------|---------|
| 1970 | 49.4 | – | – |
| 1976 | – | – | 51.7 |
| 1985 | 18.4 | – | 40.0 |
| 1987 | – | 35.3 | 24.7 |
| 1988 | 17.0 | – | – |
| 1990 | 15.0 | 34.0 | 21.0 |
| 1993 | 8.0 | – | – |

*Sources*: Rigg 1991b: 117, Booth A. 1992a: 640, King and Jawan 1996, Brookfield *et al.* 1995: 221

*Table 3.1* continued

**PHILIPPINES**
Family Income and Expenditure Surveys: percentage in poverty (head count)

|  | *Total* | *Rural* | *Urban* |
|------|------|------|------|
| 1961 | 59.3 | 64.1 | 50.5 |
| 1965 | 51.5 | 55.2 | 43.2 |
| 1971 | 52.2 | 57.3 | 40.6 |
| 1985 | 53.9 | 59.4 | 45.2 |
| 1988 | 44.2 | 50.2 | 34.5 |
| 1991 | 44.6 | 52.4 | 36.7 |

*Source:* Balisacan 1994: 122

Family Income and Expenditure Surveys: percentage in poverty (head count)

|  | *Total* | *Rural* | *Urban* |
|------|------|------|------|
| 1961 | 75.0 | 80.2 | 65.0 |
| 1965 | 67.1 | 71.1 | 57.4 |
| 1971 | 61.1 | 66.1 | 51.3 |
| 1985 | 59.7 | 63.3 | 52.0 |
| 1988 | 59.7 | 54.1 | 40.0 |

*Source:* Balisacan 1992: 135

seem to indicate that poverty is largely a rural phenomenon, requiring that development efforts focus on rural areas. By contrast, the BPS figures indicate that it is as much a feature of urban as rural areas, and its amelioration therefore requires a more balanced approach. There are also marked differences in the estimates of the balance of poverty between the different regions of Indonesia. According to the BPS figures, Java, with its relatively high urbanization rate, still supports the great rump (some 66 per cent) of poor people; the World Bank figures, by contrast, indicate that poverty in the Outer Islands is the more intractable problem. These two discrepancies are a function, largely, of the differentials utilized by each to calculate the incidence of poverty in Java and the Outer Islands, and in rural and urban areas.

One example of an attempt to compare different definitions of the poor has been undertaken by Paul Glewwe and Jacques van der Gaag, both economists at the World Bank. They were interested in investigating whether different definitions would identify the same poor, and used the Republic of the Côte d'Ivoire as a case study. The Côte d'Ivoire Living Standards Survey (CILSS) provided a bench mark and the authors selected ten other definitions of 'poor', including, for example, per capita income, per capita consumption, education levels of adult household members, and per capita agricultural land. Perhaps unsurprisingly, the study came to the conclusion that the 'message from this exercise is clear: different definitions of poverty select different population groups as poor' (Glewwe and van der Gaag 1990: 803). The study emphasizes

*Plate 3.1a* Rural transformation: a typical village house in Northeastern Thailand, 1981

*Plate 3.1b* Rural transformation: houses in the same village in Northeastern Thailand, 1994

*Box 3.1* Identifying the poor in Indonesia

Most poverty lines for Indonesia rely on data derived from the Household Expenditure Survey (HES or the *Susenas* survey) undertaken by the *Biro Pusat Statistik* (BPS – Indonesia's Central Bureau of Statistics). This is currently undertaken at three-yearly intervals, and includes questions on expenditure but not, at least until recently, on household income. Various criticisms have been levelled at the accuracy of the HES data (it is based on a sample of just 50,000 households), but the fact that the data have been collated in a systematic fashion across the archipelago from late 1963 is seen to more than make up for any deficiencies in the survey.[a]

In 1975 Professor Sajogyo argued that poor families in rural Indonesia should be identified as those whose annual expenditure was below that sufficient to buy 240 kg per capita of milled rice equivalent (MRE). Later this was refined to apply to the very poor, with the poor being defined as those whose annual expenditure was between 240 and 320 kg of MRE. The focus on rice reflects the centrality of rice in subsistence at that time – to have 'sufficiency' (*cukupan*) in terms of rice was to have sufficiency more generally. Since 1975, subsistence has broadened so that rice no longer assumes the central role it once did, and indeed is becoming increasingly peripheral. As a consequence, some scholars have argued that the Sajogyo's rice-based poverty line has outlived its usefulness.

The BPS has also produced its own figures for poverty since 1976. These are based on the cost of 2,100 calories for rural and urban areas, which is then increased to take account of other, non-food, basic needs like housing, fuel and health provision. Critics of the BPS poverty line have questioned whether using the 'cost' of calories is sensible given the varying costs of calories in different food sources,[b] and have also raised doubts over the low costs allocated to non-food basic needs. In 1986, Professor Hendra Esmara proposed an alternative poverty line based, like the BPS line, on a bundle of basic needs. This bundle also included food, clothing, housing, education and health but in addition incorporated the assumption that these basic needs would change over time as well as being different between rural and urban areas. Perhaps the principal objection to the Esmara line concerns the degree to which his dynamic view of basic needs would disguise real reductions in poverty. If a large proportion of the poor became significantly better off, and this resulted in greater expenditure as recorded by the HES, then the dynamic poverty line would simply rise to 'pull' those families back into poverty. A second difficulty concerns the assumption that just two defined bundles of basic needs (a rural and an urban bundle) are adequate to account for differences between regions, over time, and between differing qualities of a particular good (fish) or service (education).

The World Bank has also entered the Indonesian poverty fray. One World Bank study undertaken in 1984 simply took the poor to be those households

continued . . .

in the bottom 20 per cent in terms of expenditure as recorded in the HES. In this calculation poverty is interpreted as a function of inequality. Not only is it highly arbitrary, but it also suffers, like the Esmara line, from a tendency to disguise real improvements in living among the poor. The poor, to coin a phrase, are always with us in this measure, no matter what happens to real standards of living. Recognizing the deficiencies in this approach, the World Bank subsequently undertook another excercise based on 'basic food expenditure' multiplied by 1.25 to take account of other food expenditure. This figure was then divided by the share of food expenditure in total expenditure. The main criticism here is that the resultant poverty lines tend to favour those areas where food represents a smaller proportion of total expenditure – i.e. wealthier provinces. In addition, the use of variations in rice prices between regions to calculate wider price differentials is dubious because rice prices are controlled and exhibit little interprovincial variation.

The very large population of 'near' poor in Indonesia also means that small changes in the poverty line can have a very significant impact on the calculated numbers of poor. For example, raising the BPS poverty line for 1987 by just 10 per cent (and bearing in mind many commentators' view that the line is set too low) increases the number of poor by a third from 30.0 million to 41.0 million.

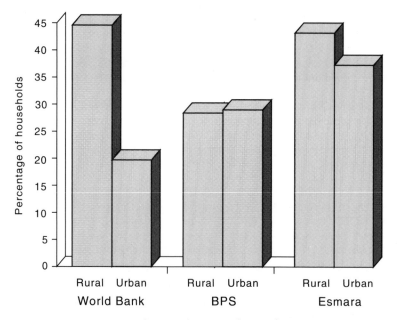

*Figure B3.1* Comparison of poverty lines in Indonesia for 1980

continued . . .

*Sources*: the above account is condensed from Booth A. 1993, also using Booth A. 1992a, 1992b and Tjondronegoro *et al.* 1992. For a local study of poverty in Bali see Firdausy and Tisdell 1992.

*Notes*

[a] Tjondronegoro *et al.* maintain that the Susenas survey tends to underestimate the wealth of the rich and also underestimate the degree of poverty (1992: 73).

[b] In rural areas, it is common for the poor to consume maize and cassava rather than rice because, calorie for calorie, maize and cassava are much the cheaper food sources.

that definitions should be selected according to the task in hand and the policies that are under consideration. To assume that one definition of poverty will satisfactorily fulfil all eventualities is unlikely.

Even more problematic than the arrival at a satisfactory poverty line for a single country, is that of comparison between countries. The incidence of poverty in the Philippines is usually quoted as 40 per cent; that in Indonesia at less than half this rate (Table 3.1). Yet, as Hal Hill observes, it would be unwarranted to take this to mean that there are (relatively) more than twice as many poor people in the Philippines as in Indonesia (1993: 65). Yet despite the apparently obvious shortcomings of such cross-country comparisons, there are instances of commentators being seduced by the apparent comparability of the data. In 1993, for example, a World Bank report unfavourably compared poverty levels in Thailand with those in other Asian countries to make the point that the Kingdom had achieved less in terms of poverty reduction: 'in 1990 Thailand's incidence [of poverty at 16 per cent] was as high as Indonesia's [of 15 per cent], even though its average GNP per capita was 2.5 times higher' (quoted in *Economist* 1993a: 81).[2] Booth, for one, has criticized the poverty lines used by the World Bank for Indonesia, claiming that they are significantly too low, and should be increased by at least 25 per cent (also see Box 3.1) (Booth A. 1992a: 637). This would clearly alter substantially the figures used by the World Bank to argue the case that Thailand has done markedly worse than Indonesia in tackling poverty given the wealth that each economy generates. Indeed, in a later paper, Booth writes that the basis of the World Bank's comparison between Thailand and Indonesia is 'highly questionable' and 'extremely misleading' in that it 'induced the World Bank to form quite erroneous conclusions about the extent of poverty in the two countries' (Booth A., forthcoming). Such debates, as noted above, are not spurious, nor should they be taken lightly as mere statistical games. They can, and do, have an effect on policy, and therefore on people.

As Box 3.1 makes clear, measures of poverty increasingly make use of data on income and expenditure to calculate whether individuals or households are able to purchase a bundle of 'basic' goods and services. While two decades ago food was the key component of these bundles – indeed in some cases effectively the only component – it is becoming increasingly less significant as non-food goods (clothing, electricity) and services (transport, health care) are incorporated to a greater degree. In other words, measures of poverty are becoming increasingly commoditized. Not only does this allow greater room for error and a divergence of views concerning what comprises a basket of 'needs'; it also highlights an issue, raised in Chapters 1 and 2 and dealt with in more detail in Chapter 4, about whether development is about economic growth or, more particularly, whether poverty is about a lack of goods and services assessed in terms of a lack of income or consumption (see page 111). The need to raise poverty lines as development proceeds and expectations grow also comes up against the desire to produce sets of comparable statistics. In Thailand, the poverty line used today is derived from a study undertaken by the World Bank in 1980 (Medhi Krongkaew 1995). Although the lines have been adjusted to take account of changes in the Consumer Price Index, there has been no attempt to adjust the lines to reflect either growing commodification nor rising expectations. As a result, the estimated figures for 1995 (calculated on the basis of extrapolated trends) reveal that the incidence of poverty in the Bangkok Metropolitan and Central regions is 0.0 per cent – a figure which common sense would lead one intuitively to reject (see Table 3.1).

## THE STATISTICAL RECORD ON POVERTY REDUCTION

> The relationship between growth and poverty alleviation is strong and positive. In the four growth economies the incidence of poverty has fallen consistently and rapidly ... [which] underlines the importance of rapid growth as the key to poverty alleviation.
>
> (Hill 1993: 63–4)

Overall, the statistical record on poverty alleviation, as Hill summarizes in the above quote, is good for the four growth economies of Southeast Asia. However for the Philippines, and more particularly for Cambodia, Laos, Vietnam and Myanmar, it is less good. This general picture, however, is not as clear as the national statistics indicate. First of all, there are significant areas, even in the growth economies, where poverty appears entrenched and standards of living have improved only marginally during a period of rapid economic growth. Second, the assumption that the countries of Indochina, Myanmar and the Philippines have done markedly less well than the growth economies requires further investigation. And third, there are significant questions to be posed regarding the very assumptions that underlie the statistical record – notwithstanding the foregoing discussion of their accuracy and comparability. This third issue will addressed in the following chapter.

## Falling through the gaps: missing out on rapid growth

It is common to read or hear the refrain that in the growth economies of Southeast Asia 'the rich are getting richer and the poor are getting poorer'. This has been referred to as the Talents Effect, drawing on the Parable of the Talents in St Matthew's gospel: 'For to every person who has something, even more will be given, and he will have more than enough; but the person who has nothing, even the little that he has will be taken away from him' (25: 29). The significance of the quote is that it implies that not only are the poor positively disadvantaged by capitalist growth, but that the wealth of some begets the poverty of others. This 'populist' view of the unequal pattern of economic growth, at least using the generally accepted (economic) measures of wealth and poverty, is hard to sustain.[3] The picture from the growth economies of Southeast Asia, albeit with a handful of exceptions, is one of increasing prosperity for rich, poor and middle alike.[4] The debate should really be framed in terms of who is benefiting most. Clearly, if it is the rich, then inequalities in economy and society will widen; if it is the poor, then inequalities will narrow.

As with most issues of development, its impact on the relative status of different income groups is highly varied. The best work of this type has been undertaken in Indonesia and much of the following discussion will draw upon this material. The diversity of results though, even within Indonesia, illustrates the difficulty of arriving at any firm, general statements about the effects of growth on different groups – except to say that few families are experiencing a real (as opposed to relative) deterioration in their livelihoods. It is accepted, though, that many commentators would object, at the outset, to the gauging of development in 'real economic' terms. In their view, a detailed critique of different poverty studies and their respective advantages and shortcomings is misguided as it is based on the erroneous assumption that income/consumption measures well-being.

### Inter-personal equity

The two countries with the most varied experience of evolving inter-personal equity are Indonesia and Thailand. In Indonesia, rapid economic growth appears not to have led to the widening of inequalities between the best and worst off, and indeed the figures seem to show a slight narrowing of differentials (Table 3.2).[5] By contrast, in Thailand, the poorest quintile of the population has seen its share of income decline from 7.9 per cent in 1962/63 to 4.5 per cent in 1988/89, to an estimated 3.4 per cent in 1995 (Akin Rabibhadana 1993: 62; Medhi Krongkaew 1995). Medhi Krongkaew *et al.* contend that nowhere in Southeast Asia has the problem of the trade-off between growth and equity been starker than in Thailand (1992: 199).

This difference in experience is usually linked to the contrasting developmental thrusts of each country. In Indonesia, the government has specifically

*Table 3.2* Per capita income distribution: Indonesia and Thailand

### INDONESIA

| | Lowest 40% | Total Middle 40% | Highest 20% | Lowest 40% | Rural Middle 40% | Highest 20% | Lowest 40% | Urban Middle 40% | Highest 20% |
|---|---|---|---|---|---|---|---|---|---|
| 1976 | 19.6 | 38.0 | 42.5 | – | – | – | – | – | – |
| 1980 | 19.5 | 38.2 | 42.3 | 21.2 | 39.0 | 39.8 | 18.7 | 37.8 | 43.5 |
| 1981 | 20.4 | 37.5 | 42.1 | 22.8 | 39.4 | 37.8 | 20.8 | 37.2 | 41.9 |
| 1984 | 20.7 | 37.3 | 42.0 | 22.3 | 39.8 | 37.8 | 20.6 | 38.2 | 41.1 |
| 1987 | 20.9 | 37.5 | 41.6 | 24.3 | 39.3 | 36.4 | 21.5 | 38.0 | 40.5 |

### THAILAND

| | Lowest 10% | Lowest 20% | Lowest 40% | Middle 40% | Highest 20% | Gini Coefficient* |
|---|---|---|---|---|---|---|
| 1962/63 | – | 7.9 | 16.5 | 33.7 | 49.8 | – |
| 1975/76 | 2.4 | 6.1 | 15.8 | 35.0 | 49.3 | 0.426 |
| 1980/81 | 2.1 | 5.4 | 14.5 | 34.0 | 51.5 | 0.479 |
| 1985/86 | 1.8 | 4.6 | 12.4 | 32.0 | 55.6 | – |
| 1988/89 | 1.8 | 4.5 | 12.5 | 32.5 | 55.0 | 0.485 |
| 1990 | – | 4.2 | 11.6 | 30.8 | 57.7 | 0.522 |
| 1992 | – | 3.9 | 11.0 | 30.0 | 59.0 | 0.536 |
| 1995e | – | 3.4 | 9.5 | 27.4 | 63.1 | 0.578 |

*Sources*: Tjondronegoro *et al.* 1992: 71 (Indonesia); Akin Rabibhadana 1993: 62, Suganya Hutaserani 1990 and Medhi Krongkaew 1995 (Thailand). Note that the 1995 figures for Thailand are estimated on the basis of the 1988 to 1992 trend.
*Notes*
\* 0 = absolute equality; 1 = absolute inequality. This trend shows inequality increasing between 1975 and 1995.

directed resources at rural areas and agricultural development. Much of this effort went into raising rice productivity in Java, mainly by promoting and supporting the dissemination of the technology of the Green Revolution. From the 1960s through to the mid-1980s, rice was at the heart of Indonesian politics, and achieving rice sufficiency was accorded the highest priority (Bresnan 1993). Though attempts to boost production through the new rice technology were, initially, faltering, by the mid-1980s Indonesia had achieved the goal of self-sufficiency. Setting a floor price, subsidizing chemical inputs (US$500 million on fertilizers and pesticides in 1987/88 alone),[6] investing in rural roads, supporting a nationwide marketing structure, expanding the irrigation infrastructure, and promoting agricultural research, all contributed to this singular achievement.[7] Revenue from the oil boom was also invested in rural roads, health centres and primary schools, all of which had a generally favourable effect in helping to distribute wealth to poorer groups.

In Thailand, by contrast, the general view is that farmers have been taxed through a series of implicit and explicit policies which have biased development in favour of urban areas and industry. In particular, the Rice Premium (a tax on every tonne of rice exported – rescinded in 1986) has been highlighted as a policy which depressed farm incomes and subsidized the urban population by lowering the price of the main staple – rice – in urban areas. Parnwell and Arghiros write that '[r]ural areas, particularly in the 1960s and 1970s, were neglected by a state élite which systematically allocated resources to urban areas, particularly Bangkok' (1996:15). Thai professor Somphob Manarangsan goes rather further in saying:

> Within five to ten years, Thai farmers will live an even poorer existence. The income gap (between farmers and non-farmers) will widen, a desperate flock of labour migrants to big cities will drastically increase, many farmers will go bankrupt and others will have no alternatives in life after the demise of their cash crops.
>
> (quoted in Choice 1995: 39; see also Pawadee Tonguthai 1987)

But two, rather different, forms of 'bias' are being alluded to here. One bias is the preferential allocation of resources to urban areas and industry. The second, is the extraction of resources from rural/agricultural areas to support urban/industrial development. Concerning the second, Muscat believes that there is 'no prima facie case for putting Thailand into the category of countries that have financed modern industrial development through large net transfers of a surplus out of the agricultural sector' (Muscat 1994: 245). Instead Muscat highlights a number of policies, including a regressive tax structure, which did not so much discriminate against agriculture, but discriminated against the less-well-off in rural areas – who happen largely to be farmers. In particular he mentions the uneven and halting expansion of health and education systems in the more peripheral regions. This constrained the ability of the poor to raise their human capital, thereby limiting their chances of joining the faster growing and more productive sectors of the economy. But, significantly, Muscat does not believe the government could have afforded the massive resource transfers that would have been necessary to prevent this uneven pattern of growth, given the implications that such transfers would have had for aggregate growth:

> In the face of the differences in the initial conditions respecting the potentialities for growth in productivity, the economic geography that has focused industrial development in the Bangkok area, the virtually unavoidable reinforcing patterns of distribution of human capital formation, and the inequalities in flexibility within the agricultural sector, one can conclude that an increase in *income inequality has been an inevitable consequence of Thailand's economic development thus far*.
>
> (Muscat 1994: 248 [emphasis added])

For Muscat, the Thai government simply failed to give the poor in rural areas sufficient help to narrow the income gap; for Somphob Manarangsan, and many others, the government promoted urban/industrial development by exploiting the rural poor (see Bell 1996).

Thailand and Indonesia, despite their undoubted differences, have both experienced rapid economic growth. The Philippines, by contrast, is often characterized as the economic 'laggard' of the five original members of Asean.[8] As Balisacan baldly states, 'overall growth performance in 1965–1986 was dismal' (1992: 127). In 1950, per capita GNP in the Philippines was twice as high as Thailand (US$150 versus US$80 at 1952–54 prices) – higher even than Taiwan and South Korea (Balisacan 1992: 127). By 1993, after years of stagnation and economic crisis, the situation had become reversed. Per capita GNP in the Philippines was US$850, while in Thailand it stood at US$2,110 (World Bank 1995).

Like Indonesia, economists working on the Philippines have access to a long-run series of consumption surveys undertaken from 1961, the Family Income and Expenditure Surveys (FIES). These give a good picture of trends in poverty and income inequality. While poverty has declined, the degree of decline – unsurprisingly given the country's economic record – has been less marked than in either Thailand or Indonesia (see Figure 3.1 and Table 3.1). Income inequality, quite pronounced to begin with in part due to the highly skewed distribution of landholdings, has also remained largely unchanged over the last three decades. But although the reduction in the incidence of poverty in the Philippines has been modest compared with its Southeast Asian neighbours, Balisacan takes issue with those scholars who argue that the incidence and depth of poverty have widened and deepened over the last thirty years: 'the oft-repeated argument that post-World War II economic growth [in the Philippines] completely by-passed the poor is not supported by the figures' (1992: 134). Miranda, one of those with whom Balisacan takes issue, would retort that the figures on which such a view is based are merely indicative of the 'poverty of economics', not of the real 'economics of poverty' (Miranda 1988).

The evidence from Thailand, Indonesia and the Philippines discussed above, and from the four tigers, indicates that although there may be a link between economic growth and 'well-being' (but see below), there is no such obvious link between economic growth and trends in inequality.[9] Rather, it seems that it is the *type of growth*, and the *policies which underpin that growth*, which are critical in determining whether rapid economic expansion is accompanied by a widening or narrowing in income distributions.[10]

## Intra-personal equity

The above discussion of interpersonal income distribution appears to indicate that the poor are becoming relatively more entrenched in Thailand. In Indonesia, although there may be some (but debatable) evidence of a slight narrowing of differentials, even so this still leaves a large rump of poor people controlling only a fraction of total wealth. However these figures, and most studies, do not show the degree to which there is intra-personal movement of the poor and non-poor. One of the few studies in Southeast Asia which has followed individuals over time is that undertaken by Edmundson (1994).

Edmundson and his colleagues undertook a survey of forty-six families in three East Javanese village over a period of three years and six months in the two decades between 1971 and 1991 (Edmundson 1994). Their conclusions regarding the distributional effects of growth can be summarized as:

- The majority of families who were defined as poor in 1971 have become relatively richer, while the majority of families defined as rich have become relatively poorer.
- The poor, whether defined as a cohort or as a percentile group, have raised their income in real terms[11] (Figures 3.2a and 3.2b)
- In terms of cohorts, inequality between rich and poor narrowed between 1971 and 1991 (Figure 3.2a)
- In terms of percentile groups, inequality between rich and poor widened between 1971 and 1991 (Figure 3.2b)

As Edmundson writes, based on the conclusions above there are two answers to the question whether the rich are getting richer and the poor, poorer. One is that the original poor are relatively richer. The second is that the poor of 1991 are relatively poorer than the poor of 1971. There is also a third answer to the question. And that is that in real, rather than relative, terms both the rich and the poor are getting richer – whether they are defined in terms of cohorts or percentile groups. There are clearly some difficulties with the work, and the research team had to make a series of debatable decisions about, for example, how to measure inflation and therefore to gauge the real buying power of money in 1971 and 1991, and how to measure 'wealth' and therefore ascertain which families deserved the titles 'rich', 'middle' and 'poor'. In addition, the evidence from a sample of just forty-six families in three villages in East Java cannot be taken as indicative of processes elsewhere in Java, let alone in Indonesia or the wider Southeast Asian region. None the less, what the work does reveal is the mobility of people in economic terms. As Edmundson writes, '[t]he rich and poor may be "with us always" but the individuals who comprise these groups appear to be always changing' (1994: 140). Importantly, there was no entrenched core of permanent poor; all families showed upward mobility and improving real incomes. As the next chapter will argue, to some extent being poor is a transitory status associated with the life cycle. It is for this reason that studies refer to the 'striving poor', as opposed to the 'entrenched poor' (see page 144).[12]

## Geographical exclusion: balancing the space economy

regional disparities in per capita consumption expenditures [in Indonesia] ... widened between 1980 and 1987. ... Some provinces outside Java actually experienced negative real growth in consumption expenditure over these years.

(Booth A. 1992b: 353)

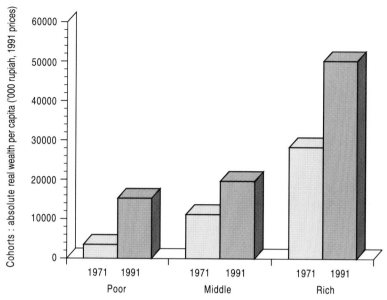

*Figure 3.2a* Per capita income in East Java based on cohort groups (1971 and 1991)

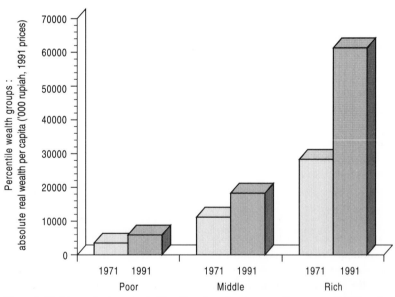

*Figure 3.2b* Per capita income in East Java based on percentile groups (1971 and 1991)

Three decades of statesponsored, reasonably coordinated planning in Thailand have come up with lasting solutions to few, if any, of the really pressing problems of the country's rural areas and peripheral regions.

(Parnwell 1990: 1)

Although the incidence of poverty is declining in all the growth economies of Southeast Asia, and real standards of living are improving on a broad front, the national statistics do reveal continuing areas of concern. Among these, one of the most significant is the geographically uneven distribution of wealth. Levels of poverty, household income, gross regional product, consumption, the incidence of diseases, mortality rates and educational attainment all vary markedly between areas. Every country has its 'lagging' regions – with the exception of Singapore – as well as its lagging people. This inequality is sometimes expressed in terms of a rural–urban or core–periphery dichotomy – by which it is also meant an agriculture–industry dichotomy – sometimes in terms of a regional imbalance, and often both. While governments in every country have instituted development policies to target these backward regions it is fair to say that attempts at regional development have failed significantly to create any greater 'balance' in the space economy. In some instances, most notably Thailand, space inequalities have become even further accentuated as development and modernization have proceeded. For reasons of both human equity and political stability, governments have become increasingly concerned at the apparent failure to effect spatially even growth.

## The administration of space

Measures of regional inequality are usually based on large administrative units. Statistics collected at the regional level are used to construct an image of (relative) backwardness (see, for example, Figure 3.3). Gross regional products, the incidence of poverty, or consumption data are employed to argue the case that a region is lagging when compared with other regions, or with the country as a whole. Such a crude approach to inequality has a tendency to ascribe human characteristics to (often arbitrary) spatial designations. It is people who suffer from poverty and low levels of consumption, and intra-regional inequalities are often more pronounced than inter-regional ones. Thus, the emphasis on spatial inequality can serve to disguise deep-seated structural and social inequalities. To put it simply, poor farmers in East Nusa Tenggara (Indonesia), Northeastern Thailand, Mindanao (Philippines), or East Malaysia often have more in common with poor farmers in West Java, Central Thailand, Central Luzon or West Malaysia than they do with skilled factory workers or teachers in their own 'backward' regions. Yet regional development has a tendency to regard regions as uniformly disadvantaged. As a result, the practice of regional development often brings additional benefits to individuals who were not disadvantaged to start with – they just happened to live in disadvantaged regions.

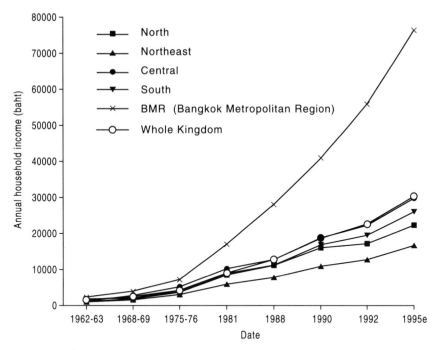

*Figure 3.3* Thailand: annual household income, by region (1962–95)

*Source*: Medhi Krongkaew 1995

*Note*: Based on current prices, baht per person per year. The figures from 1975/76 are based on the two-yearly household Socio-economic Survey undertaken by the National Statistical Office. Earlier figures are derived from other sources.

There are a series of possible explanations to account for why, when poverty overall is falling rapidly, the relative backwardness of certain areas is becoming more pronounced. To begin with, it has been consistently argued that spatial development policies are flawed because they are formulated in a vacuum. Aspects of the political economy and sectoral policies often work against regional initiatives. In Thailand, for example, the government was 'taking' money from farmers in the poorest region, the Northeast, through the Rice Premium (see page 89) at the same time as it was 'giving' money through various dry season employment projects. Although it is not possible to 'tot up' the balance of all policies and programmes with respect to the Northeastern region, it is tempting to see regional initiatives as essentially an exercise in politics, rather than development.[13] It is notable that at the time when regional development became a central tenet of faith in Thai planning circles (essentially from the Third Five-Year Economic and Social Development Plan [1972–76] but becoming particularly pronounced during the Fourth Plan [1977–81]), the Communist Party of Thailand (CPT) was growing in influence and one of its main bases of support was in the disadvanataged Northeast. With a Thai

Deputy Prime Minister stating on radio that 'if stomachs are full people do not turn to Communism' it is easy to assume that regional development was politically driven (Caldwell 1974). It is significant that two key programmes in the development of the Northeast – the ARD (Accelerated Rural Development) and MDU (Mobile Development Units) programmes – were funded by the United States which at that time was actively engaged in fighting Communism across mainland Southeast Asia, and operated under the auspices of the Thai military. Their emphasis was as much on securing peripheral areas through expanding the communications infrastructure than it was on promoting human development (see Muscat 1994: 139–42). Now that the CPT has been vanquished, it might further be argued that from the political and security perspective at least, regional development has been an enormous success.[14]

Like Northeast Thailand, Indonesian government investments in East Timor (particularly) and Irian Jaya reflect the need to secure these areas militarily just as much – if not more – than they reflect a recognition that the areas represent poor provinces requiring regional development assistance. On a per capita basis, East Timor has been the country's largest recipient of central government funds and is highly dependent on these transfers (Hill and Weidemann 1991). Yet it is significant that a large proportion of the funds has gone towards such things as improving and expanding roads, funding the security forces and building police stations. Soesastro states that the province's 'deep-seated (and long-standing) problems of poverty, the circumstances surrounding the civil war and integration into Indonesia, the security situation, and continuing problems of apathy and low morale in some quarters all contribute to making East Timor a test case for the Indonesian government's commitment to equitable regional development' (Soesastro 1991: 228). Yet, though East Timor may be a 'test case' for regional development, it remains poor despite the sums of money invested in the province.[15] The same is true of the other provinces that constitute Eastern Indonesia which, despite receiving more than their 'fair share' – in population terms – of INPRES grants from the centre, continue to lag behind the rest of the country. In 1993–94, the four provinces of East and West Nusa Tenggara, Maluku and East Timor received some 11 per cent of INPRES grants, yet supported 5 per cent of the country's population and 9 per cent of its poor population. Booth suggests that this is because the three pillars of Indonesian economic development during the New Order period – import substitution industrialization, labour-intensive export-oriented industrialization, and resource-based industrialization – have all been centred elsewhere in the archipelago.[16] In infrastructural terms there is a serious imbalance between Eastern and Western Indonesia and 'as long as infrastructural imbalance of these orders of magnitude remain, the longer run development prospects for Eastern Indonesia must remain uncertain' (Booth A. 1995b: 117).

Although the political and security rationale for promoting peripheral regions is compelling, there is also a significant case that regional development often fails because it is wrongly conceived, insufficiently supported, and inadequately implemented. Many economists doubt whether the development plans produced by the Thai National Economic and Social Development Board (NESDB), for example, amount to much more than paper plans, or at best to a series of good intentions. As a confidential World Bank report revealingly observed in 1978:

> There is . . . little evidence that Thailand's development plans systematically guide or govern the actions of departments or, for that matter, the cabinet itself, in the day-to-day conduct of government affairs. Although national development plans should never be treated in mixed economies as binding and inflexible statements of government intentions, the frequency and extent to which development plans appear to be disregarded in the allocation of administrative and financial resources and in the introduction of new policies, programs and projects is indicative of a lack of full commitment to the concept of development planning. In recent years it has become increasingly difficult to discern a sense of direction and purpose in public sector behavior that is in any way comparable to its stated intentions and objectives.
>
> (World Bank 1978: 28)

Why successive governments in Thailand have, on the one hand, embraced the concept of economic planning with such enthusiasm and yet, at the same time, have not made use of the resulting plans in any systematic manner seems to be linked to two issues. First, the operation of the Thai political economy makes coherent implementation difficult. The NESDB does not have cabinet status and the role of the NESDB in decision-making largely depends on the particular relationship between the secretary-general of the NESDB and the prime minister of the time (Muscat 1994: 178–179). In addition, ministries operate as 'kingdoms' with a high degree of autonomy, while influential interest groups within and outside the cabinet are often in a position to circumvent or amend plans if they desire (see Demaine 1986). Second, and more fundamentally, there is the question of whether planning, except in a loose sense, makes sense in free-wheeling and volatile market economies. Although some might argue that 'indicative' – rather than 'guidance' – planning is possible, it is significant that Singapore is the only country in Asean which does not produce national development plans. As Dr Goh Keng Swee, a key figure in the management of Singapore's economic success, observed:

> Actually when we [the PAP] first won the elections in 1959, we had no plans at all. We produced a formal document called the First Four-Year Plan in 1960, only because the World Bank wanted a plan. We cooked it up during a long weekend. I have very little confidence in economic planning.
>
> (cited in Toh Mun Heng and Low 1988: 23)

In the formulation of its Eighth Five-Year Development Plan (1997–2001), Thailand's NESDB tried to make the document more relevant by letting citizens' groups and non-governmental organizations participate in its formulation. Partly as a result, the central theme is not the attainment of economic targets, but the achievement of 'human development'. None the less an academic involved in drawing up the document was quoted as remarking that '[t]he plan will be nice as a historical document, a snap-shot of Thailand's main concerns at the present', but added that 'if the NESDB had the power to implement these plans, Thailand wouldn't be in the mess it is in today'. Even an NESDB director, Witit Rachatatanun, admitted that 'it is hard to get the government to implement many things in the plan' (Bardacke 1995).

On both the Left and the Right there is increasing doubt concerning the efficacy of state-led efforts to promote regional development in a globalizing economy. None the less, it does seem that some academics continue to set considerable store by national development plans, perhaps because they are usually the most accessible statements of government development intentions and are often also published in English. They are used as baselines from which success or failure can be gauged. Yet while academics may be using them in a prescriptive fashion, governments in the region appear to accept that they are not founded on rigorous econometric analysis and are happy to treat them in a relatively cavalier manner. The assumptions on which they are based are, in any case, often overtaken by events (Hill 1993: 58–9).

## Rural–urban inequalities

Critics of regional development observe that spatial inequalities are usually a function of rural–urban inequalities, and not regional inequality *per se*. Those regions where the population are most disadvantaged tend to be rural areas dominated by agriculture. At the same time, those in such lagging regions who suffer from poverty and underdevelopment most acutely, tend to live in the countryside and engage in farming. Thus, for example, in Thailand, the incidence of poverty in municipal (urban) areas in the lagging Northeast is considerably below the national average (19 per cent versus 24 per cent), while for rural areas in the Northeast it is substantially above the average (40 per cent) (see Table 3.3). It is also notable that the incidence of poverty in Bangkok – in many scholars' eyes the proto-typical primate city – is a mere 3 per cent (or zero, if the 1995 projected figures are to be believed). In addition, between 1975/76 and 1988/89 urban poverty declined to a much greater degree than did rural poverty (see Table 3.3).

It is, of course, this observation – that regional underdevelopment is really all about rural underdevelopment – which has provided the impetus to the wide-ranging and long-standing debate on urban bias in development. But – and this is an issue which is discussed in more detail in Chapters 7 and 8 (see pages 239 and 276) – there are serious questions to be posed concerning whether it is

*Table 3.3* Regional and rural–urban incidences of poverty in Thailand (1975/76, 1988/89 and 1992)

| | 1975/76 (%) | 1988/89 (%) | 1992 (%) |
|---|---|---|---|
| **North** | 33 | 23 | 14 |
| Villages | 36 | 25 | |
| Sanitary districts[a] | 19 | 19 | |
| Municipal areas | 17 | 11 | |
| **Northeast** | 44 | 37 | 22 |
| Villages | 49 | 40 | |
| Sanitary districts | 25 | 20 | |
| Municipal areas | 21 | 19 | |
| **Central** | 13 | 16 | 6 |
| Villages | 14 | 19 | |
| Sanitary districts | 8 | 6 | |
| Municipal areas | 11 | 8 | |
| **South** | 31 | 22 | 12 |
| Villages | 34 | 24 | |
| Sanitary districts | 18 | 11 | |
| Municipal areas | 21 | 12 | |
| **Bangkok (BMR[b])** | 8 | 3 | 1 |
| **Whole country** | 30 | 24 | 13 |
| Villages | 36 | 29 | |
| Sanitary districts | 14 | 13 | |
| Municipal areas | 13 | 7 | |

*Sources*: Suganya Hutaserani 1990: 16, Medhi Krongkaew 1995
*Notes*
[a] 'Sanitary districts' are areas which achieve a certain minimum population size and density and which embody some urban characteristics. Scholars have been divided as to how to treat them, partly because the government has been slow in reclassifying villages which have attained sanitary district status in terms of their characteristics. For example, Suganya Hutaserani (1990) lumps 'sanitary districts' along with 'villages' in the category 'rural', while Medhi Krongkaew prefers to treat them as 'urban'.
[b] BMR = Bangkok Metropolitan Region

Note that there is some discrepancy between these figures and those in Table 3.1

correct to conceive of discrete 'rural' and 'urban' worlds in competition, one with the other. Not only are many of the poor in urban areas officially defined as rural inhabitants because of deficiencies in registration methods (see page 161), but such is the degree of circulation between rural and urban areas (and between agricultural and non-agricultural work) that it is often impossible to pigeonhole households or individuals as 'belonging' and therefore owing 'allegiance' to a geographically defined category such as that of 'rural' or 'urban'. None the less, the force of the apparent statistical support for there being a rural–urban imbalance is such that there are few studies of Southeast Asia's development which do not make it a central block in the edifice of 'uneven' development (e.g. Parnwell and Arghiros 1996: 14–15).

## The poverty (and wealth) of remoteness

While regional development has received considerable scholarly and planning attention, many would wish to disassociate themselves from regional development *per se*, and instead re-formulate spatial inequalities in terms of peripheral areas and the geography of remoteness. This allows a rather more flexible view of geographical marginality which is not tied to particular regional designations, nor to rural areas in their entirety.

In Indonesia, some of the wealthiest provinces are, at the same time, some of the most remote and the poorest (Table 3.4). Irian Jaya and East Kalimantan, for example, had above average per capita gross regional products (GRPs) in 1989, but also suffered from some of the highest provincial levels of poverty. There are two reasons for this seeming mismatch between economic wealth and human well-being. First, the high GRP figures for Irian Jaya and East Kalimantan, as well as for Aceh, Riau and South Sumatra, are linked to the presence of extractive industries, namely timber in the first two cases, and oil and gas in the latter three. The wealth that flows from these 'enclave' industries tends not to translate into higher standards of living for the mass of the local

*Table 3.4* Inter-regional wealth and poverty: selected Indonesian provinces

| | Index of per capita GRP (1989)[c] | Incidence of poverty (1987) | |
|---|---|---|---|
| | | Rural | Urban |
| Riau[a] | 524 (89)[d] | 16.0 | 8.8[e] |
| East Kalimantan[a,b] | 503 (203[d] | 34.8 | 34.4 |
| Aceh[a] | 298 (83)[d] | 16.3 | 13.7[e] |
| DKI Jakarta | 217 | n.a. | 1.4 |
| South Sumatra[a] | 133 | 14.0 | 23.6[e] |
| Irian Jaya[b] | 110 | 67.0 | 33.0 |
| **Indonesia** | **100** | **17.2** | **14.7** |
| South-east Sulawesi | 61 | 40.7 | 19.2 |
| South Sulawesi | 60 | 25.2 | 22.3 |
| North Sulawesi | 57 | 11.4 | 14.4 |
| Central Sulawesi | 52 | 21.4 | 13.0 |
| Bengkulu | 41 | 9.9 | 15.0 |
| East Timor | 30 | n.a. | n.a. |

*Sources*: Booth A. 1992b; Tisdell and Firdausy, unpublished
[a] oil/LNG-rich provinces
[b] timber-rich provinces
[c] 100 = GDP per capita for the whole country
[d] Figures in brackets are GRDP per capita figures excluding mining
[e] Figures are 1990

population living in those provinces. Thus the main beneficiaries of Indonesia's oil and timber booms have been industries and élites living elsewhere, and particularly in Java (see Box 3.2).

Second, even when mining is discounted, this does not mean that GRP figures are any more useful as an indicator of standard of living. This is largely because they are a measure of production, not of consumption, and much of the wealth derived from economic activity over-and-above that from extractive industries such as oil and timber is also transferred out of the provinces concerned. Just 'as cream rises to the top of the milk', Castles writes with reference to Indonesia's capital, Jakarta, '[so] surpluses from whatever industries are currently flourishing tend to gravitate to the metropolis' (Castles 1991: 234). For Irian Jaya in 1987, a remarkable 55 per cent of GDP was neither invested nor consumed in the province, but syphoned off, and since then this has declined only slowly to 37 per cent in 1991 (Booth A. 1992a: 640 and 1995b: 118). During the 1980s, Booth estimates, around 50 per cent of Irian Jaya's generated wealth has been remitted to other provinces. If concentrations of regional poverty and differences in living standards are to be gauged more accurately then consumption data – such as those collected by the Susenas survey (see Box 3.1, 'Identifying the poor in Indonesia') – must be analysed. In addition, if peripheral provinces like Irian Jaya and East Kalimantan are to develop, then governments must ensure that a greater proportion of wealth generated in such provinces be invested there (Booth A. 1992a: 640–41 and 1992b: 333). There is evidence that the central government's long-term policy of draining wealth from resource-rich, but human development-poor provinces like Irian Jaya and East Kalimantan is leading to growing dissatisfaction and potential unrest, turning economic relations between the centre and the periphery into an increasingly vexed political issue.

---

*Box 3.2* Freeport Indonesia: local wealth, local poverty

The most striking illustration of the spatial coincidence of immense wealth set against stubborn poverty is Freeport Indonesia, a gold and copper mine near Tembagapura in the southern highlands of Irian Jaya. The mine is probably the richest in the world: the gold and copper reserves have been valued at US$38 billion (1994 prices). Around 100,000 tonnes of ore are moved each day along a 118-km long conveyer belt, a town of 25,000 near Timika and named Kuala Kencana (River of Gold) exists to support the mine, and it is Indonesia's fourth largest tax payer (US$212 million in 1993). The area around Tembagapura is also a centre of activity by the Organisasi Papua Merdeka (OPM or Free Papua Movement) fighting for Irian Jaya's independence from Indonesia. The Indonesian military have been accused of human rights abuses in the area and the OPM has also attacked Freeport

continued . . .

---

vehicles using the road linking Timika with Tembagapura (*Inside Indonesia* 1995). Inevitably, Freeport has been drawn into the wider debate over the distribution of benefits in Indonesia and the OPM's political demands. In a report released in September 1995 the Indonesian Commission on Human Rights stated that Freeport had a responsibility to provide environmental, social, cultural and economic support for the surrounding area. How far, and how directly, the mine can be associated with the reported human rights abuses (not to mention the environmental costs of the massive operation) is a contested point. None the less it is clear that the Indonesian military have a hightened presence in the area largely because of the mine and the strategic role that it fills. In an agreement – known as the January Agreement – signed in 1974 the Amungme, the local tribal group, effectively signed away their rights to the Grasberg mine. The Amungme were largely relocated to Timika and although they have received some benefits, such as free medical care, their physical relocation has led to social dislocation and cultural alienation (Menembu 1995). In March 1996 tribespeople (mostly Dani, 4,500 of whom have settled in the area) rioted in Timika and Tembagapura leading to the temporary closure of the mining operation. Their demand: that they should receive more of the economic benefits from the operation. They suggested a figure of 1 per cent of the mine's gross revenue or US$17.5 million per year.

Sources: *Inside Indonesia* 1995; McBeth 1996b; Menembu 1995

## The urban poor

Many studies of poverty emphasize that most of the poor live in rural areas and therefore that poverty is essentially a rural phenomenon. Yet this association of poverty with the countryside is unsatisfactory on a number of counts, and more to the point is becoming increasingly so as modernization proceeds. To begin with the data, imperfect though they may be, seem to indicate that urban poverty is becoming relatively more prominent over time (Table 3.5). In Indonesia, for example, while in 1976 the urban poor represented just 18 per cent of the total poor population in the country, by 1990 this had almost doubled to 35 per cent.[17] Although the percentage of poor in the total urban population may be decreasing (i.e. the incidence of urban poverty is declining), this needs to be set against a rapid rate of urbanization. As a result, the numbers of urban poor have remained stagnant during a period when deep inroads have, apparently, been made into rural poverty.

It would be tempting to see this increasing prominence of urban poverty as a reflection of government policies which have perceived poverty as almost entirely a problem of rural areas and agriculture, thus leading to an over-concentration on the countryside and a concomitant neglect of urban poverty. But it could be argued that this interpretation misconstrues the roots of urban

*Table 3.5* Urban poverty in total poverty: Indonesia, Thailand and the Philippines

| | | Rural:urban poverty ratio* | |
|---|---|---|---|
| | *Indonesia* | *Thailand* | *Philippines* |
| 1961 | – | – | 70:30 |
| 1965 | – | – | 74:26 |
| 1971 | – | – | 76:24 |
| 1976 | 82:18 | – | – |
| 1978 | 82:18 | – | – |
| 1980 | 78:22 | – | – |
| 1981 | 77:23 | 83:17 | – |
| 1984 | 73:27 | – | – |
| 1985 | – | – | 68:32 |
| 1987 | 65:35 | – | – |
| 1988 | – | 81:19 | 70:30 |
| 1990 | 65:35 | – | – |
| 1991 | – | – | 59:41 |

*Sources:* Thailand: Warr 1993: 46, Amara Ponsapich *et al.* 1993: 9 and Somchai Ratanakomut *et al.* 1994. Indonesia: Booth, A. 1993: 65. Philippines: Balisacan 1994: 122
* The number of poor people in rural and urban areas expressed as a ratio.

poverty for, to a significant extent, urban poverty is, itself, a manifestation of rural poverty. The poor in rural areas – often landless – are forced to migrate to cities and towns as a strategy of survival.[18] Thus the rural poor, through geographical dislocation and relocation, become redefined as 'urban' poor (Balisacan 1992: 125). Further, and arguably even more significant in the Southeast Asian context, many of the rural poor can simultaneously be numbered among the urban poor. Such is the degree of interaction between rural and urban areas that families embrace livelihood strategies which ensure they have a foot in both the rural and urban worlds (see Chapters 5, 6 and 7). Invariably these poor households and individuals are classified as 'rural', even though their lives and livelihoods may be just as rooted in urban areas and urban work. As a result, it is likely that urban poverty is seriously underestimated in Southeast Asia – even should one discount the difficulties connected with setting urban poverty lines and the changing definition of what consitutes an urban area in each country.[19]

But it is important to emphasize that the problem is deeper than mere underestimation, for poverty statistics ignore the structural causes of poverty and the critical links that exist between urban and rural areas. To a significant degree, rural and urban poverty are not discrete and different, but are mirror images of each other (see *Environment and Urbanization* 1995; Wratten 1995: 21–2). There are also parallel dangers of assuming that urban life has nothing to do with agriculture. Statistics released by the Biro Pusat Statistik (BPS) in Indonesia, for example, reveal that 24 per cent of the urban poor are principally engaged in agricultural pursuits (Firdausy 1994: 77). Cities and towns support a sizeable urban farming sector geared to providing the urban population with

*Plate 3.2* The urban poor: a woman drying fish in the squalor of Jakarta's Pasar Ikan

vegetables, fruit, eggs, milk and some meat and many of those who work in this sector can be counted among the urban poor. Conversely, of course, there are many rural poor who earn the bulk of their income from non-agricultural activities.

## POVERTY AND ECONOMIC REFORM IN INDOCHINA (AND MYANMAR)

The material and data on poverty in the countries of Indochina and Myanmar are far less plentiful than on the Asean nations. There are no long-run poverty statistics giving an impression of change over time; nor is there the wealth of micro-level studies that exist for the other countries of the region; nor, for that matter, is there much in the way of regional (sub-national) data. The first expenditure and consumption survey in Laos, for example, was only undertaken in 1993 and the results released in July 1995. This survey showed that while Laos is egalitarian – almost two-thirds of households have a level of consumption within 50 per cent of the median – it is also uniformly poor with almost three-quarters of households allotting 60 per cent or more of their consumption on food (see Box 3.3). However although the degrees of poverty and inequality may be substantially different in the countries of Indochina and Myanmar to those of the growth economies of Asean, the material and data that are available show that there is a preoccupation with similar themes, and in particular the widening of inequalities in the wake of economic growth. As Kerkvliet and Porter remark, 'Vietnam has dilemmas that are also common to neighbouring countries', specifically mentioning equity, distribution, quality and range of public services, and environmental degradation and sustainability (1995: 31). Although there is still a tendency to talk as if there are two Southeast Asias, increasingly their experience of development is similar (see Chapter 1). Indeed there is a case that the two Southeast Asias were never as different, at least in terms of economic structure and function, as the formerly deep political and ideological divisions might lead one to assume (see page 19). Because the experiences of the countries of Southeast Asia are increasingly similar – at least in kind, if not in degree – many of the issues highlighted above with reference to the growth economies apply equally to the countries of Indochina and, though less obviously, to Myanmar. The problem is that the general paucity of data makes it difficult to examine poverty in nearly the same level of detail.[20]

### Regional inequality in Indochina

Although there may be a grave shortage of sub-national data, Vietnam, Laos and Cambodia should prove to be some of the most interesting – and complex – countries to study in terms of regional (spatial) inequalities. Each country has suffered a destructive and divisive war, and each has made – and is making – the transition to the market.

*Box 3.3* Consumption in Laos: poverty and equality

In 1990 Laos received assistance from the United Nations Development Programme to carry out the country's first expenditure and consumption survey. The survey, undertaken between March 1992 and February 1993, covered 2,937 households (19,574 people) drawn from a sample of 147 villages from every province of the country. The results of the Lao Expenditure and Consumption Survey (LECS) were released in July 1995. They show, as one might have expected, a country where levels of consumption are uniformly low. Of total consumption, 62 per cent is allocated to food, 13 per cent to housing, and 5 per cent to personal care and recreation (Figure B3.3). Average annual household consumption in 1992–3 was just under 85,000 kip, or less than US$100. By some international standards, if food accounts for 60 per cent or more of total household consumption then a household should be regarded as poor, and if it is above 80 per cent then its members should be designated as living in severe poverty. Using these measures, the LECS revealed that 72 per cent of households were poor, and 44 per cent severely poor (Figure B3.4). On this basis, Laos is very poor even when compared with countries like Indonesia.

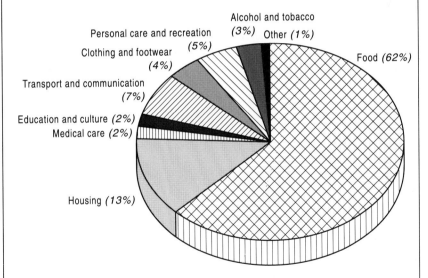

*Figure B3.2* Level and structure of household consumption, Laos (1992–93)

However if we take poverty to be a function of inequality then the picture is rather different. Almost two-thirds of households fall around the median level of consumption, defined as consumption within 50 per cent of the

continued . . .

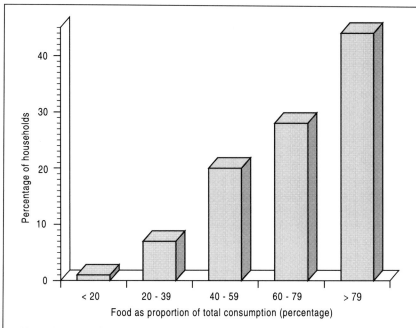

*Figure B3.3* Food as a proportion of total consumption, Laos (1992–93)

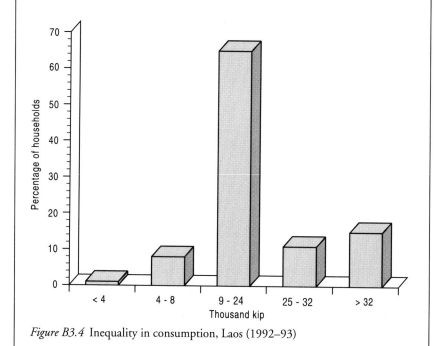

*Figure B3.4* Inequality in consumption, Laos (1992–93)

median (between 8,000 and 24,000 kip per 'consumption unit', with a median of 16,000 kip). One in four households have a consumption level above 24,000 kip while less than one in ten have a level below 8,000 kip (Figure B3.5). These figures on interpersonal inequality can be usefully compared with those for Thailand (see Table 3.2).

In Vietnam the complexities are even greater because while the northern half of the country was undergoing a transition to socialism between 1954 and 1975, the South was experiencing a deepening of capitalist relations. Further, the conflict also led to the regional dispersal of industry across the North as the leadership attempted to shield factories from the American bombing campaign. Individual provinces were allowed considerable autonomy as the exigencies of the war, rather than Marxist theory, drove industrial policy. Provinces were seen as self-sufficient production units and the North's urban population was relocated to the countryside as a protective measure. Hanoi's population by the end of 1967 was perhaps just 250,000 – a quarter of the pre-war figure. In the South, the war also had far-reaching economic implications, leading to massive infusions of American assistance and causing (in contradistinction to the North) the drift of millions from the countryside to the relative safety of the cities. Between 1964 and 1974/75, the population of Saigon (Ho Chi Minh City) grew from 2.4 million to perhaps as much as 4.5 million (Thrift and Forbes 1986: 124). The population of Da Nang, following the Tet Offensive of 1968, expanded by 58,000, and by a further 200,000 after the 1972 offensive (Desbarats 1987: 47). In 1974 the city supported an estimated 500–600,000 inhabitants, while ten years earlier in 1964 its population was just 149,000 (Forbes 1996: 26). Since the early 1980s, and following reunification in 1975/76, the leadership in Hanoi has embraced wide-ranging market reforms (see Chapter 1). The south, with its better-developed entrepreneurial culture and infrastructure, and links into a wealthy overseas Vietnamese community, has been far better placed to take advantage of these reforms, and though the government may not have explicitly favoured the south, the effect of the reform process has been to allow the area around Ho Chi Minh City to outpace the rest of the country (Figure 3.3 and Table 3.6). The same pattern is evident in Laos where the capital Vientiane and its environs, and towns along the valley of the Mekong River have experienced significant economic gains from the reforms, while geographically more marginal areas have failed (so far) to see much increase in prosperity (see Rigg forthcoming, and Jerndal and Rigg forthcoming).

A Living Standards Measurement Survey (LSMS) was undertaken by the Vietnamese government in 1993 to ascertain standards of living in the country.[21] This, as most prior observations had indicated, showed that while GDP per capita in the northern mountains and midlands, and in the north central coast region averaged less than US$80 (<350,000 dong at 1989 prices), in the Mekong

*Plate 3.3* Booming Ho Chi Minh City. The south of Vietnam, and especially Ho Chi Minh City and its environs, has the highest incomes and a large proportion of foreign investment is concentrated there

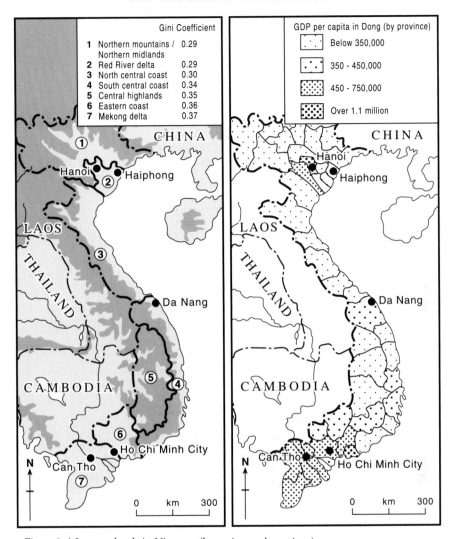

*Figure 3.4* Income levels in Vietnam (by region and province)

Delta, as well as around Hanoi and Haiphong in the north, it ranged between US$100 and US$275 (450,000–1,237,500 dong). Such data as are available all indicate that the south is growing faster than the north, and that the northern mountain provinces, particularly, are lagging behind the rest of the country. The emergence of such uneven spatial development was a particular topic of debate – and concern – at the Seventh Party Congress in 1992 (Beresford and McFarlane 1995: 60) (see Figure 3.4).

However, although regional differentials exist, there is not the available statistical evidence to say for certain whether such differentials pre-date the reforms in

*Table 3.6* Poverty in Vietnam, by selected province and region (1989–92)

| Province (region) | Households living in poverty (%) | | Gini Coefficient | |
|---|---|---|---|---|
| | 1989 | 1992 | 1989 | 1992 |
| Yen Bai (Northern Mountains) | 31.9 | 13.2 | 0.287 | 0.254 |
| Nam Ha (Red River Delta) | 25.4 | 13.1 | 0.202 | 0.307 |
| Binh Dinh (South Central Coast) | 27.0 | 12.9 | 0.217 | 0.314 |
| Dac Lac (Central Highlands) | 30.1 | 12.6 | 0.343 | 0.325 |
| Can Tho (Mekong Delta) | 12.1 | 8.6 | 0.267 | 0.291 |
| Average | 25.3 | 12.1 | 0.263 | 0.298 |

*Source*: Dao The Tuan 1995: 143. Based on household surveys carried out by the General Statistics Office. The poverty line is defined as income equivalent to 20 kg of rice per month per person.

Vietnam (see Fforde and Sénèque 1995: 116). Those who have travelled between regions tend to report the impression that inequalities are becoming more pronounced, but such observations should be treated with caution. One household survey conducted between 1989 and 1993, for example, revealed no indication that inter-regional differentiation had deepened following the introduction of the highly significant land reform measures in 1988 (so-called Resolution 10) in which household property rights were strengthened (Dao The Tuan 1995: 144). The same survey did, however, uncover some evidence that intra-regional inequalities had widened. Beresford and McFarlane, on the other hand, quote figures to indicate that the North was highly egalitarian prior to reunification and the economic reforms. These figures are for 1970 and are based on retail markets (taken as a proxy for consumer incomes and spending) and relative population in the north's three main regions (1995: 57). They show:

> Northern midlands and highlands 0.9
> Red River delta 1.1
> Central coastal provinces 0.8
>
> (where 1 = equality)

But, as they point out, 'these figures probably indicate nothing more than what Vietnamese commentators have often pointed out . . . that equal incomes meant shared poverty' (Beresford and McFarlane 1995: 57). Like the growth economies of Asia then, Vietnam may be making the transition from shared poverty to unequal prosperity. 'Official egalitarianism, in so far as it existed, has been abandoned' (Beresford and McFarlane 1995: 58).

In rural areas of Vietnam and Laos, the agricultural reforms have undermined the apparent uniformity of the past, contributing to a significant increase in inequality. The income spread in rural areas of Vietnam in 1990 was said to have widened to forty to one and individual land holdings in some rice-growing areas had reached 100 hectares, and in upland areas over 500 hectares (Kerkvliet and Porter 1995: 16, Kerkvliet 1995a: 73).[22] One poor farmer in Ban Un, a village in northern Vietnam, for example, observed that poor households, because they own no buffalo, cannot get manure to apply to their fields, and therefore are constrained in their attempts to raise yields (ActionAid Vietnam 1995). This means that they are effectively barred from taking advantage of the reforms. The same picture emerges in work on Laos where Trankell reports that while villagers welcomed the additional opportunities for economic enterprise created by the reforms, were at the same time quite aware 'that the new system would probably have at least as many losers as winners' (1993: 92). This coincides with Håkangård's findings from rural Laos where the reforms had created numerous new money-making opportunities for richer households, but poorer households were too busy surviving to take advantage of them (Håkangård 1992).

## Categorizing the poor

It is often argued that before they can tackle poverty, governments must begin by identifying the poor (e.g. Quibria 1991: 105). As a result, many studies of poverty which glean their information from national surveys, like those mentioned above, turn to the issue of 'characterizing' the poor. This sometimes involves running regression analyses on the data set, thereby arriving at a series of defining characteristics of poor households which build up a picture of the 'anatomy' of poverty (e.g. Balisacan 1994; Mason 1996 and see also Suganya Hutaserani 1990). Often, the 'characteristic' poor family will apparently show similar features whatever its nationality. Its members live in a rural area, often in a lagging and remote region. They will be engaged in and dependent on farming, but will tend to either work smaller than average landholdings – on which the family are more likely to be tenants – while having a higher than average component of subsistence production and consumption, or they will be landless agricultural labourers. The family will be larger than average, have a young head of household[23] with a low level of educational attainment, and will have more restricted access to social services like health and education, and to financial capital (see Table 3.7).

Critics of this approach to poverty analysis contend that it conceals the diversity and dynamism that must be recognized if a full understanding of poverty is to be achieved. It stereotypes and pigeon-holes the poor, which is only a short step from assuming that the experience of poverty is uniform and therefore that the solutions to poverty are similarly uniform.

This chapter offers a statistical and reductionist view of poverty: Who are the poor and the non-poor? Where do they live? How successful has development been in reducing the incidence of poverty? This approach to poverty and the poor is about categorizing and measuring. Critics of what might be termed the 'economics of income poverty' would argue that it tells us little or nothing

*Table 3.7* Characterizing and categorizing the poor in Indonesia, the Philippines and Thailand

|  | Indonesia | Philippines | Thailand |
|---|---|---|---|
| Rural-based | ✓ | ✓ | ✓ |
| Dependent on farming | ✓ | ✓ | ✓ |
| Concentrated in peripheral regions | ✓ | ✓ | ✓ |
| Smaller than average land-holdings | ✓ | ✓ | |
| More likely to be tenants or landless agricultural wage labourers | ✓ | ✓ | |
| Larger than average element of subsistence production and consumption | | | ✓ |
| Larger than average family | ✓ | ✓ | ✓ |
| Young household head | | ✓ | ✓ |
| Household head with a low level of education | ✓ | ✓ | ✓ |
| Low level of access of social services such as health and education facilities | ✓ | ✓ | ✓ |
| High level of underemployment | | ✓ | |

*Sources*: Thailand: Medhi Krongkaew 1993 and Medhi Krongkaew et al. 1992; Philippines: Balisacan 1992; Indonesia: Brookfield *et al.* 1995, Booth, A. 1993, Mason 1996

about *why* people are poor – except to banally observe, for instance, that most live in rural areas and are engaged in agriculture. The analysis of poverty, in these approaches, is about quantification and not about understanding. Time and again, micro-studies emphasize that poverty is complex, differentiated and dynamic (Chambers 1995: 173). It has multiple causes, is manifested in multiple ways, is experienced differently, varies over space, and changes through time. But to be fair, these critics, for argument's sake, usually discount or play down the degree to which economists engaged in such work *are* aware of the complexities of poverty as well as the social (as well as economic) roots and manifestations of poverty. The mantra that poverty is about 'more than income' is attractive to other social scientists attempting to carve out a niche distinct from that occupied by economics, but this should not be taken to mean that economists have been so blinkered as to have entirely ignored the issue. The difference, perhaps, is that many economists continue to believe that it is critical to set down markers (or poverty lines) and to assess wealth and poverty (through consumption surveys) if policies are to be formulated to assist the poor. Governments, they might argue, simply cannot operate using a multitude of micro-level social studies which lack any common base lines and produce highly nuanced interpretations of poverty, often at the intra-household level. Further, economists continue to maintain that poverty *is* about economic growth – and they would point to the experience of the growth economies to support their case:

> It was not caprice that made economic growth central to thinking about development, that put GNP on the throne and has, in the face of all the

onslaughts and pinpricks, firmly kept it there. . . . Economic growth
. . . was the only effective way of making the people of the Third World
materially better off, of raising consumption standards.

(Arndt 1987: 173)

## Notes

1  For Southeast Asia, examples include: Dixon, Chris and Drakakis-Smith, David
   (forthcoming) (eds) *Uneven development in South East Asia*, Aldershot: Avebury;
   Parnwell, Michael J.G. (1996) (ed.) *Thailand: uneven development*, Aldershot:
   Avebury. More generally, we can point to Amartya Sen's (1992) *Inequality re-
   examined*, Oxford: Clarendon Press.
2  These figures have subsequently been quoted in other, scholarly, publications – for
   example Schmidt does so in his paper entitled 'Paternalism and planning in
   Thailand: facilitating growth without social benefits' (1996: 79).
3  The important issue of whether the 'accepted' economic measures are acceptable
   measures is discussed in the next chapter.
4  Ji Ungphakorn takes the authors of a series of articles in an issue of the *Thai
   Development Newsletter* (*TDN* 1995b) to task for suggesting that the poor are worse
   off today than ten or twenty years ago. He writes: 'We do the poor no service by not
   starting out with the facts. Numerous studies have been made as to the effects of
   economic growth on the well-being of the Thai population and there is every reason
   to believe that the standard of living of the majority of Thais has improved over the
   last 20 years' (1995b: 54).
5  It should be stressed that some scholars maintain that inequalities in Indonesia *are*
   widening despite the fact that the BPS figures suggest the reverse (e.g. Tjondronegoro
   *et al.* 1992: 88–89). This is because the statistical base on which this view is founded
   is so narrow (see Chapter 5).
6  In 1968 the *rumus tani* or farmer's formula was introduced, explicity linking the
   price of rice with the cost of fertilizers – with the aim of maintaining 'incentive
   parity'.
7  Indonesia's modern rice economy has an extensive literature. See Booth A. 1985 and
   1988; Bresnan 1993; Fox 1991; and numerous papers in the *Bulletin of Indonesian
   Economic Studies*. Note that a shift into non-rice crops, supported by the government, has raised the possibility that Indonesia may return to rice deficit.
8  The original members were: Indonesia, Malaysia, the Philippines, Singapore and
   Thailand. Brunei and Vietnam joined the Association in 1984 and 1995 respectively. Laos and Cambodia are scheduled to join in 1997 and possibly Myanmar too.
9  Simon Kuznets famously hypothesized that income distribution would widen in the
   early stages of economic growth and then narrow as countries attained greater wealth
   – resulting in an inverted 'U'.
10 Medhi Krongkaew observes with regard to Thailand's development-equity experience: 'The blatant, continuing income gap may be looked on as an indication of
   unsuccessul economic management' (1995: 64).
11 A 'cohort' comprises the same group of families, measuring their progress over time.
   Cohorts are based on the composition of the various percentile groups in 1971.
   'Percentile groups' refers to those families who comprise a particular group at any
   point in time. Thus, the members of a percentile group will change over time as
   some families improve themselves (relatively) and others fall back (relatively).
   Cohort membership, though, remains unchanged.
12 The Indonesian Population Ministry in a December 1994 report referred to this
   group of near-poor as 'pre-prosperous' (Cohen 1996: 28). This not only raises a

great opportunity for punning, but also brings to mind the definition of the difference between an optimist and a pessimist as being those people who refer to either a 'half full' or a 'half empty' glass.

13 For papers on the Northeastern region of Thailand, see: Demaine (1986); Dixon (1978); Parnwell (1988 and 1992); Parnwell and Rigg (1996); and Rigg (1986).

14 There is a strong case, though, that factors beyond these regional development initiatives were also instrumental in leading to the demise of the CPT. In particular, splits within the ranks of the CPT between pro-Beijing/Phnom Penh and pro-Moscow/Hanoi factions, former Prime Minister Prem Tinsulanond's political amnesty for members of the CPT, and Beijing's rapprochement with Bangkok.

15 In this sense at least, it shares the experience of Northeast Thailand. What makes East Timor different from Northeast Thailand is that it remains a source of political instability and a threat to the unity of the nation. Regional development, therefore, has been unsuccessful in the province in both economic *and* security terms.

16 Irian Jaya is the exception with its strong mining and timber sectors. Yet it still has a high incidence of poverty, possibly the highest in the country – see Box 3.2.

17 These estimates are based on figures which were questioned earlier in the chapter and should be treated with some caution.

18 Assigning single motivations to migration decisions is highly suspect. It is likely that migrants make the decision to move on the basis of several factors, forming a web of motivation, and to assume that economic factors are either the only, or even the major influence is dubious. None the less, it does seem that, overall, the desire to increase income is a highly important factor driving migration streams in the region.

19 It has been argued that urban poverty lines are set too low and that they fail to take into account the higher costs of living (food, rent, transport, etc.) in towns and cities (*Environment and Urbanization* 1995). Two examples of how changing definitions of urban areas can distort figures, come from Thailand and the Philippines. In 1991, the total population of Thailand's 'municipal' areas was 10.0 million. In 1992, with an expansion of municipal status to towns which previously had not warranted this classification, the population of Thailand's municipalities almost trebled to 29.0 million (NSO 1993: 18). In the Philippines, the ratio of the number of rural to urban poor increased between 1961 and 1971, from 70:30 to 76:24. Since then the ratio has declined and in 1991 stood at 59:41 (Table 3.5). This decline, Balisacan contends, is due to the redefinition of extensive 'rural' areas as urban between 1980 and 1990 (1994:123–24). The same difficulty has been highlighted with respect to the East Malaysian state of Sarawak. Here urban poverty in 1990 was estimated to be just 5 per cent. Yet a study of poverty in the urban sprawl that lies beyond the administrative boundaries of the township of Bintulu, yet is clearly part of the urban fabric, revealed an incidence of poverty of 21 per cent (quoted in Brookfield *et al.* 1995: 222).

20 The discussion here draws disproportionately on work conducted in Vietnam. There has been almost no work recent work on Myanmar, little on Cambodia, and just an emerging literature on poverty in Laos.

21 See page 111 for a critique on the manner in which the poor are being 'discovered' in Vietnam.

22 Hoang Thi Thanh Nhan states that the income gap between rich and poor has widened from 1:2 in the early 1960s, to 1:3–4 in the 1970s and 1980s, and had reached 1:40–100 by the early 1990s (1995: 19–20).

23 Implying that the classic 'poor' household is fairly 'young' in terms of the household life cycle with dependent children.

# 4

# THE EXPERIENCE OF EXCLUSION

## Introduction

'Exclusion' in the preceding chapter was largely used in the sense of 'Excluded from development' or rather 'Excluded from modernization'. The excluded, in these terms, are usually identified according to income-poverty, and then characterized geographically and economically. While it might be argued that the poor are therefore those who have 'missed out' on development, the gauging of progress against such a measure makes the determination of success and failure intimately linked to notions of economic growth and commercialization. There is therefore a critical basic assumption built into the geography of exclusion – as 'viewed from above': that we can gauge accurately the poor and the non-poor, or the excluded and the included, according to surveys which base their results on estimates of income/consumption. And following on from this, that income/consumption is, in itself, an appropriate measure of exclusion or poverty.

It is widely acknowledged that deprivation is much more than a lack of income or consumption. Yet most measures of deprivation use income-poverty, often calculated on the basis of consumption patterns, as their central determining variable. In this way, poverty tends to become narrowly defined, despite protestations that such an approach is inadequate. Chambers writes of this tendency:

> It is then but a short step to treating what has not been measured as not really real. Patterns of dominance are then reinforced: of the material over the experiential; of the physical over the social; of the measured and the measurable over the unmeasured and the unmeasurable; of economic over social values; of economists over disciplines concerned with people as people. It then becomes the reductionism of normal economics, not the experience of the poor, that defines poverty.
>
> (Chambers 1995: 180)

Escobar, in writing of exclusion, contends that the 'most important exclusion ... was and continues to be what development was supposed to be all about: people' (1995a: 44). This resonates with Chambers' view. Development, as a

109

technocratic exercise in modernization, obscures the real effects of the process on individuals, their lives and livelihoods. It is clear that people are not excluded from development merely because they lack income. Nor is exclusion from development necessarily the key issue – although it is traditionally presented as such. The implication is that achieving high economic growth rates, though perhaps a necessary condition for progress on a broad and inclusive front, is not a sufficient one. Even in such miracle economies as Singapore, Thailand and Malaysia, there are groups who, for an assortment of reasons, have found themselves excluded or marginalized. In addition, even those who have acquired some of the fruits of economic growth, may in the process have sold their 'souls'. An example discussed in more detail below is prostitution, where the economic benefits must clearly be set against the resulting social stigma and health costs. The modernization 'bargain' for this group at least is very much a Faustian one.

The discussion that follows is an attempt to shed some light on the people that lie behind the aggregate statistics. Who is excluded – in the widest sense – why are they excluded, and how is this manifested and experienced. The various sub-groups are clearly not mutually exclusive categories and for many it is their multiple exclusion which consigns them to an existence of poverty, powerlessness, stigma and disapprobation. It is also true that what may, superficially, appear to be a 'geography' of exclusion can have its origins in powerful underlying political or economic structures of exclusion. A subsistence farmer in the Indonesian province of Irian Jaya is poor/excluded, arguably, because he is a Papuan; because he is a farmer; because he has no formal education; because he lives in a remote area of Indonesia; because his commitment to Indonesia is subject to doubt; and because he is viewed as 'primitive'. It may also, of course, be because he is a rotten farmer. But although this notional Papuan subsistence farmer may be excluded, the experience of exclusion is not set in stone. In part it may be associated with the progress of the life cycle; with retirement, for example. Nor are people consigned to their fate; they have the opportunity, even if it is a constrained one, to escape and to build new lives (see page 240 for a discussion of the poor as victims). Inclusion and exclusion are therefore multidimensional, overlapping and dynamic.

There is a tension, and to some extent a contradiction, in discussions of exclusion. To some, the study of exclusion is important because it highlights those groups and individuals who would be overlooked in simple income-poverty studies. It illustrates and emphasizes the point that because development is multidimensional, so measures or identifiers of development–underdevelopment must also be multidimensional. However in this schema, the rationale and emphasis remain embedded within the modernization ethos. It represents not so much a rejection of income-poverty, but a fine-tuning. Other studies of exclusion apparently reject any attempt to link progress with modernization, or vice versa. But if such perspectives do not imply excluded from development, then what exclusion(s) are they referring to? The difficulty, as noted below, is that those who would wish to challenge development and its economistic slant are forced to do so in developmental – and often economistic – terms. The tyranny of the modernization discourse leaves little choice; just as it leaves governments and people in the so-called developing world with little choice.

## Poverty and exclusion

*Poverty = lack of modernization*

> As goods increase,
> so do those who consume them.
> And what benefit are they to the owner
> except to feast his eyes on them?

*Ecclesiastes* 5: 10–11

> Both Kings and ordinary people must die in the midst of want,
> never reaching an end to desire and craving.

*Buddhist scriptures*

Poverty was 'discovered' at the end of the nineteenth century in the developed world and on a global scale in the middle of the twentieth century (see Esteva 1992). Virtually overnight, scholars, politicians and journalists could shock their readers, listeners and constituents with claims that 'two billion people live in absolute poverty'. Figures of truly biblical proportions were – and still are – bandied about to emphasize the necessity for development. It was as if merely thinking about poverty could transform ordinary people into 'poor' people requiring massive help and assistance (i.e. development). This change comes at the tail end of a longer-term transformation which successively reconfigured the unknown world (*terra incognita* – 'here be dragons') into the primitive and pagan world; the primitive into the colonial world; and the colonial into the post-colonial and underdeveloped Third World (see Peet and Watts 1993: 236). In parts of Asia at least, the Third World is now entering its latest reincarnation as the miracle world.

The assumption that poverty is about exclusion from development can be seen reflected in how the lives and livelihoods of tribal peoples are characterized. Sahlins, for example, writes that: 'The world's most primitive people have few possessions, *but they are not poor*. Poverty is not a certain small amount of goods, nor is it just a relation between people. Poverty is a social status. As such it is an invention of civilisation' (quoted in Griffin 1989: 13). In contrast to Sahlins, Prime Minister Mahathir Mohamad of Malaysia in replying to the environmental activist Bruno Manser and the latter's defence of the hunter-gathering Penan of East Malaysia, contends that: 'it is the height of arrogance for you [Bruno Manser] to advocate that the Penans live on maggots and monkeys in their miserable huts, subjected to all kinds of diseases. . . . Have they no right to a better way of life?' (*FEER* 1992: 9).

The poor are still being discovered. In Vietnam the years since the late 1980s has seen a spate of poverty surveys, revealing that anywhere from 12 per cent (Dao The Tuan 1995: 147) to over 50 per cent (Schwarz 1995) of the population are living in poverty. In 1992–93 the first consumption survey was undertaken in Laos revealing, by one measure, that 72 per cent of households are poor (see Box 3.3). What resonates in these surveys with the views of the post-developmentalists is that the poor are being conjured into existence by applying modernization criteria to their lives. In reviewing poverty in Vietnam's countryside, Dao The Tuan identifies the main task of rural development as 'help[ing] peasants change from subsistence to commercial farming' (1995: 147–9). Another Vietnamese economist, Hoang Thi Thanh Nhan, interprets the problem (and therefore the solution) of poverty in the same manner when he observes that poverty 'hits' those exclusively engaged in agriculture. These households, he seems to imply, are by definition 'under-employed' and poverty is an 'inevitable' outcome of such a condition (1995: 19).

The overwhelming tendency to determine development in terms of material (economic) progress is, some post-developmentalists would argue, the single greatest victory of the development discourse (see page 32). In gaining the high ground, the discourse can set the field of debate and is therefore in a position to determine the game that will be played. That game is the race to modernity and the boundaries of the debate are set in advance, ensuring that there is little fundamental questioning of what is being promoted. Even those scholars who apparently applaud the anti-developmentalist thinking that would wish to reject the assumed rationality of development and modernization, at the same time point to the 'appalling spectacle' of the polarization of global wealth as it is detailed in each year's *World Development Report* (e.g. Peet and Watts 1993: 238).

### Stand up and be counted: letting the poor identify themselves

It is usual for the determination of poverty to be based on criteria selected and weighted by economists. Although these economists are increasingly nationals of the countries concerned they are still, largely, Western-trained and form part of an urban-based and Westernized élite. As such, they are unrepresentative of the population as a whole. It is because of the perceived bias inherent in such determinations of poverty that alternative development workers – usually working with NGOs – and scholars have begun to construct rankings of wealth based on a much wider range of criteria, often drawing on the views and opinions of the respondents themselves.[1] The poor (as well as the middle and rich) set their own standards of poverty, deprivation or exclusion – creating, in the process, emic poverty 'lines' or wealth rankings.

Robert Chambers, for one, sees considerable advantages in using participatory methods that draw upon the poor's knowledge of themselves:

The new [participatory] methods enable poor people to analyze and express what they know, experience, need and want. They bring to light

many dimensions of deprivation, ill-being and well-being, and the values
and priorities of poor people.

(Chambers 1995: 185)

What is notable about the results of such studies is not just that they bring (in
some senses) greater precision to our understanding of poverty, given that
conceptions of deprivation are locally rooted, highly differentiated and dynamic.
But also that they appear to arrive at different visions of what is important to
poor people. Such studies have shown, in some instances, that contrary to the
normal economist's 'reality', poor people are less concerned with income *per se*
and more concerned with issues such as self-respect, joys and sorrows, status,
education, and health and well-being (Chambers 1995: 186–7) (Table 4.1). The
familiar refrain that poor people can only enjoy the luxury of such concerns
when they have become non-poor is refuted by such studies.[2] Porter's work on
Vietnam (1995a) emphasizes this point that the poor there are not concerned
with income, so much as 'endowments' and 'risks to livelihood'. Their fears
revolve around morbidity and mortality and an inability to save, while their
perception of the problem tends to focus on environmental marginality, remoteness from markets and social infrastructure (health and educational facilities
particularly), and the particular difficulties that confront female-headed households. Taken together these endowments create marginality and, in turn, this
constrains choice and amplifies risk.

However asking poor people to count themselves, though it may be in line
with current methodological fashion, does have shortcomings.[3] To begin with, it
should not be assumed that in the process the 'invisible' poor will necessarily
become visible. If the very poorest (or ultra poor) are excluded from village
meetings and not regarded as true and full community members then their voices
will not be heard. They may well remain invisible even to survey teams ostensibly

*Table 4.1* Views of deprivation

| Deprivation: an economist's view | Deprivation: a people's view |
| --- | --- |
| Income | Vulnerability |
| Consumption | Stability |
| Resources | Regularity of employment |
| Employment | Dependency |
| | Self-respect |
| | Food deficit (or being forced to eat staples other than rice) |
| | Lack of a tiled roof or a brick house |
| | Female-headed households |

*Sources*: this table is drawn from numerous papers on wealth rankings published largely in *PLA
Notes* (*Participatory Learning and Action Notes*), formerly *RRA Notes* (*Rapid Rural Appraisal Notes*).
*Note*: this is a composite table drawn from work conducted in numerous countries. Not every
person in every place will identify the criteria listed in the right-hand column. See note 2 for a
discussion of whether such non-economic factors are more important.

adopting a people-oriented approach (see Lindberg *et al.* 1995: 12–14). Action-Aid Vietnam (AAV) found in 1990 that when they first asked villagers to rank their fellow villagers according to wealth criteria defined by the ranker they were 'extremely reticent on the subject of relative wealth and poverty, shy because of the continual presence of local officials' (Turk 1995: 37). Although AAV found two years later – by which time the official presence in the area had much reduced – that villagers were far more forthcoming with their opinions, a general feature of such participatory surveys seems to be that local people are often reluctant to offer an opinion on the relative wealth of their fellow villagers (see Carter *et al.* 1993). It should be added that participatory approaches are also highly dependent on the skills and sensitivities of the practitioners.

Furthermore, just because poverty surveys are conducted in a participatory manner, it does not automatically follow that local people will be any more accepting of the results. AAV found in Mai Son District, Son La Province in the north of Vietnam, that there was always the possibility of rapid change in a household's status. If a member fell seriously ill, then even a comparatively wealthy household might find itself in a 'rapid downward spiral' (Turk 1995: 38). This emphasizes the high degree of dynamism and mobility between the poor and non-poor in any given population, reiterating the point raised in the discussion of Edmundson's work in Java in the previous chapter (see page 85). However, AAV discovered that this was also an issue which threatened to undermine their own work, as it made the identification of the poor and non-poor, particularly at the margins, extremely arbitrary. It sowed confusion among the population as to why some households were included in the programme, and others excluded. As a result, 'there was considerable [community] resistance to the exclusion of households from [AAV's] programme activities' (Turk 1995: 41).

There are also semantic difficulties: what do outsiders and insiders mean by such fuzzy concepts as 'wealth', 'poverty' and 'deprivation'? The emphasis, as in the standard income-poverty methods discussed in the last chapter, is still on material assets – particularly with regard to the term 'wealth'. Not only is it a case of insiders and outsiders having different perceptions; there are also likely to be marked differences between insiders. Between men and women, the elderly and the young, the poor and the non-poor, between different ethnic groups, and so on. This importance of semantics is clear in Sjafri Sairin's paper on plantation labour in North Sumatra. The author writes of the importance of the cultural notion of *tentrem* (*tenteram*) to workers. The dictionary defines the word as meaning 'calm, peaceful, safe and tranquil' (Echols and Shadily 1987). Yet Sairin notes that among plantation workers it is defined significantly differently by various people:

> For some it is an ideal situation in which everything is going well. . . . For others, *tentrem* is having enough food to eat and dwelling in decent housing, with no crimes in society and no personal conflicts. Yet for some others, *tentrem* may suggest a condition of stability.
>
> (Sairin 1996: 12)

It seems that *tentrem* encompasses two different aspects of life. It can be used to refer to sufficiency within the economic realm – the fulfilment of basic daily needs – and to stability and tranquility in terms of social relations. Only when both of these elements have been fulfilled can a person be said to have achieved *tentrem batin*, or 'peace in one's heart'. This paper is interesting because it also illustrates that economic security remains highly important. Plantation work is viewed, in some quarters, as particularly dehumanizing. Yet 'from the [Javanese] workers' point of view, employment in a plantation provides a genuine sense of security' – *cekap sandang, pangan lan papan* or 'enough clothing, food and housing' (Sairin 1996: 11). Sairin links this with James Scott's subsistence ethic (1976). In this instance, though, the subsistence ethic is being met in an economic system where wages are low and which many scholars decry as particularly exploitative.

Wealth rankings of the type described here and used, on the ground, by many NGOs – like ActionAid Vietnam – are very hard to standardize and are not conducive to scaling-up for application at a wider level (see Leach 1994: 53; Baulch 1996: 8). The information is locally gleaned, probably on the basis of just a single survey, or surveys spread over a comparatively short period. It is for these reasons that governments, among others, operating from the centre, often still continue to favour nationwide consumption surveys. Because they are standardized they are amenable to wider application, and because they have been conducted over a number of years there are the longitudinal data available that allow trends over time to be investigated.

So although this approach to wealth ranking and the assessment of poverty avoids some of the problems associated with the more standard, economic approaches discussed in the last chapter it is, by no means, an answer. Some scholars would further suggest that it still does not escape the difficulty of assuming that a lack of development is a problem requiring rectification. With most of Southeast Asia firmly aboard the modernization bandwagon there are few people, even in marginal villages, who would not see lack of modernization as a key indicator of deprivation.

## EXPERIENCES OF EXCLUSION

### Minority groups and marginal peoples

#### Of tribes, hill peoples and natives

In all the countries of the region, policies – whether explicitly, implicity, or merely through their unintended operation – have marginalized particular groups and widened ethnic divisions. In mainland Southeast Asia the focus has tended to be on the hill peoples of Thailand, Myanmar, Laos and Vietnam – the so-called hill 'tribes'. In the archipelago, the emphasis has been on forest-dwelling tribal groups, especially those of Borneo and the Indonesian province

115

*Box 4.1* Developing the villages that have been 'left behind' in Indonesia

On 1 April 1994, the Indonesian government launched the three-year Presidential Instruction Programme for Less Developed Villages, better known as *Inpres Desa Tertinggal* (IDT). Despite its title, this effort was an attempt to move away from the programmatic, top-heavy and overly-bureaucratized approaches to poverty alleviation attempted in the past. Prior to the programme, the National Development Planning Board under-took a survey of all the country's villages and arrived at a map designating 20,633 so-called *desa tertinggal* or 'villages that have been left behind', communities where the 27 million Indonesians classified as 'poor' were concentrated.[a] The National Development Planning Minister then took the unusual step of consulting with NGOs in the formulation of a policy specifi-cally to address the problems of these 27 million poor. The result was a programme which empowered self-help village groups in each *desa tertinggal* to identify their own needs and priorities. Funds, set at 20 million rupiahs (over US$9,000) per village, were then directly allocated to the self-help groups.

Although a comprehesive survey of the programme is still to be published, some problems have already emerged. It has been found in some communities that influential villagers have dominated proceedings, effectively excluding the poor from gaining access to the self-help groups through which funds are allocated. Their exclusion is often made on the pretext that their inclusion would slow repayment rates, increase the chances of fines being imposed for late repayment, and therefore jeopar-dize other group members. Women, it seems, are also excluded in some instances, with the village head of Tegal Alur in West Java explaining: 'According to our cultural norms, it's the husband who takes responsibility. Though the wife might be smarter . . . that's the way it goes' (Cohen 1994: 58). One informal survey of ten villages across the archipelago revealed that villagers were setting stringent repayment schedules for themselves – far more stringent than strictly necessary given the terms of the loans. 'Ironically, it is the villagers themselves who are setting the payback targets [and therefore] putting a tight squeeze on the poorest of the poor' (Cohen 1995: 24). It also seems that although NGOs may have been involved in the initial design of the programme, they have been excluded from the later implementation phase. In Tegal Alur for example, district officials did not invite NGOs to be involved in the training programmes. Further, all projects had to be approved by local officials who were also available to assist villagers in the process of highlighting their needs and priorities.

The experience of Indonesia's *Inpres Desa Tertinggal* illustrates the difficulty – some might say the impossibility – of the State designing and implementing truly grassroots, locally-driven development programmes. To expect the poor to play a determining – or even an equal – role in fractious and hierarchical communities is expecting a great deal; while to envisage

continued . . .

that local government will not play a significant role in such programmes when officialdom is so pervasive and intrusive is equally unrealistic. As John Friedmann remarks with reference to alternative development in general:

> Like it or not, the state continues to be a major player. It may need to be made more accountable to poor people and more responsive to their claims. But without the state's collaboration, the lot of the poor cannot be significantly improved. Local empowering action requires a strong state.
>
> (Friedmann 1992: 7)

### Note

a The very means by which villages are defined as 'backward' is, arguably, highly programmatic. Twenty-seven indicators of backwardness like accessibility, village facilities and mortality rates were used to categorize villages as 'non-poor', 'poor' and 'very poor' (Brookfield *et al.* 1995: 224–5). This runs against the more nuanced determinations of poverty that many analysts would like to see being utilized. For a more detailed discussion of the semantics of this programme see page 49.

of Irian Jaya. There are, though, many other minorities who have faced either active persecution – like the Rohingya Muslims of Myanmar's Arakan state and the Christian East Timorese of Indonesia – or have found, as a group, that development has seemingly passed them by – for instance the Tamil community of Malaysia and the Lao population of Northeast Thailand. It is difficult to draw out generalizations from the wide range of experiences, except once more to emphasize that these groups often find themselves marginalized in multiple and overlapping ways (Table 4.2). Taken together, such multiple marginality has often created an obstinate degree of exclusion. In addition, the fact that they are marginal and only partially visible has meant that governments have often been able to deny that any problem exists, or at least to ignore it. When such minority groups do become a problem requiring action, it is usually because they have become viewed as a security risk or political threat.

Thailand is often regarded as a homogeneous country in religious, linguistic and ethnic terms with over 90 per cent of the population classified as Thai-speaking Theravada Buddhists. Perhaps it has been this façade of unity which has enabled the authorities to construct policies which either ignore, marginalize or malign the Kingdom's minorities. Nowhere does this view have greater resonance than among the hill peoples of the North. Numbering almost 750,000 or 1.3 per cent of the total population of Thailand, and divided (officially) into nine sub-groups, they live on and at the margins of the Thai state (Table 4.3). They occupy the highland borderlands abutting Myanmar and Laos, are often Christian or animist rather than Theravada Buddhist, embrace cultures which set

*Table 4.2* The multiple dimensions of minorities' marginality

**Geographical marginality**
- Sensitive borderlands
- Remote hill regions
- Inaccessible forest areas
- Dispersed population

**Political marginality**
- Lack *de facto* citizenship
- Lack *de jure* citizenship
- Excluded from mainstream political engagement

**Cultural and social marginality**
- Are distinctive in terms of material and non-material cultures
- Regarded as 'primitive' by the majority
- Regarded as 'other' by the majority
- Lack necessary education to engage with the modern economic and political systems
- Embrace minority religions and beliefs
- Speak minority languages

**Economic marginality**
- 'Primitive' in the sense of not commercial and therefore not modern
- Embrace distinctive livelihood strategies, often regarded as unproductive

**Environmental marginality**
- Regarded by élites as the cause of environmental degradation
- Inhabit, and suffer from, degraded environments often caused by others
- Especially vulnerable to 'natural' hazards

*Table 4.3* Hill peoples of Thailand

| Hill group | Population in Thailand | Origins | Approximate date of arrival in Thailand |
|---|---|---|---|
| Karen | 292,814 | Myanmar | 18th century |
| Hmong (Meo) | 91,537 | China | Late 19th century |
| Lahu (Mussur) | 57,144 | Yunnan (China) | Late 19th century |
| Mien (Yao) | 34,545 | southern China | Mid 19th century |
| Akha (Kaw) | 32,041 | Yunnan (China) | Early 20th century |
| Htin | 25,613 | – | – |
| Lisu | 25,051 | | |
| Lawa | 8,227 | – | – |
| Khmu | 8,705 | – | – |
| Total hill people population | 575,677 | | |

*Source:* Schmidt-Vogt 1995: 49; *TDN* 1993: 20
Note: estimates of the population of the hill people of Thailand should be treated with caution. The above population figures are taken from a 1992 report of the Tribal Research Institute in Chiang Mai (northern Thailand). A more recent study in 1993 gave a total hill people population of 749,353. There are other tribal groups – like the Mlabri (at last count numbering just 139, and falling) and the Sakai or Semang (not more than 500 individuals) – who are not included in most population estimates.

them apart from the majority 'Tai' population, often lack citizenship,[4] and adopt distinct livelihood strategies which further mark them out as 'different'. This multiple separation from the mainstream of Thai life has resulted in their multiple exclusion. As the influential Thai Development Research Institute (TDRI) has stated, 'for the Royal Thai government, the hill tribes pose a series of profound political, social and ecological problems' (TDRI 1987: 80; and see Tapp 1989).

Successive Thai governments have used the supposedly environmentally destructive character of shifting cultivation as a reason to settle the hill peoples in *nikhom* (resettlement villages) and actively to evict them from forest reserve land traditionally regarded as a communal resource, in the process undermining their livelihoods. It is common for officials and official publications to place the blame for many of the problems of the North, especially deforestation and associated environmental degradation, squarely on the hill peoples. Security concerns are also prominent in the designation of the hill tribe 'problem'. Thailand's National Security Council (NSC) for example has played a central role in the administration of those areas inhabited by ethnic minorities and its assimilationist policies are driven in large part by security fears such as separatism, communist insurgency, drug trafficking and opium cultivation, and conflicts with neighbouring countries. This narrowness of thinking is reflected in the NSC's Master Plan for Community and Environmental Development and Narcotic Plantation Control in Highland Areas (1992–96). The plan's identification of the problems facing highland areas is almost exclusively linked to the activities of the hill peoples. That other agents, like logging concessionaires and agribusinesses, might also play a role is not considered. In the process, of course, the authorities are able neatly to deflect attention from the equally, if not more pertinent contribution of corrupt officials, incompetent management, lowland agricultural and business interests, and ineffective policymaking, in creating the conditions that exist in the Kingdom's highlands (Rigg and Stott, forthcoming).

The Akha village of Ban Mae Poen in Chiang Rai province is an example of such multiple exclusion. Of its 300 inhabitants, only 30 had received Thai citizenship by the late 1980s. The remainder lived in a state of insecurity, unable to travel beyond the district and therefore denied the opportunity to search for work, at risk of resettlement and the economic and social dislocation that that might bring, and open to abuse by the police and army. Sanitsuda Ekachai wrote at the time that '[t]hey live in a state of inferiority to lowlanders and of frustration at being blamed for the problems of the North' (1990: 182).

The same dominant – and domineering – attitude of the state towards minority groups is seen in Borneo.[5] King, in his study of the Dayak peoples of the island, believes that their future is bleak.[6] The process of modernization will transform them into marginal peasants, estate workers and urban wage labourers. More to the point, this destiny is not just a sad but unintended consequence of the operation of economic forces, but one 'which governments probably more consciously than unconsciously have marked out for them' (King 1993a: 302).

119

The operation of *politik pembangunan*, or the politics of development, ensures that strategies of development stress modernization and this, in turn, results in the logging of the forests for foreign exchange, the conversion of logged land to settled agriculture, and the marginalization, at least, of the Dayak way of life (King 1993b).[7] In another, more recent paper, King summarizes the 'Iban dilemma' – namely their economic and political marginality – as being a consequence of a series of environmental and historical factors which have meant that the Iban lack political influence commensurate with their numbers, and partly as a consequence lack the ability to mould economic policies to their best advantage (King and Jawan 1996). In the state elections of 1991, the Iban-based Parti Pansa Dayak Sarawak or PBDS managed to win just seven seats on a ticket of 'Dayakism'. This constituted 12.5 per cent of the 56 seats in the state parliament, while the Iban represent around 30 per cent of the population, and all Dayak groups some 45 per cent (Table 4.4) (Chin 1996; Mason 1995). 'The clear loser [in the election]', Chin writes, 'is the Iban/Dayak community . . . [who] . . . by remaining disunited and splitting their vote . . . cannot hope to promote their political and economic interests significantly' (1996: 40). The Iban find themselves classified as *bumiputras* – the putative beneficiaries of Malaysia's policies of ethnic restructuring – and yet because they are lumped in together with the dominant Muslim Malays often find that these policies are inappropriate.[8] The same ethos, and the same pressures, are evident in other areas of the region.

Most studies of the hill peoples of Thailand, Laos, Myanmar and Vietnam, as well as the native peoples of Malaysia and Indonesia, emphasize the degree to which their multidimensional distinctiveness has led to their being excluded in

*Table 4.4* Population and poverty in Sarawak

| | Population | | Percentage of 1995 total | Poverty (incidence, 1990) |
| | 1990 | 1995 | | |
| --- | --- | --- | --- | --- |
| Iban | 493,000 | 510,000 | 30 | |
| Bidayuh | 84,000 | 136,000 | 8 | |
| Other Indigenous | 51,000 | 119,000 | 7 | |
| *Total indigenous* | 628,000 | 765,000 | 45 | |
| Malay | 181,000 | } 442,000 | 26 | |
| Melanau | 53,000 | | | |
| *Total bumiputra* | 862,000 | 1,207,000 | 71 | 29% |
| Chinese | 294,000 | 493,000 | 29 | 4% |
| Others | 10,000 | – | | |
| *TOTAL* | 1,166,000 | 1,700,000 | | 21% |

*Sources*: King and Jawan 1996; Brookfield *et al.* 1995; Mason, R. 1995

various ways. Yet government policy in the region is uniformly directed at the inclusion of such people into the mainstream of political, economic and cultural life. Development policies aim, for example, to incorporate such peoples through infrastructural developments, through education and the promotion of the national language, through the encouragement of missionary work, and through resettlement. These policies, though, are predicated on the belief that minority groups should become 'more like us' – i.e. more like the majority. It is through bowing to the collective economic, political and cultural will of the majority that such groups can partially throw off the discriminatory label of 'minority' and become part of the 'majority'. Indonesia seems, at one level, to revel in the diversity of cultures and peoples that constitute the Indonesian nation. The country's motto, *Bhinneka Tungal Ika*, is usually loosely translated as 'Unity in Diversity'. Yet this slogan, along with the state ideology of *Pancasila*, were not created so that the newly-independent nation could celebrate diversity, but rather to maintain unity in the face of diversity.[9] The national language Bahasa Indonesia, for example, is not the native language of the majority Javanese population but a hybrid language, neutral in ethnic terms, that draws largely on the trading *lingua franca* of the archipelago. This Malay language was selected to deflect likely accusations of 'Javanization'. As time went by, people began to call the language Bahasa Indonesia (the Indonesian language) rather than Malay and, as Ricklefs puts it '[t]he linguistic vehicle of national unity was thereby born' (1981: 176).

Although governments have had considerable success in pursuing this process of incorporation, arguably far more important – and more insidious – has been the homogenizing effect of the modernization ethic.[10] Writing of the Akha village of Ban Mae Poen noted above, Sanitsuda Ekachai comments that the effect of children receiving a Thai education is not just that they 'speak Thai fluently' but also that 'their dream is to live like the people they see on television' (1990: 185–6). Government policy, coupled with commercial propoganda, has resulted in the reincarnation of the Akha youth in the clothes of Thai modernity. Apparently, the authorities do not just want the Akha, and the other hill peoples, to be Thai, but also to be T'ai.[11] The same is true of many other minority groups in the region.

Appearances, though, can be deceptive. Toyota, in her study of an Akha village in Northern Thailand, questions whether we can draw any conclusions about a 'weakening' of ethnic identity from, for example, a decline in the prevalence of traditional costumes, dances and cermonies (1996: 237). The Akha, like most other 'tribal' groups, have always been in contact with the wider world (see the section below), and have always been open to outside influences: 'it should be recognised that, since ethnic identity is a product of adaptation, resistance and compromise, it may be renewed, modified and remade in each generation' (1996: 239). At the same time as we can question whether cultural 'erosion' is an appropriate term to use when culture and identity are always in a state of flux, so it is also true that people and regional and national governments have recognized that there is money in cultural distinctiveness. In the Baliem Valley of Irian Jaya, a statue of a Dani warrior resplendent in *koteka* (penis sheath) and little else, graces the capital of Wamena, and the government supports the production of local tribal handicrafts for the tourist market (James 1993). Not too many years ago, in the 1970s, the government was trying to encourage the Dani to wear less 'primitive' attire and seemed intent on obliterating local distinctiveness (see note 9). Outside Kuching in Sarawak, East Malaysia, the Sarawak Cultural Village is a 'living museum' designed to preserve the arts and cultures of the native Dayaks. In these instances, culture is being sold in the name of development, in the process transforming culture into a commodity. It has little to do with preserving culture for its own sake and its own terms, and it can tell us little about the inherent vigour of any particular culture.[12]

### *Temporal exclusion: the denial of common time*

The hegemony of the development discourse can be seen reflected in the way 'tribal', 'indigenous', or 'fourth world', peoples are portrayed as 'backward' and 'living in the past'.[13] Development is regarded as necessary in order that these people can 'catch up' with the rest of the population. In this way, such marginal groups are denied 'common time' with those modern groups at the centre, a tendency which both reflects their position as excluded peoples as well as continuing to justify their treatment as different, distinct and, to a large extent, inferior in terms of their level of 'progress'. This view is reflected in the extract from Prime Minister Mahathir Mohamad's letter to Bruno Manser quoted in the introduction to this chapter, but it has also been present in the work of scholars. Economic constructions such as that of the 'dual' economy inherently depict some groups as living in a backward state, isolated from the advanced, modern economy.

The portrayal of tribal peoples as distinct and isolated has tended to obscure the degree to which such peoples have always played a role in the wider economy and have articulated – socially as well as economically – with the 'core'. The dual economy was often more imagined than real. Relationships between the interior (tribal) and coastal (Malay) peoples of Borneo, for example, were mediated by the river. This is reflected in the terms used to describe interior populations as *hulu* (upriver) and coastal settlers as *hilir* (downriver). Aromatic resins and rare woods, and forest products like animal skins and hornbill ivory were exchanged for salt, metal goods and cloth (Jay 1996: 90). But this relationship was an unequal one. Just as the existence and under-development of the 'third' world can only be fully understood in terms of its relationship with the first world, so the pattern of relations between the fourth world and, say, the dominant lowland *Khon Müang* of Thailand or the Malays of Indonesia and Malaysia is critical in forging an understanding of those minorities that constitute the Fourth World. In others words, far from living in a different time, the Fourth World is sustained by, linked to, and a reflection of relations with the dominant group.

*Ethnicity, class and citizenship among the Chinese diaspora*

The Chinese community in Southeast Asia represents an interesting counter-point to the more usual discussions of excluded minorities. It is well-known that the Chinese diaspora in the region, who number perhaps 25–30 million, are economically powerful and highly influential (Table 4.5). It is partly because of their very economic power that their role, position and identity is contentious – so much so that in Indonesia, for example, the issue has traditionally been termed the *masalah Cina* or the Chinese 'problem'. This 'problem' is a multiple one. It has its roots, perhaps, in the economic power of the Chinese, but extends to include their political loyalties and ('alien') cultural identity. Should, in short, the Chinese be allowed to integrate into Southeast Asian society but maintain their cultural distintiveness (an argument for multiculturalism), or should they be encouraged to assimilate? Governments in the region have approached this issue in different ways, and within individual countries policies have changed over time. In Indonesia, an assimilationist perspective has characterized government policy since 1965 (see Tan 1991). In Malaysia, where the Chinese community represents a much larger (close to 30 per cent) segment of the population, the Malay-dominated government has been forced to accede to the wish of the Chinese community to maintain its distinctiveness, even though certain education and language policies have a Malay accent and emphasis (see Rigg 1991b). Sino-Thais, by contrast, have embarked upon something of a cultural awakening, learning Mandarin, and rediscovering their Chinese origins and cultural roots (see Vatikiotis 1996).

The Chinese in Thailand (until recently) and Indonesia have tended to disguise their Chinese origins by taking local names, learning the local language and

*Table 4.5* Ethnic Chinese in Southeast Asia

|  | Total (millions) | Percentage of population |
| --- | --- | --- |
| Indonesia | 7.2 | 3.8 |
| Thailand | 5.8 | 10 |
| Malaysia | 5.2 | 28 |
| Singapore | 2.0 | 70 |
| Myanmar | 1.5 | 3.4 |
| Vietnam | 0.96* | 1.4 |
| Philippines | 0.8 | 1.3 |
| Laos | 0.008 | 0.2 |
| Cambodia | 0.33 | 4.0 |
| Brunei | 0.075 | 25 |

*Sources:* various
* In 1975, just before reunification, the ethnic Chinese population of Vietnam was around 1,500,000 or 3 per cent of the total. By the time of the 1989 census the Chinese population had fallen to 961,702. The great majority have always been concentrated in the south of the country (Tran Khanh 1993).

embracing the local religion in an attempt to merge with the majority – to assimilate and disappear.[14] Sometimes this suppression of Chinese identity has been forced, and sometimes encouraged. None the less, even in countries where the Chinese have apparently been most seamlessly integrated, like Thailand, neighbours and business associates of Sino-Thais will be quite aware of their ethnic origins. Thai ID cards indicate ethnicity along with citizenship. But at what point a Thai of Chinese extraction becomes sufficiently Thai to be identified as 'Thai' rather than 'Chinese' appears entirely arbitrary (Smalley 1994: 322). At a popular level to be truly Thai is associated with three symbols: King, Nation and Buddhism. Thus the Muslims of the South are, by association, only half-baked Thais and do not warrant the label Thai *thae* or 'real' or 'genuine' Thai. Terminology becomes important. Sino-Thais – i.e. Thais of Chinese extraction – are usually referred to locally as Overseas Chinese. While in Indonesia, when the local media refer to Sino-Indonesian conglomerates investing in China, they write of 'Chinese' investments in China.[15] In these two examples, the stress is on Chinese ethnic identity, not on the actors' Thai or Indonesian citizenship.

As a group and comparatively, the ethnic Chinese in Southeast Asia are rich. Yet this, as Wang Gungwu has put it, is 'wealth without power' (quoted in Blussé 1991: 2). In Indonesia in 1993 about 5 per cent of ethnic Chinese – some 300,000 people – were of indeterminate citizenship and like Thailand the ID cards of ethnic Chinese continue to have a code signifying the cardholder's ethnic origins. In Malaysia, the New Economic Policy (NEP, 1971–90) represented a far-reaching programme of positive discrimination in favour of Malays which is continuing, albeit in a slightly watered-down version, in the replacement New Development Policy (NDP, 1991–). In Vietnam, there occurred the effective expulsion of about 40 per cent of ethnic Chinese or *Hoa* living in the country between reunification in 1976 and the census of 1989 as the Chinese population there became perceived as a political and socio-economic 'problem' (Tran Khanh 1993: 3).

It is clearly difficult to generalize about the position of the Chinese in the region. In economic terms they often operate at the core of the system; indeed, in some respects and in some countries, they are the core of the system. Taking income poverty as an indicator of exclusion, it would be hard to characterize the Chinese as an obviously excluded group. In the East Malaysian states of Sarawak and Sabah for example, while the incidence of poverty was 29 and 41 per cent respectively for the *bumiputra* population, for the Chinese it was just 4 per cent in both states (Brookfield *et al.* 1995: 222).[16] Yet in other senses, their position is marginal. Politically they may be denied citizenship or find that the terms on which they are granted citizenship are circumscribed. And culturally they may feel uncomfortable about openly expressing their ethnic origins, and may be encouraged – even forced – to assimilate. Even economically the Chinese are dependent on the state, and yet at the same time cannot guarantee that the state will protect their interests. Nor, for that matter, is it obvious that discussing the 'Chinese' as a group has great utility. As immigrants they were scarcely culturally homogeneous. And today their economic positions are at least as diverse as those of the majority population, while the line dividing Chinese from indigenous is becoming increasingly blurred. Few would write of the majority as an undifferentiated mass, yet somehow five million Chinese Malaysians, for example, are regarded as amenable to generalization. This, though, is not just a scholarly tendency; it is also a reflection of intranational visions of the Chinese population which have an equal tendency to eschew more nuanced interpretation in favour of ethnic stereotyping.

Work on the economic position of the Chinese in Southeast Asia tends to take one of two positions. First, there are those accounts which stress cultural and historical factors (the Confucian work ethic or Asian values argument, coupled with the role of the Chinese as recent immigrants); and second, those that emphasize structural and class-based interpretations. Mackie writes of Sino-Indonesian businessmen:

> the most distinctive feature of the emergence of the big Sino-Indonesian conglomerates . . . has been the extent to which they have been able to benefit from deviations from free-market principles by taking advantage of privileged access to resources (particularly subsidized loans), quasi monopoly situations, and rent seeking opportunities. They have been able to leverage the huge profits generated from them into enormous capital gains. . . . Political connections and protection have been an almost essential condition of economic success for business of all races since early in the New Order [1967–].
>
> (Mackie 1991: 91–2)

In Vietnam, the persecution of the Chinese might be seen as driven not so much by their ethnicity, as by their membership of the compradore bourgeoisie which the government in Hanoi, after the liberation of the South in 1975, was committed to expunging (Tran Khanh 1993).[17] In these terms, the fact that most of the compardore bourgeoisie were ethnic Chinese was merely coincidental. A similar line of argument has been presented to explain the so-called 'double dualism' that characterized the Malaysian economy, where the division between the modern and traditional sectors of the economy was overlain by an ethnic division where the Chinese were overwhelmingly concentrated in the modern sector, and Malays in the traditional. Malaysia's NEP, it has been argued, served to widen intra-ethnic inequalities (Shari and Jomo 1984). The preoccupation with ethnicity has disguised deep-seated class divisions. The real struggle, then, is not between races but between classes; it is just that, to date, this class-based conflict has been fought out on an ethnic landscape (see Brennan 1985; Jomo 1984; Rigg 1991b).[18]

Though such a class-based interpretation of the Southeast Asian political economy is attractive, it is forced to deal with a political reality that often ignores class. Even in Vietnam immediately after 1975 when visions of the future were apparently ideologically driven, the Chinese found themselves persecuted not so much because they were branded class enemies, but because of their ethnic origins. (This became even more acute leading up to and following China's incursion into north Vietnamese territory in 1979.) Chinese *bang* (clan or dialect) associations, newspapers and hospitals in Saigon (Ho Chi Minh City) were either closed down or taken over by the communist authorities; 320,000 ethnic Chinese were forcibly sent to a new and harsh life in the frontier New Economic Zones (NEZs) between 1975 and 1978; and in 1979 alone over 270,000 'boat' people and overland escapees fled the country, most of them Chinese, or some one million between 1975 and 1989 (Tran Khanh 1993: 81–7). Directive 10 of the Vietnamese Communist Party issued on 17 November 1982, although it stipulated that people of Chinese ethnicity living in Vietnam would be considered as citizens, also stated that Chinese citizens were not allowed, for example, to become officers in the People's Army. Politburo Act 14 of 13 September 1983 went even further in excluding Chinese citizens from full participation in economy and society, by expressly forbidding 'Chinese participation in commerce, transport, printing, processing, cultural business, information, opening of schools, etc.' (quoted in Tran Khanh 1993: 90). It should be added that since 1986, with the introduction of the economic reforms, the Chinese have found their roles far less circumscribed by the state (see page 10). None the less, the experience of the Chinese emphasizes the difficulty of separating issues of class from issues of ethnicity and identity.

An interesting counterpoint to the experiences of the Chinese in Vietnam and Malaysia concerns the difficulties that the Communist Party of Thailand (CPT) had in mobilizing support on the basis of class struggle. The CPT reached its ascendancy during the mid- to late-1970s when it had some 12,000–14,000 armed insurgents under its control. The *Journal of Contemporary Asia* in 1978 baldly stated in an editorial predicting imminent civil war, that a 'level of class struggle has been reached almost unimaginable until a few years ago when the effects of the integration of Thailand into the world capitalist economy since the 1950s began to take form. . . . Armed opposition now pervades the greater part of the country' (*JCA* 1978: 3–4). Yet the CPT was never able to escape from the tyranny of ethnicity: most of those who joined up, though they may have been poor, did so because they were Lao (in Northeast Thailand), belonged to one of the hill 'tribes' of the North (especially Hmong and Karen), or were Malay Muslims (in the South). Again, as in Vietnam, class was articulated through an ethnic lens (see Parnwell and Rigg 1996; Keyes 1989: 107–9).

### 'Guest' workers and refugees: invited and uninvited

Southeast Asia is rapidly becoming a regional human resource economy (Figure 4.1). The factors that lie behind this increase in international mobility – both legal and illegal – are diverse. To begin with, borders have become more permeable and legal and physical barriers less of an impediment to movement. In addition, labour shortages in countries like Malaysia, Singapore and Brunei,

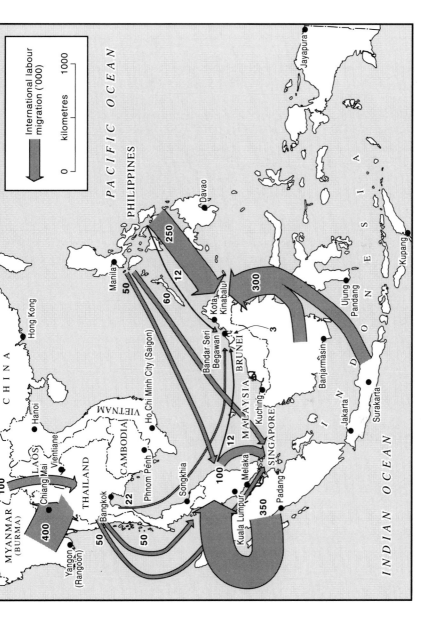

*Figure 4.1* Labour migrants in Southeast Asia

coupled with a widening in wage differentials between countries and between sectors within countries, have amplified the economic incentive for movement. This has been assisted by the economic and political rapprochement in the region since the late 1980s. What is interesting is that the dynamic of development in Southeast Asia has led to a cascade of international mobility: labourers, both male and female, enter Thailand from Myanmar, Cambodia, Laos and China to work on building sites, in warehouses and in the sex industry;[19] Thai agricultural workers cross the southern border of the Kingdom into Malaysia to harvest rice on the fields of Kedah; Malaysian industrial workers, in their turn, take the causeway to Singapore to take up occupations there. In each case, neo-classical migration theory would lead one to identify wage differentials – and therefore economic factors – as the primary explanatory cause (see Hugo 1993 with reference to Indonesians working in Malaysia). Although recent research indicates that migrants move for a multiplicity of reasons, and to isolate just one factor is highly reductionist, micro-studies none the less tend to bear out such a perspective.

Legal labour migrants face poor working conditions and low wages in comparison to the nationals of the country where they are guests. In 1994, Singapore's Ministry of Labour recorded that there were (officially) 200,000 foreign workers in the island republic (quoted in Davidson and Drakakis-Smith 1995: 22).[20] These workers live in shacks on building sites or are crammed into old Housing Development Board (HDB) blocks which are rented by companies from the government. The government stipulates that they not be paid more than S$1,500 per month, so that they remain in the low-skilled and low-paid sectors of the economy. Many receive much less than this in any case (see Davidson and Drakakis-Smith 1995: 22). For illegal labour migrants conditions are potentially far worse. Writing of migrants – both male (gay) and female – from Myanmar working as prostitutes in Thailand, Fairclough comments:

> As illegal immigrants working in an illegal industry, people like Maung Htay [a 17-year-old male prostitute working in a gay bar in Chiang Mai] and Wan [a female prostitute working in a brothel in Mae Sai] are especially vulnerable. They cannot go to the authorities for help, since they would risk imprisonment and deportation. They also end up working in the least remunerative and most dangerous places. And because they don't know as much about Aids prevention as their Thai counterparts, they have much higher infection rates.
>
> (Fairclough 1995a: 28)

Koy Suon, a Cambodian working also working illegally in Thailand described to Christine Chaumeau how he paid a middleman 1,500 baht (US$60) to smuggle him across the border. He found work on a construction site but rather than receiving the promised fortnightly wage of 1,950 baht (US$78) he was paid only 820 baht (US$32). After 27 days' work Thai workers on the site reported him to the police. He was arrested and spent five months in Suan Phlu

Detention Centre in Bangkok having been sentenced for just one and a half months. 'During the five months, I never saw the sun's rays. We didn't have a mosquito net . . . [and] when we got a fever, we didn't have any medicine. We tried to ask why we stayed more than the one and a half months, but the guards wouldn't answer us' (Chaumeau 1996: 5). It is stories like Suon's which tend to make the news, but there are, in all likelihood many hundreds more who escape the authorities and the more rapacious labour recruitment agencies and middlemen. Ra, one of Suon's fellow villagers from Chom Nom in Battambang, worked for three months in a warehouse in Bangkok, earned 10,000 baht (US$400), and returned safely home. Ra is sure he will return to Thailand. 'Here', he commented, 'even if we work hard, we do not earn enough of a living for our children'.

To some extent the Burmese working in Thailand as prostitutes and construction labourers, the Thai agricultural workers toiling in the fields of Kedah, and the Malaysians helping Singapore achieve its Second Industrial Revolution, are economic refugees. In recent history there have also been many hundreds of thousands of political refugees, the majority of who have sought refuge in Thailand (Figure 4.2).[21] Refugees suffer from manifold dislocations and, probably more than any other single group, can be classed as excluded. They are usually stateless and families will often be divided, whether through death or spatial separation. Uprooted by war or civil strife, most arrive with few possessions and little or no money, and sometimes suffering from malnutrition, disease or physical injury. The experience of dislocation also brings with it severe mental and psychological tensions. Having 'escaped', refugees are then confined to refugee camps with limited physical (shelter, water) and social (schools, clinics) infrastructures, little political security, possibly no chance of employment, dependent on the charity of their host national government and various voluntary and multilateral bodies, and facing an uncertain future. As Gnem Im, a 33-year-old Cambodian refugee, physically weak and with a paralysed leg explained to two journalists: 'We are hanging in the balance. Life was better at home, but we still don't know if we have to go forward or backward because our cattle, land and house are gone' (Livingston and Ker Munthit 1993: 2).

Monzel's account of Hmong refugees from Laos describes three women whose lives are characterized, more than anything, by a 'lack of control'. The war in Laos was appalling enough and they were also born into a patriarchal society. Because many Hmong fought for the Americans, the victory of the communist Pathet Lao in 1975 led to widespread persecution.[22] Thousands fled to Thailand, and many more died in the process. At the Ban Winai refugee camp in Thailand's Northeastern region[23] not far from the town of Loei the refugees' lives were tightly controlled. 'Everything' as Kue put it 'was distributed to us. Everything was controlled'. They were allotted one gallon of drinking water each per day, food was short, medical care rudimentary, sanitation primitive and they were robbed periodically by the Thai. The Hmong would beg for water from Thai farmers in the dry season, some committed suicide so distressed were they

*Figure 4.2* Refugees in Southeast Asia

at leaving their homes and land. Even their resettlement in the United States was a harrowing and disorientating experience. Kue described her first winter in Syracuse:

> It was so cold! All the trees outside looked dead so I went out to collect firewood. Then I realized that the wood was green. My hair was still wet from the shower, and it turned to icicles at the ends. . . . I was afraid to touch the snow, afraid that something might happen to my hands. We didn't touch the snow until the next winter.

> (Monzel 1993: 129)

Now settled in Syracuse the women are happy. Xai admits that 'it doesn't matter that my children and grandchildren have lost the Hmong ways. I am happy that they have been given another chance' (ibid.: 130).

130

## Excluded lives and livelihoods

The discussion above was on groups who are excluded for *who they are*. This is an exclusion which is rooted in the 'differentness' of people and the tendency for majority populations to discriminate against minorities, especially if they are regarded as in some way inferior. It emphasizes that identities, spaces of citizenship, and relations between groups are both internally generated by the individuals that constitute any local or national population, but are also reflections of wider (and, perhaps 'deeper') political, economic and social structures. The following pages, though, focus on groups who are excluded for *what they do* – their livelihoods. These two populations are clearly not mutually exclusive. Many prostitutes in Bangkok, for example, are hill tribe girls from the Northern region or economic migrants from Myanmar who have been enticed or sold into the sex industry. They are therefore members of minority groups employed in an 'excluded' occupation.

### *Prostitutes and 'deviancy'*

Prostitution is the most obvious occupation where participation stigmatizes the worker and causes her (or him) to be excluded from the mainstream of society. Officialdom tends to view prostitution as a pathological deviancy to be controlled where possible, even if it cannot be stamped out. The same was formerly true, if to a lesser degree, of informal sector occupations more generally. Alison Murray summarizes one supposedly scientific work on prostitution in Indonesia by Kartono which 'claims that women become prostitutes because they are nymphomaniacs; have been deserted by their husbands; are too lazy to work; have no morals; or are just stupid' (1991: 107).

The economic and societal forces behind the growth of prostitution in the region, and whether (and in what ways) sex workers can be regarded as 'excluded', are varied. McIlwaine, in discussing sex work in the Visayan Islands in the Philippines, observes that it 'perhaps represents the most explicit example of the feminisation of export-oriented employment' (1995: 10). The origins of the 'hospitality' industry there are associated with the presence of US military bases in the country, and the expansion of the R&R ('Rest & Recreation') industry, particularly during the conflict in Vietnam.[24] But even before the closure of the bases in 1992, and especially since, the sex industry had begun to take advantage of the growth of international tourism. Thus the roots, growth and persistence of sex work in the country can be seen to be export-oriented.

Prostitution in Thailand, however, cannot be so clearly linked to foreign influences. Reid reports that in the 1680s an official was granted the licence to run the prostitution monopoly in Ayutthaya (the capital of the Siamese (Thai) kingdom of Ayutthaya, 1350–1767), using 600 women who had been enslaved or captured (Reid 1988: 156). A League of Nations report recorded that in 1930 there were 151 registered brothels in Thailand while the Fox Report of 1957 concluded there were 20,000 prostitutes in the Kingdom, 10,000 in Bangkok (Vitit Muntarbhorn 1986: 410). All these estimates indicate that there was a flourishing prostitution industry prior to the entry of the USA into the conflict in Vietnam and the growth of international tourism.[25] Yet despite the evidence that prositution has a long history in Thailand, there is a tendency to equate the rise of the sex industry with the expansion of international tourism and, more broadly, with the incorporation of the Kingdom into the global capitalist economy:

> We are witnesses to a sad period in the history of humanity. With the implementation of certain modern economic development scenarios there has been a parallel explosion in the prostitution industry in third world countries. . . . It is ironic that while Thailand is claiming to be a NIC (Newly Industrialized Country), it is also a PIC (Prostitution Industrialized Country).
>
> (Prawase Wasi 1991: 26)

That poverty and a lack of opportunity are critical impetuses to the growth of the sex industry in the region is broadly accepted. Sanitsuda Ekachai, writing about a village in Phayao province in Northern Thailand, recounts:

> Why do they [the villagers] let their girls, as young as 13, be sexually violated at the whim of strangers? Is it hunger? Poverty? Or is it greed? All such questions have to be swallowed when coming face-to-face with Pon Chaitep, a peasant who let his daughter go to work as a prostitute in Sungai Kolok, a town on the Malaysian border, in exchange for 15,000 baht [about US$600]. . . . 'I didn't sell my daughter,' mumbles the father of eight apologetically. 'She saw me suffer, she saw the family suffer. And she wanted to help'.
>
> (Sanitsuda Ekachai 1990: 169–70)

Similarly, Murray sees prostitution in Jakarta, Indonesia as a logical and rational choice for women who have few other opportunities. It enables them not only to survive, but to live comparatively well and satisfy their consumerist aspirations. They have, as she puts it, 'bought' into the system, while the poor have been 'sold in the name of development' (Murray 1991: 125). While many prostitutes may feel, in Murray's terms, that they have made a rational choice given their situation, McIlwaine would prefer to describe their actions as being driven by 'constrained choice'. The industry exploits, and depends upon, entrenched gender stereotypes which reflect patriarchal hierarchies and gender inequality. Women become commodities to be exploited by men in an industry which, ultimately, is also controlled by men. In Thai, the colloquial term for a pimp is *maeng da*, the horseshoe crab, because, like this prehistoric beast, the male survives by clinging onto, and being carried by, the female.

## *Prostitutes: excluded through their inclusion*

Prostitution, perhaps more than any other occupation, illustrates the difficulty of talking about the 'excluded' in one-dimensional terms. As noted above, many women enter the sex industry because they or their families have few other opportunities; their exclusion from the mainstream of development encourages them to become prostitutes. By entering the sex industry their bodies become commodities to be bought and sold, part of the system of market relations. In the case of the Philippines this can also be linked to the demands of international tourism and therefore to the broader export-oriented growth strategies being pursued in the country. But at the same time as these women become part of the landscape of modernization, they may also exclude themselves from the mainstream of society. In Indonesia and Malaysia they become 'deviants'; in the Philippines and Thailand they become part of a separate social order, and only when they retire from prostitution can they re-enter 'normal' society.[26]

However, even this economic–social dichotomy is not sufficiently nuanced to account for the multiple ways in which a prostitute (and no two experiences are the same) articulates with the world around her. In Southeast Asia, entering employment in the sex industry does not, usually, denote an attempt to 'flee' from poverty and oppression at home. Thai cultural norms, for example, ensure that daughters remain in contact with their families and remit money on a regular basis. Pasuk Phongpaichit's study of 50 'masseuses' records that 49 maintained links with their families (1984: 254). Family obligations are strong. So becoming a prostitute may be seen as a means to drag one's family – not just oneself – out of poverty. The money accrued is most obviously spent in building, extending or improving the family residence. 'The [new and modern] houses, paid for by their girls' remittances, are evidence of a daughter's virtue: her readiness to sacrifice herself, her gratitude to her parents and, more importantly, her success' (Sanitsuda Ekachai 1990: 171). Nor does this increased income merely translate into greater buying and spending power; it may also result in greater status for the prostitute's family. Sanitsuda Ekachai goes on to suggest that:

> Riam [a prostitute] has given her father more than a house, a television set, a refrigerator and a stereo. She has made him someone in the village. He was once a landless peasant, one of those who sat in the back row at village meetings. Now he sits at the front. . . . The old man was recently appointed supervisor of the temple's funds, a position given only to the village's most trustworthy person.
>
> (Sanitsuda Ekachai 1990: 173)

In this way, in some villages in the North and Northeast of Thailand, entering into prositution is regarded as a respectable choice (even if it is a constrained one) for a daughter to make when a family is experiencing economic hardship. Prostitution has become an accepted strategy of survival, consolidation or accumulation in an otherwise bleak working environment.[27] But, and this is an important caveat, why taking up prostitution should be viewed as an honourable, even a virtuous, decision for a girl to make must be seen in the light of the existing structure of gender relations.[28] Prostitutes in Thailand are sometimes termed *phuuying sia* – women who have been 'lost' or 'gone rotten' (Harrison 1995: 129). Thus although daughters may be virtuous in sacrificing themselves for their families, they still cannot escape the stigma of being a prostitute. Some scholars would also wish to link prostitution in the country with Buddhist philosophy, where women, having accumulated bad deeds in a previous life, are one step below men on the cycle of birth and rebirth. Villages where large numbers of girls have 'gone south' in many cases are endowed with lavishly decorated *wats* (monasteries). It is perhaps this coincidence of acceptance (even encouragement) and denial that explains why villagers are often reluctant to admit what everyone knows and use euphemisms like daughters who have 'gone south' when talking of prostitution.

The previous paragraph, drawing on work conducted in Thailand, character-izes prostitutes as dutiful daughters encouraged or driven into prostitution as part of a household livelihood strategy. In the Philippines, by contrast, the increased earning power of such an occupation gives prostitutes a degree of economic autonomy, permitting them to create female-headed or female-only households, and freeing them, in many instances, from domineering husbands and fathers. 'This, in turn, allows women an arena for female solidarity and a potential base from which to challenge wider gender inequalities' (McIlwaine 1995: 12). But while providing a degree of autonomy, this is only achieved by becoming part of the system which perpetuates gender stereotyping and the unequal position of women. As such, the higher incomes and degree of economic independence must be seen as a hollow victory in the context of the wider war. It has also be pointed out that by entering the sex industry, women exclude themselves from 'normal' society, and as a result lose their ability to challenge patriarchy in the 'real' world (Chant and McIlwaine 1995a: 302).

Of course the fact that prostitutes and their families talk of poverty and a lack of opportunities being critical motivating factors, as well as the manner in which the income derived is spent – much on consumer goods – is merely indicative, so post-developmentalists might argue, of the hegemony of the development discourse. Further, the incorporation of Myanmar, Laos and Yunnan (south-west China) into a regional sex industry centred on Bangkok is evidence, not so much of extreme poverty in those countries, but of the insidious permeation of the modernization ethic.[29] People without televisions, radios, modern houses and so on have, through their gradual incorporation, become 'poor', and this has driven their engagement in prostitution.

### Deviancy and non-conformity

Prostitutes, especially in Malaysia, Singapore, Indonesia and Brunei, are often viewed as 'deviants'. However, deviancy alludes to a much wider range of occupations and lifestyles than just those engaged in the sex industry. Deviants, in practice, have become loosely defined as those members of society who differ markedly from the norm; who fail to conform; or who are engaged in occupations and activities that are at, or beyond the margins of the socially acceptable. They are commonly regarded as 'sick' or 'pathological' cases.

The country in Southeast Asia where the state has been most active in controlling so-styled 'deviancy' is Singapore. Chapter 2 briefly reviewed the elements that comprise the Republic's promotion of Asian values, and central to them are the overlapping notions of conformity, corporatism, consensus and harmony. This has allowed the state sharply to constrain critical debate and to deal harshly with opposition politicians that might undermine the integrity and stability of the country.[30] It has also, arguably, allowed the state to limit individualism by restricting self-expression and to map out and promote what it regards as 'normal' and acceptable lifestyles.[31] In Singapore, single parents are not eligible for income support through Public Assistance (PA) or the Rent and Utilities Assistance Scheme (RUAS), nor are they eligible to buy Housing Development Board (HBD) appartments. They also find that child care places, for example, are highly restricted. The reason for their exclusion is they do not meet the government-stipulated requirement of representing the 'norm' (i.e. a married couple with children) (Davidson and Drakakis-Smith 1995: 15). The government's policy towards single parents is driven by a perceived need to discourage mothers from having children out of wedlock. This has led Davidson and Drakakis-Smith to state that '[s]ingle parent families have been abandoned and ignored by the government' (1995: 21). As Liew Kim Siong expressed in the *Business Times*, the PAP appears to view welfare not as an entitlement but as a privilege 'doled out at the discretion of a paternalistic government that expects gratitude from the citizenry' (Liew Kim Siong 1994: 54).

### *Inside out: the garbage pickers of Smoky Mountain*

Garbage picking is another occupation which marks people out as different, even from those working in other low-wage occupations. The garbage pickers of the *barrios* of Magdaragat and Looban, popularly known as 'Smoky Mountain', outside Manila in the Philippines, recognize this when they talk of people on the 'outside' (*tagalabas*). 'Our aspiration', one resident of Smoky Mountain remarked to Alex Brillantes, 'is to one day be regarded as no different from them. Now, if they find out that you live in Smoky Mountain, they immediately look down on you as "one step down" because you live on and make a living out of garbage' (Brillantes 1991: 187). In the early 1990s there were some 3,000 families (20,000 people) making a living rummaging through, sorting, and selling garbage on Smoky Mountain. Everything has a value from broken glass (0.10 pesos/kilo), to rubber (1.5 pesos/kg), bones (0.5 pesos/kg) and scrap copper (9 pesos/kg). Scavenging for 12 hours a day can earn a picker US$0.75–1.25. Popular conception within the Philippines has it that the garbage pickers of Smoky Mountain

---

*Box 4.2* AIDS and mainland Southeast Asia's regional prostitution economy

Increasingly the sex industry in Thailand is using prostitutes from neighbouring countries, especially Myanmar but also Cambodia, Laos and Yunnan (south-west China). Thailand also has the most serious HIV/AIDS epidemic in the region. By December 1993 30 per cent of sex workers in Thailand were HIV infected and at the end of the following year 840,000 Thais were thought to be HIV positive, or 1.4 per cent of the population. An estimated 46,000 Thais died of AIDS in 1995. In Myanmar in 1991, 11 per cent of sex workers were HIV positive while the figure for Cambodia in 1994 was 69 per cent (up from 9 per cent in 1992). In all the countries neighbouring Thailand, AIDS is believed to be either 'increasing' or 'rapidly increasing' – although the information on which projections are made is sometimes thin. It would be inaccurate to write that AIDS in Thailand's neighbours is solely due to the emergence of this regional prostitution economy. However there can be little doubt that the regional economic integration of these formerly isolated countries is a major factor driving the epidemiology of AIDS, and that this, in turn, is intimately linked to the occupational patterns of the migrants.

*Sources*: Brown and Xenos 1994; Brown and Werasit Sittitrai 1995a and 1995b

---

represent the ultra poor – the epitome of the excluded. When a foreign film crew wishes to capture on film a vision of extreme poverty they make for Smoky Mountain. The inescapable message of the images, of course, is that the life portrayed is one of misery and degradation. Yet some residents would rather that they be held up as an example of successful struggle against the odds: 'It is better to make a living among the garbage dumps than to steal. Others see only the hardship and poverty. . . . Whatever others might say, we manage, however difficult, to have some dignity' (Brillantes 1991: 191). In 1983 (during the Marcos era) the Philippine National Housing Authority forcibly relocated the squatters of Barrio Magdaragat to a new settlement 30 kilometres outside Manila and provided each family with a small plot of land, four walls and a toilet. They were not, though, provided with a livelihood, and as one resident rhetorically commented to Alex Brillantes, 'what use is the toilet bowl if you have nothing to defecate?' (1991: 193). At first, those who had been relocated commuted to Smoky Mountain to work; in time they drifted back to their old haunt and rebuilt their shacks.

But although the residents of Smoky Mountain may have a livelihood, their status as squatters means that they cannot benefit from local social services. They are excluded from most welfare schemes and, in the words of a local Catholic priest, '[n]ational events are nothing to them . . . Smoky Mountain is like a ghetto, an island' (Brillantes 1991: 195). The spaces of citizenship for these people is incomplete. None the less, while their short-lived relocation in 1983

*Plate 4.1* AIDS comes to Vietnam: a billboard in Hué informs English-speaking residents and foreigners of the dangers of the disease

*Plate 4.2* A garbage picker in the Sumatran city of Pekanbaru. Pekanbaru is the capital of the oil-rich province of Riau

brought access to social services and welfare provision – making their citizenship complete – the feeling among those who were relocated was not one of gratitude, but rather of betrayal. In late 1995, the residents of Smoky Mountain were once again facing eviction, this time under the Presidency of Fidel Ramos. And once again some residents mounted a campaign to resist their resettlement. Having learnt from the last attempts at relocation, on this occasion the selected site was centrally located and fragrant with the promise of jobs (*Economist* 1995a). The garbage pickers of Smoky Mountain illustrate, yet again, how people may choose to be excluded in one sense, in order to be included in another. The greatest irony, though, is that the livelihoods of these people is squarely based on the rejected scraps of modernization: garbage.

## The dynamics of exclusion

So far in this chapter, exclusion has been portrayed largely as an embedded characteristic, most clearly seen in the exclusion of minority ethnic groups like the Chinese and various 'tribal' peoples. However, there is also a dynamic built into exclusion. Government policies may change over time – as they did *vis-à-vis* the Chinese in Vietnam for example – and individuals may find their place in society changing as they pass through the life cycle. In other words, exclusion is not set in stone. Just two facets of this dynamic are addressed below. First, the extent to which the economic reforms in Vietnam have undermined the presumed securities of the past; and second, the generational angle to exclusion.

### Deconstructing the iron rice bowl in Vietnam?

Early pronouncements on the economic reform process in Vietnam had little to say about equity and welfare implications. Economists subscribed to the 'trickledown' theory of development and were led by the assumption, apparently, that economic stagnation was most severely felt by the poor. Therefore, reform and economic growth could only but be beneficial to the poor (Kerkvliet and Porter 1995: 13). More recently however, local economists and the leadership in Vietnam, as well as Laos, have begun to appreciate that liberalization can have pernicious effects on the poor and on general welfare; indeed, the dangers of over-zealous and uncontrolled reform have become key themes in recent plenary sessions of the VCP (see Thayer 1995: 41–2).[32] Some of these effects have already been touched on in the discussion in Chapter 3 on poverty in Indochina (see page 98).

It is usually assumed that the cooperative system, though it may have had shortcomings in terms of production, at least provided a universal safety net for poorer peasants and for those living in poorer regions (e.g. Kolko 1995: 24–5) (see Box 4.3). By contrast, the reforms are often characterized as intrinsically divisive, creating a system of unequal relations, worsening the pattern of income distribution and creating a class society (see Kolko 1995: 32–3 on the effects of

*Box 4.3* Subsidy and survival in Phuc Loi, Son La Province, Vietnam

There is considerable doubt as to the extent to which the commune system in Vietnam ever provided an egalitarian safety net for the poor. In Phuc Loi village in Son La Province, in the far north of Vietnam, the cooperative system did provide some subsidies to poorer families. For example, the elderly who were unable to work were provided with special rations while those with household members receiving medical treatment could buy rice at subsidized prices from the cooperative. The same was true during periods of harvest failure. In terms of medical facilities, villagers had to pay for treatment at the commune clinic. Education, though, was free and the state also subsidized the cost of building and maintaining school buildings.

The economic reforms of the 1980s and 1990s have had the greatest impact on access to education. One respondent reported that subsidized fees were abolished in 1981, a decision which had a catastrophic effect on educational standards in Phuc Loi. In 1987 villagers were once more exempted from paying fees, although equipment (stationery, school uniforms, etc.) costs were still met by parents. The system of subsidies during the cooperative period hardly provided a comprehensive safety net for the poor. As a result, poorer households did not regret their removal, reporting that the major increases in yields resulting from the economic reforms, and which they also enjoyed, had more than compensated. The report concludes: 'Except in the field of education, the abolition of "subsidised" services has had little effect on households, as in reality such services were never heavily subsidised' (ActionAid Vietnam 1995: 25).

*Source*: ActionAid Vietnam 1995

the reforms in Vietnam). Adam Fforde and Steve Sénèque suggest that there are three key perspectives on the interplay between the economic reforms and well-being in the countryside. One is that social differentiation driven by the reform process is not the cause of impoverishment, but allows the more efficient allocation of resources and so assists in raising production and standards of living on a broad front. In this perspective, state policies are seen as at best un-important, at worse a hindrance. The second perspective holds that given the tensions of reform it is necessary for the state to intervene to control markets and preferentially to allocate some resources to those areas and people who would otherwise be excluded. The third perspective holds that the reforms are causing social and economic polarization and are creating a large mass of excluded people (Fforde and Sénèque 1995: 98–9).

Which of the above three scenarios is selected depends, initially at least, on how the foregoing cooperative or command period is interpreted. In the first chapter it was argued that the market was never obliterated, even in North Vietnam, and that cooperatives failed to deliver in a number of important respects. None the less, there is evidence that the reforms have led to greater economic differentiation, and perhaps even marginalization (see page 104). The economic reforms of the last decade have also weakened community support systems leading to increasing interpersonal inequality. 'The deterioration', Beresford writes, 'though uneven, is palpable' (1993: 41; see also Kolko 1995: 35–7). There is evidence, for example, that the disintegration of the cooperative system has led to a decline in the quality and availability of child-care in Vietnam, with a consequent rise in the incidence of child malnutrition (Kerkvliet and Porter 1995: 17).[33] Similar developments – which are recognized by the leadership of the VCP as so-called 'negative phenomena' linked to the reform process – have been reported for education and welfare services where provision has reportedly both declined and deteriorated (Thayer 1995: 41; Pettus 1995). Party leader Do Muoi warned at the VCP's Fourth Plenum in January 1993, for example, that the incidence of some diseases and the illiteracy rate were actually rising (Irvin 1995: 740). The reforms have removed public funding and introduced fees for education and health. Buildings have fallen into disrepair, there are severe shortages of equipment and supplies, and even welfare services which are ostensibly free have hidden costs. Parents of primary level pupils, for example, are expected to contribute anywhere from US$0.50–5.00 to the costs of supplies, repairs and even salaries (Pettus 1995). (See Box 4.3 'Subsidy and survival in Phuc Loi, Son La Province'.)[34] Women also seem to have lost out. Scholars have identified widening disparities during the 1980s in such areas as child survival rates, marital opportunities, childbearing preferences, employment patterns, and leadership roles (Goodkind 1995).

However, the emphasis in many studies on these 'negative phenomena' has tended to obscure the production increases that have been generated by the reforms, particularly in rural areas:

> The general picture [in rural Vietnam] is that most rural people have enough rice and a diversity of other foods; many have sufficient money to repair and rebuild their houses and buy radios, televisions, bicycles, and other consumer items; and some can accumulate a surplus with which to start businesses. Production has been increasing, cereal grains per capita figures have been rising, and land previously idle is being farmed again. Vegetables, meats, and other produce are more plentiful than people can remember. Peasants have incentives to grow more and work more energetically because the energy and money they invest will directly benefit them. . . . most, I think, would agree that conditions would have been far worse without the [reform] changes.
>
> (Kerkvliet 1995a: 71)

The reforms, then, have translated into higher standards of living for most, if not all, and as Fforde and Sénèque remark: 'that a large number of people, previously hungry for many months a year, now eat rice, is an achievement of historic significance' (1995: 108–9). One villager in Ban Un in the northern Vietnamese province of Son La, while highlighting the difficulties of the past also noted the way in which the economic reforms have opened up new live-lihood strategies for the poor:

> if the harvest is lost, there is still cassava to eat and one can go wage labouring. In the past one could only go and dig up *cu mai* tuber in the forest, while some households could go and ask for food from relatives and friends.
>
> (ActionAid Vietnam 1995: 10)

Lo Van Hoa, another farmer in Ban Un, commented that 'making a living is easier now, because activity is less regulated, less timetabled and one is not allocated tasks to do' (ActionAid Vietnam 1995: 8). The reforms have unleashed a production surge, some of it linked to under-utilized investments made prior to the initiation of the reforms. This issue of the sequencing of causes and effects is important in understanding where and how the reforms are influencing people's lives. For example, it seems that gender inequalities in Vietnam were widening before *doi moi* had been embraced by the leadership and it would be too simplistic to associate such changes purely and only as products of economic reform. Goodkind suggests that while the two decades since reunification (1976–95) have seen a worsening of the position of women, the next two decades may see an improvement as women's high literacy leads to upward social and economic mobility, as demographic changes ease the marriage squeeze associated with the skewed sex ratio (a surplus of women linked to the war), and as production increases, particularly in the countryside, reduce the need to discriminate against female children (Goodkind 1995).

There seems little question that the reforms in Vietnam have led to some measure of social and economic polarization. In addressing this issue however, the Vietnamese economist Hoang Thi Thanh Nhan, while calling for measures to ensure social justice and the alleviation of poverty and hunger, argues that '[i]t would be terribly wrong if we should solve the said polarization by blocking wealthiness [*sic*] and sharing out poverty' (1995: 20).

### The first and final circuits: young families and old age

Wealth rankings and poverty surveys indicate that there is a generational angle to destitution: those households that are more likely to be poor are young families and the elderly. Southeast Asia's population is currently a very 'young' one with only a small proportion aged 60 or over (Table 4.6). However, a significant challenge for the future will be the support and care of the elderly. With a steep decline in birth rates and associated smaller families, an increase in life expectancy, combined with the gradual erosion of the culture of filial obligation, so it is likely that there will be a large increase in the number of elderly people requiring state support (Table 4.6). This erosion of the family as a support unit and safety net is partially associated with social changes linked with moderniza-tion. However, greater mobility and the spatial dysjuncture of family members also makes formerly geographically-rooted extended families where co-residence was the norm (Table 4.7), hard to maintain. Thus a combination of economic and demographic necessity, combined with social choice, is likely to make the support of the elderly a key issue. Nowhere in the region is the onset of the ageing population so far advanced as Singapore, and nowhere has a government considered the issue so thoroughly and entertained such radical solutions.

*Table 4.6* The greying of Southeast Asia

| | *Percentage of the population 60 years or older* | | |
| | *1960* | *1990* | *2025* *(projected)* |
| --- | --- | --- | --- |
| Indonesia | 5.2 | 6.3 | 13.6 |
| Malaysia | 5.3 | 5.8 | 12.6 |
| Philippines | 4.9 | 4.9 | 10.9 |
| Singapore | 3.7 | 8.7 | 27.0 |
| Thailand | 4.5 | 6.3 | 16.8 |

| | *Total fertility rates (TFRs)* | | |
| | *1960–65* | *1970–75* | *1985–90* |
| --- | --- | --- | --- |
| Indonesia | 5.4 | 5.1 | 3.5 |
| Malaysia | 6.7 | 5.2 | 4.0 |
| Philippines | 6.6 | 5.3 | 4.3 |
| Singapore | 4.9 | 2.6 | 1.7 |
| Thailand | 6.4 | 5.0 | 2.6 |

| | *Life expectancy at birth* | | |
| | *1960–65* | *1970–75* | *1985–90* |
| --- | --- | --- | --- |
| Indonesia | 42.5 | 47.5 | 60.2 |
| Malaysia | 55.7 | 63.0 | 69.5 |
| Philippines | 54.5 | 57.9 | 63.5 |
| Singapore | 65.8 | 69.5 | 73.5 |
| Thailand | 53.9 | 59.6 | 67.3 |

*Table 4.7* Co-residence in Southeast Asia: proportion of elderly (60+) living with children

| | *Percentage* | *(Date of source/data)* |
| --- | --- | --- |
| Indonesia | 67 | (1990) |
| Malaysia | 69 | (1984) |
| Philippines | 68 | (1988) |
| Singapore | 88 | (1988) |
| Thailand | 77 | (1986) |

*Source*: Hermalin 1995: 6
*Note*: Co-residence rates in the USA in 1910 were about 60 per cent – i.e. close to those rates indicated above.

In Singapore the increase in the population over 60 years of age from 3.7 per cent to 8.5 per cent between 1960 and 1990, with projected figures of 27.0 per cent by 2025 and 29.4 per cent by 2030, has raised fears that those who have not made sufficient contributions to the Central Provident Fund may find themselves without an adequate income (Silverman 1995: 52).[35] Indeed, the emphasis on Asian values is, in some ways, an attempt to anticipate this problem: such values emphasize the enduring commitment of children to their parents. In November 1994 the Singaporean government passed the Maintenance of Parents Bill, legally requiring children to support their parents in retirement. Even so there are increasing numbers of elderly poor who have no children or who cannot count on the support of their children. Davidson and Drakakis-Smith note that currently many of these unsupported elderly are immigrants who arrived alone and did not marry, including *samsui* women (female labourers), *coolies* (male labourers) and *amahs* (domestic servants) (1995: 19). Voluntary Welfare Organizations (VWOs), part-funded by the state, exist to meet the needs of this group, and they are also eligible to apply for Public Assistance (PA) and to the Rent and Utilities Assistance Scheme (RUAS) (Davidson and Drakakis-Smith 1995: 14). Almost all of those who apply for Public Assistance are elderly. 'There is enormous pressure on all services and resources, especially senior citizens' clubs and nursing homes, clearly reflecting not only the immediate problems but those of the future' (Davidson and Drarkakis-Smith 1995: 20).

Governments in the region look at the costs of welfare programmes for the elderly in the West with some alarm and are beginning to search for solutions than might minimize these costs. Many are likely to be founded on at least some of the core Asian values noted in Chapter 2. There is even the possibility that such radical solutions might influence Western approaches (Hermalin 1995: 2).[36]

Work on the position of young families is not as widespread as that on the elderly. None the less, there is considerable evidence to show that newly established or 'young' families are more likely to be counted as poor (see page 95). This is partly because they have not had the opportunity to accumulate wealth, and are unlikely (yet) to have inherited it. In addition, they may well have young children to support. Wealth rankings undertaken in 1994 in Ban Bong and Sam Ta, two villages in Son La province, north Vietnam, supported this contention, revealing that those at the foot of the wealth rankings consisted mostly of newly established families. But these poor households were separately identified as the 'striving poor' – to distinguish them from the mainstream poor. The term indicates that their categorization as poor was not expected to be permanent, and that as they became established families so they would move out of poverty (see page 85) (ActionAid Vietnam 1995).

## Including the excluded: new social movements

The inadequacies and insufficiencies of class-based interpretations of struggle and identity led, from the 1980s, to a growth of scholarly concern for 'social movements' and particularly, 'new' social movements.[37] In many cases they aimed to represent and further the interests of excluded groups in society. They were a means to raise the profile and power of specific population groups who, hitherto, had been ignored, persecuted or who had suffered from discrimination.

As one might expect then, new social movements are disparate and highly heterogeneous groups and they lie outside traditional class-based divisions of society. They may be based on geographic criteria, on employment categories, on gender-based segmentations, on religious affiliations, or on language and ethnic criteria. 'A multiplicity of groups independent of traditional trade unions and political parties, [including] squatter movements and neighbourhood councils, baselevel communities within the Catholic church, indiginist associations, women's associations, human rights' committees, youth meetings, educational and artisitic activities, coalitions for the defense of regional traditions and interests, and self-help groups among unemployed and poor people created a new social reality' (Peet and Watts 1993: 245). Much of the scholarly literature on new social movements in the developing world has originated in Latin America, although Scott's, and Turton and Tanabe's work on 'everyday forms of peasant resistance' in Malaysia and Thailand could, loosely-speaking, be placed within the same broad category (Scott 1985, Turton and Tanabe 1984). It is also true that NGOs and grassroots organizations, the very stuff of such new social movements, have played an increasingly critical role in the development discourse in the region (this is particularly true of the Philippines, Malaysia and Thailand). As was noted in Chapter 2, we are once again at an odd dysjuncture: NGOs and other such organizations are at the forefront of the development frontier in Southeast Asia, reflecting developments elsewhere in the South, and yet local scholars, in general, are strangely silent on the key issues which fill the pages of academic publications devoted to such things in other parts of the world.

New social movements and patterns of resistance may have their roots in deeper structures of relations – even in the processes of globalization – but the focus is on how these are manifested and resisted at the micro-level. The movements are products, on the one hand, of existing relations yet at the same time embody the purpose of challenging and/or neutralizing the effects of those systems of relations. Escobar would have us believe that these groups are engaged in the rejection of the entire development paradigm.[38] New social movements hold the potential of becoming the bases – the nuclei – for a new discourse of development. Only in this way, he believes, can the current crisis of development be reversed (Escobar 1995a: 216–17; see also Escobar 1995b; Watts 1995: 58–60).

The assumption – or the hope – seems to be that these new social movements (and perhaps some revitalized old ones too) will represent the interests of the excluded, will challenge exisiting entrenched power structures, and will lie largely outside the context of the dominant development discourse. On all three counts there is good reason to be sceptical. First, new social movements are usually, themselves, exclusive. Their membership is often limited by geography, gender, age, culture, ethnicity and/or employment; and they are also, in many instances, narrowly based in terms of their objectives (many are single-issue groups). They rarely therefore represent the interests of the excluded; just a very small sub-set of the potentially excluded. Second, it is by no means a safe assumption that these groups are in the process of challenging the status quo. In Singapore, for example, the government there has made highly effective use of supposedly independent groups as sounding boards for its policies. In a Green Paper presented to Parliament in 1988, Goh Chok Tong stated that 'the ties between elected leaders and the people . . . must be constantly nurtured through continual discussion, feedback and explanation' (quoted in Tremewan 1994: 161). It might seem reasonable to reject such groups as not truly representative of new social movements, but given their extreme heterogeneity it is hard to identify where resistance ends and compliance begins. Porter notes how 'intriguing' it is that new social movements are being shaped by the dominant development discourse itself, 'being legitimated by natural science, and wrapped in a bold new programme which it is morally imperative we adopt' (Porter D. 1995b: 83). This echoes Schuurman's observation that social movements 'are not expressions of resistance against modernity; rather, they are demands for access to it' (1993: 27). Many of the excluded find new social movements enticing not because they espouse an alternative development, but because they embody an alternative path to development. In other words, they offer an alternative means to the same end. In Thailand, there has been a flowering of new religious (social) movements. Two of the best known are Santi Asoke and Thammakai. Both promote a socially-engaged Buddhism that might be seen as Thailand's equivalent to Liberation Theology. Yet in critical ways they promote visions of Buddhism's role in society that are at variance. Thammakai endorses modernization and consumerism and has drawn much of its support from the urban, educated middle classes, often using modern media techniques to get its message across. Santi Asoke meanwhile espouses simplicity and the rejection of consumerist culture (Schober 1995). While the former articulates global issues such as human rights, gender equality and environmentalism, the latter takes a fundamentalist position where the aim is to remould the world in the light of the Buddhist scriptures. The third reason why we should be sceptical of the ability of new social movements to change the developmental landscape in Southeast Asia is because they are usually engaged in an attempt to ensure that the exisiting development 'cake' is apportioned more fairly and with less damage to people and the environment; comparatively few appear to recommend that a new cake be baked.

## The interface of exclusion–inclusion

Exclusion is an unfortunate term as it implies a certain detachment. As has been argued with respect to prostitutes, some tribal peoples and the garbage pickers of Smoky Mountain, their marginality is driven by their inclusion in development, not by their exclusion from it. Just as the 1980s saw the gradual realization that those engaged in informal sector pursuits were a critical and integral part of the modern economy – and were far from being a marginal enclave – so, more recently, there has emerged the recognition that some of those viewed as 'excluded' are in fact intimately engaged in development and modernization. It is the means and method of their inclusion which, somewhat paradoxically, permits scholars to talk of their exclusion. Inclusion and exclusion are, therefore, two sides of the same coin. A person's articulation with the modern economy may lead to their being marginalized in social terms. At the same time, social deviancy may mean that welfare entitlements and full inclusion in the economy are denied.

It is easy enough to identify groups in economy and society who particularly warrant the label 'excluded'. The assumption seems to be that these people represent exceptions to the rule. There is also, though, a body of work which maintains that exclusion is not an exception, and that workers in general deserve such a categorization. This is linked to the belief that Southeast Asia's development has been driven by 'competitive austerity' – or, to use Alice Amsden's phrase introduced in Chapter 1, by a neoclassical market-friendly growth strategy which is 'not-so-friendly-to-labour'(see page 26). The success of the region is predicated on keeping salaries low (or competitive), retarding or institutionalizing unions, resisting the development of a welfare state, and maintaining an investment-friendly environment. Schmidt argues, on the basis of this, that '[s]ocial exclusion and marginalization is the *strategic result* of Southeast Asian leaders using economic success to boost their political legitimacy and justify authoritarian regimes' (Schmidt, forthcoming [emphasis added]). This perspective makes the working classes of Southeast Asia the victims of a lattice of economic and political relations that, in broad terms, determine their lives. Human agency, in other words, is circumscribed. Exclusion here might be better described as a combination of constraint and coercion.

Part III, 'Change and interaction in the rural and urban worlds', rejects the appeal of such structural interpretations for explanations which are rooted in the uniqueness of people and places. It explores in greater depth the difficulties of ascertaining whether people's actions, in the face of modernization, are evidence of marginalization or exclusion. Does mechanization in agriculture, for example, displace landless and land poor wage labourers? Does it, in the process, selectively displace women? Is the rapid expansion of industrial (factory) employment, much of it geared to export generation, therefore a reflection of poverty and lack of opportunity in rural areas and in agriculture? The evidence for all these questions, and many others, is far from clear and is open to multiple interpretations.

'Exclusion' is often used in an all-embracing way to highlight the mal-effects of modernization. Yet when it comes to specifics, the word loses its usefulness.[39] It is a catch-all phrase – now much beloved of politicians possibly for the very reason that it is so woolly – which does not do justice to the delicate and shifting balancing act that most individuals must undertake in the management of their lives. It categorizes people and fails to explore adequately the ways in which people struggle against, and resist, such depictions.

## Notes

1 It has been suggested that Socio-Economic Dimension Ranking (SEDR) is a more accurate term, avoiding the overt economic connotations connected with the word 'wealth' (Gujit 1992: 10).

2 Common sense would also lead one to question the answers to such participatory surveys. In the developed world it is common to hear that 'money isn't everything' and many (most?) people would state that happiness, a stable family life, and self-respect, for instance, are far more important than wealth. Yet the money machine continues to dictate life and to be the single greatest driving force in many (most?) people's lives. Bearing this in mind, the wealth rankings derived from such work perhaps define what *ought* to be, and not what *is*. They are normative.

3 As little work of this nature has been conducted in Southeast Asia the discussion draws largely on work conducted in Africa and South Asia.

4 In 1985, only 49.9 per cent of hill peoples living in Thailand enjoyed citizenship (Kanok Rerkasem and Benjavan Rerkasem 1994: 6).

5 The island of Borneo comprises the East Malaysian states of Sarawak and Sabah, the sultanate of Brunei, and the Indonesian provinces of South, Central, East and West Kalimantan.

6 Dayak is the collective term for the native (often called 'tribal') peoples of Borneo including the Iban, Bidayuh and Orang Ulu (the smaller non-Muslim and non-Malay groups).

7 On the basis of the observation that 'the central problem is simply stated . . . the forest people of Borneo and the [Malay] Peninsula are a small minority whose demand for land extends over a disproportionately large area', Brookfield *et al.* suggest that it was not unreasonable for the Malaysian and Indonesian governments to target these areas for logging and land settlement (1995: 138).

8 King argues: 'Iban underdevelopment is part of Malaysian inter-ethnic relations and politics' (1996: 212).

9 Pancasila, or the 'five principles', were advocated by Sukarno and embraced by the nationalist leaders in 1945. They later became the state ideology, or philosophy, of the independent nation. Pancasila incorporates: belief in one supreme God, the unity of the nation, just and civilized humanity, social justice for all the people of Indonesia, and democracy guided by the inner wisdom of unanimity.

10 One of the more amusing examples of policy failure was Operasi Koteka – or Operation Penis Gourd – in the Indonesian province of Irian Jaya. The programme, initiated by the army in the early 1970s, aimed to encourage tribal peoples to dispense with their 'primitive' penis gourds by airlifting jogging shorts and dresses into the jungled moutains. As it turned out the men found the shorts more becoming as hats while the women discovered that the dresses could be usefully converted into holdalls (see James 1993; Eliot 1995a: 1269).

11 In other words, not just to gain Thai citizenship but also to embrace Tai-ness.

12 Michel Picard's study of culture and tourism in Bali emphasizes the slippery nature of 'traditional/authentic' culture. In Bali, dances invented for the first tourists in the 1930s have now become key markers of Balinese, and in some cases of Indonesian, culture. In the process, the Balinese have become 'self-conscious spectators of their own culture – taking the growing touristification-cum-Indonesianization of their culture as the very proof of its "renaissance"' (Picard 1993, also see Vickers 1989).

13 All the terms employed are problematic in one way or another. 'Tribal' has connotations of 'primitive' and has been so loosely used over the years as to have lost all precision. 'Fourth World' implies that these people occupy a different geographical space, reinforcing the illusion that Fourth World peoples are different and inferior. The term was also originally used to draw a distinction between local/indigenous and (First World) immigrant populations in Canada. Clearly such a distinction does not apply to Southeast Asia. Finally 'indigenous' simply does not apply in some instances. The hill (tribal) peoples of Thailand are, in many cases, recent arrivals in the Kingdom and are therefore less indigenous than the majority lowland population. See Walker (1995) for a comprehensive discussion of work on Thailand and Peninsular Malaysia.

14 However, the motivation for this appears to have been different. In Thailand assimilation was largely voluntary, and possibly related to the inclusive and flexible nature of Thai Buddhist culture. In Indonesia it seems to have been a response to fears of being branded pariahs.

15 This, in turn, has led to the widely-held impression in Indonesia that the Chinese community are moving their capital off-shore, abandoning Indonesia in favour of their 'homeland'. The Sino-Indonesian investors, however, would argue that they are like other Indonesian businessmen and should be viewed as exploiting investment opportunities in China for the ultimate benefit of Indonesia.

16 *Bumiputra* or 'princes' or 'sons of the soil' includes all Malays as well as indigenous native groups (mostly Dayaks).

17 The Provisional Revolutionary Government of the South issued its economic policy statement in September 1975 stating: 'All properties of compradore capitalists . . . whether they have fled abroad or remained at home is placed under the control of the state and confiscated in whole or in part depending on the nature and extent of their offences' (quoted in Tran Khanh 1993: 81).

18 It is notable that in Malaysia the ethnic group to have gained least (as opposed to suffered most) during the course of the twenty-year NEP have been the Indians. In 1970 they controlled 1.1 per cent of wealth; by 1992 this had decreased to 1.0 per cent and it has been estimated that two-thirds of Indians live in poverty. Without the political power of the Malays nor the economic influence of the Chinese they found themselves sidelined and their particular plight ignored. In many respects the most 'excluded' group in Malaysia are the ethnic Indians – and not the Chinese, Malays nor the indigenous 'tribal' groups who receive the bulk of attention. In late 1995 Transport Minister Datuk Ling Liong Sik suggested that '[t]he time may come for an NEP for the Indians' (Jayasankaran 1995: 26).

19 One estimate put the number of Burmese in Thailand at well over 300,000 (Fairclough 1995a), while the Thai authorities believe there are 100,000 Cambodians working illegally in the Kingdom (Chaumeau 1996).

20 Most commentators believe that there are many thousands more illegal immigrant workers in Singapore. In 1992, 2,000 illegal workers were detained and then deported (Davidson and Drakakis-Smith 1995: 22).

21 The difficulties with defining who is and who is not a refugee are well known (e.g. Black 1993). Even the most widely-used definition derived from the 1951 UN Convention on Refugees which defines refugees as those people who are 'outside

their own country, owing to a well-founded fear of persecution for reasons of race, religion, nationality, membership of a particular social group, or political opinion' allows for a wide range of interpretations. This was been clear in the long-running debate over whether Vietnamese 'refugees' in Hong Kong should be forcibly repatriated.

22 Monzel states that they were singled out for 'extermination' (1993: 119) by the Pathet Lao, but on the basis of most of the available evidence this is unlikely.

23 The camp has since closed.

24 Before they were closed in 1992, Subic Bay and Clark Air Field were two of the largest US military facilities outside the continental United States. Subic Bay alone covered over 20,000 hectares.

25 In 1967 an agreement was signed between the Thai and US governments, allowing US soldiers on tours of duty in Vietnam to visit Thailand on R&R (Truong 1990: 161). In the late 1960s there were 40,000 US servicemen stationed in Thailand and many more visiting the country on R&R.

26 Pasuk records in her study of Thailand that returning former prostitutes do not seem to have a problem finding a husband and reintegrating into village life – indeed, she suggests that former prositutes often have a surfeit of suitors because of the savings they have managed to accrue (Pasuk Phongpaichit 1984: 255).

27 This view is, to some extent, obsolete, although it held true throughout the 1970s and 1980s. Many of the prostitutes entering the industry in Thailand today are from neighbouring countries, especially Myanmar. Rising wage rates in the Kingdom, the availability of factory work, and the greater awareness of the dangers of AIDS have combined to transform prostituion from a national into a regional (international) industry. But the impetus has not changed, it has merely shifted: Burmese, Chinese and Lao prostitutes come for the same reasons – poverty – and maintain links with their families, remitting money to support their families (see Fairclough 1995a).

28 Rachel Harrison's paper on prostitutes and prostitution as depicted in Thai short stories provides a fascinating perspective on Thai literary visions of the industry (Harrison 1995).

29 It has been estimated that there are 20,000–30,000 women from Myanmar working as prostitutes in Thailand (Fairclough 1995a: 27).

30 Critics would say that these policies are not designed to protect the country, but rather to protect the interest of the ruling People's Action Party (PAP).

31 The finest example of this concerns the transvestites of Bugis Street. This infamous street was, until it was demolished in 1985 to make way for the Mass Rapid Transit system, an international tourist attraction. A street of stalls and hawkers by day, as night fell it became a place where transvestites would strut their wares. Its demolition seemed to reflect the government's distaste for such tacky non-conformity. But in 1989 it was decided to re-create Bugis Street, down to the last detail ('right down to the toilet building . . . it will be back with smell and all' as the project consultant was quoted in the *Singapore Bulletin*), in a post-modern attempt at tourist promotion. The street is, indeed, a remarkable re-creation of the past, but it lacked one thing if the façade of authenticity was to be created: there were no transvestites. This affected business, and in a U-turn the government gave permission to bars and restaurants to hire transvestites as 'customer relations officers'. They would though, the *Straits Times* assured it readers, be watched by closed-circuit cameras to make certain that families could still visit the street with impunity. When the Singapore Tourist Promotion Board mounted a PR event in Hong Kong to publicize the street, and hired transvestites to help with the show, incredulous journalists were told that they were not transvestites at all, but 'female impersonators, professional artistes'. As well as being a wonderful metaphor for modern Singapore, the transvestites of Bugis

Street illustrate the state's ability to manipulate deviancy and defuse it as a presumed threat to society (from Eliot 1995a: 498–9).

32  'There is hardly any area of the overall reform program that has succeeded in the way that those who advocated it promised, if at all . . . the growing wastage and debilitation of human capital, juxtaposed against social inequities, corrupt cadres, and high-living nouveaux riches, poses the fundamental dilemma of how Marxists, even of this hybrid variety, should react as the oppressed, maltreated masses begin to respond' (Kolko 1995: 38).

33  One study published in 1992 estimated that 36 per cent of children in Vietnam were malnourished (quoted in Kerkvliet and Porter 1995: 17).

34  Trankell, for Laos, similarly reports that villagers are concerned that the reform process is undermining village support networks and communal solidarity (1993).

35  The CPF is Singapore's compulsory savings scheme set up in 1955 to provide pensions and medical care for workers. Savings can also now be used to purchase property and stocks. In 1995 employers were required to contribute 18.5 per cent of a worker's salary to the CPF and the employee a further 21.5 per cent (Silverman 1995: 51). It has been estimated that the dependency ratio in Singapore will increase from 1:8.2 in 1988 to 1:2.2 in 2030 (Davidson and Drakakis-Smith 1995: 19).

36  Britain's Labour leader Tony Blair visited Singapore at the end of 1995 to learn about the republic's Central Provident Fund.

37  'Old' social movements are such bodies as religious associations, trades' unions and farmers' groups; 'new' social movements are human rights organizations and women's environmental associations, for instance.

38  As Watts puts it, some scholars claim that new social movements are 'principally claiming territory from developmentalist states' (1995: 59).

39  An interesting piece of research would be to explore how 'exclusion' is understood and conceptualized in Southeast Asian languages.

# Part III

# CHANGE AND INTERACTION IN THE RURAL AND URBAN WORLDS

The discussion in Parts I and II, despite the attempt to provide the conceptual debate with a regional and local gloss, tended to strip places and people of their unique identities as part of an attempt to merge the conceptual and experiential landscapes. Anyone with field experience in the region, or outside it for that matter, may have found themselves muttering, 'it wasn't like that in xxxxx'. The discussion in this section is an attempt to put people back into the equation. It is drawn, in the most part, from the abundance of case study material from Southeast Asia. An important distinction between the following chapters and those that have gone before is that while the latter largely began with the conceptual and theoretical and then applied it to the Southeast Asia region, the following section will begin with the Southeast Asian experience and then link this back to the conceptual and theoretical.

In essence, the three chapters in this part of the book examine the ways in which people have responded to the challenge set by the process of development and should, ideally, be read as a single unit. The argument that links the chapters is that an understanding of key changes in rural and urban areas of Southeast Asia must be rooted in an understanding of the links and interactions between the two. Indeed, the evidence for a thorough dismembering of 'rural' and 'urban' is becoming increasingly persuasive. The notion that there are distinct and separate worlds, where agrarian change is fundamentally linked to agrarian processes, and urban change to urban and industrial processes, is shown to be deeply flawed as more and more people, and with greater frequency, cross the 'divide' between the two.

However, the chapters also reveal the extent to which, except at the most banal, it is impossible to generalize about important processes of change and their impacts on individuals and communities. The effects of mechanization on

153

people's livelihoods, the relative returns in agriculture and industry, and the impacts of non-farm employment on different classes, for example, are all shown to be highly diverse. The picture is of a process of 'change' – one is chary to use the terms 'modernization' or 'development' – which is not just complex in time and space, but also subject to multiple interpretations given different cultural and ideological backdrops.

# 5

# NEW RURAL WORLDS
## More than the soil

### Introduction

On the face of it, and despite the industrial growth of the past decades, Southeast Asia remains a region of farmers. Excluding the city state of Singapore and the oil-rich sultanate of Brunei, between 27 per cent (Malaysia) and 76 per cent (Laos) of each country's population are classified as working in the agricultural sector, and between 55 per cent and 88 per cent as living in rural areas (Table 5.1). Weighted according to population, the combined figures for the region as a whole are 58 per cent and 70 per cent respectively. Such statistics lend credence to the belief that the countries of Southeast Asia, like most developing countries, are dominated by agriculture. This view has it that the Southeast Asian world is still primarily a rural world; that this rural world is essentially an agricultural world; and, for the truly unreconstructed, that this agricultural world is, or

*Table 5.1* Population working in agriculture and living in rural areas

|  | Rural population as percentage of total, 1992 | Percentage of labour force in agriculture, 1990–92 | Total population, 1992 (millions) |
|---|---|---|---|
| Brunei | 42 | n.a. | 0.3 |
| Indonesia | 70 | 56 | 191.2 |
| Malaysia | 55 | 27 | 18.8 |
| Philippines | 56 | 45 | 65.2 |
| Singapore | 0 | 0 | 2.8 |
| Thailand | 77 | 67 | 56.1 |
| Cambodia | 88 | 74 | 8.8 |
| Laos | 80 | 76 | 4.5 |
| Myanmar | 75 | 70 | 43.7 |
| Vietnam | 80 | 67 | 69.5 |
| Total |  |  | 460.9 |
| Weighted total | 70 | 58 |  |

*Source*: UNDP 1994: 148

*Plate 5.1* The rural idyll: the popular view of the Southeast Asian countryside. A patchwork of paddy fields, field huts, coconut and banana trees, and surrounding forested higher ground, in this instance in Bali, Indonesia

should be, a peasant world. As McVey explains: 'save for the city-state of Singapore, [Southeast Asia] has long been imagined as quintessentially agrarian, its symbol the peasant toiling in his paddy-field, and plantations the only real representatives of the modern economic sector' (1992: 7). But 'rural', like 'urban', has become a metaphor for much more than open spaces and green trees: it also implies a way of life, a set of values, and a shared commitment to a certain livelihood.

From the early 1960s, studies of rural Southeast Asia have made the point that the village economy is undergoing fundamental structural change as the forces of modernization impinge on formerly isolated, inward-looking, self-sufficient and agriculturally-based communities. Howard Kaufman's community study of Bangkhuad in Bangkapi (in the Central Plains region of Thailand, but now a suburb of Bangkok) concludes that 'Bangkhuad, like most of Thailand, is in the throes of extensive socio-economic change . . . from a simple, minimum subsistence economy to one of a complex export economy geared to enlarged production' (1977: 210). Kaufman's field research was conducted in 1954 and his book first published in 1960. Many of the characteristics of the 'New Rice Economy' that he identifies are also issues with which this part of the book is concerned: the growing importance of non-farming activities, migration, mechanization and the shift into alternative crops, for example. Indeed, Kaufman was sufficiently prescient to identify a 'non-farming, urban – salaried – future' (ibid.: 211) for the village. Michael Moerman's research for his community study in Northern Thailand was conducted between 1959 and 1960. Unlike Bangkhuad, Ban Ping was truly isolated – at least in the sense that getting there was not easy.[1] Even so, extra-village links were important in defining the village itself and Moerman identifies a pattern of change not dissimilar from that of Kaufman's study. On his return to Ban Ping in 1965, the degree of 'progress' surprised the author, although he maintained that peasant life, at its core, remained redoubtable:

> It is tempting to exaggerate these changes, to suggest that Ban Ping has progressed from insularity to dependence, from status to contract, from *Gemeinschaft* to *Gesellschaft*. . . . Although Ban Ping has indeed progressed towards increasing participation in the nation and in the world market, it is still a peasant village.
>
> (Moerman 1968: 191)

This chapter is also about change, but focused on economic change. The perturbations in society, culture and politics, though important and profound, are not the main concerns here and will be discussed only insofar as they impinge and relate to economic change. Simply put, the chapter will present the evidence for fundamental change in the economies of rural Southeast Asia. Like Moerman, it will be argued that this change is, in a strict sense, one of degree rather than kind. The roots of commercialization are bedded deeply in the history of the region. Production for the market, payment in cash, the renting and sale of land, and the movement of people all, for example, pre-date modernization. In this sense, it would be wrong to write that communities have moved from subsistence to market orientation; from family to wage labour; from autonomy to dependence. Such dichotomies ignore the fact that there have always been market interactions and dependencies – and a flourishing non-farm economy (see Rigg 1994a, and below). But it might be argued that as degree augments degree, almost imperceptibly, a change of kind occurs.

The observation that rural Southeast Asia is changing is nothing new, but rarely is the breadth and depth of change in 'farming' communities in the developing world clearly set out. Nor, for that matter, are the implications of such changes sufficiently addressed. As a result perhaps, the enduring image of the village is one where a community of farmers gains a livelihood from working the land. The figures presented in Table 5.1, coupled with the apparent 'timelessness' – in visual terms – of country life, tend to support such a perspective. Perhaps it is for these reasons that Angeles-Reyes finds that she is able to write that the 'Philippine economy remains predominantly rural. . . . No fundamental structural transformation has taken place in the economy since the later 1960s' (1994: 134), while Nipon Poapongsakorn can state that 'even [in] the dry season [in Thailand] . . . the agricultural sector still accounts for almost 60 per cent of total employment' (1994: 168). However, it does not take a deep incision to cut through these layers of image to reveal the truth that the rural world has become a *new* rural world. This requires a rethinking of the rural economy and rural life, a reappraisal of policy initiatives and planning strategies, and a reformulation of theories of agricultural and rural development. It also, it should be added, emphasizes the point that the modernization ethic has become part of the Southeast Asian landscape, both mental and physical. In this sense, it could be convincingly argued with respect to Southeast Asia that the post-developmentalists are, over much of the region, fighting yesterday's battle (see Chapter 8 for further discussion).

## Village identity and identification: who is a farmer? What is a household?

### The invention of the village

Until quite recently, most studies of Southeast Asia assumed that the basic unit of rural life in the region was the village. The village represented the essential unit of identity and organization within which agricultural and other activities occurred. This view was not just restricted to Southeast Asia; Eric Wolf's designation of the village as a 'closed, corporate community' gained wide currency across the developing world (1967). However, from the early 1980s studies that re-examined the historical evidence began to question whether the village, at least as it exists today, was not, in fact, a creation of the colonial period and our understanding of its structure and function a product of those whose minds were firmly rooted in European conceptions of rural life.

From the end of the nineteenth century, colonial authorities, and in the case of Siam the national authorities, found it necessary to control people and space more closely. To achieve this, a series of administrative reforms were introduced of which the creation of the administrative village (to distinguish it from the traditional village) was one. In Siam, for example, the brilliant and influential prince Damrong Rajanubhab (1862–1943), half-brother of King Chulalongkorn (Rama V) and Minister of the Interior, reformed village administration in the early 1890s.[2] Bunnag describes the process as it was implemented near Bang Pa-In in the province of Ayutthaya, north of Bangkok:

> *Luang* Thesachitwichan asked the heads of ten households, whose houses were situated near one another, to elect a village elder [*phu yai ban*]. He then asked the elected elders of villages, the number of which was governed by the natural features of the locality, to choose one of themselves as commune [*tambon*] elder [*kamnan*]. . . . *Phraya* Sisuriyaratchawaranuwat . . . began to persuade the heads of five to twenty households to elect a village elder . . . [and] then asked approximately ten village elders to choose a commune elder.
>
> (Bunnag 1977: 110)

The picture is clearly one where people were manipulated in the interests of state control. The same appears to have been true in Java where the village has been described as a 'harnessed construction' (Breman 1980) and in Bali where villages became 'rationalized' into administrative communities (*desa perbekelan*). 'People found themselves' as Vickers puts it 'in new villages to which they previously never had any connection' (1989: 134). In a similar vein, Shamsul, writing on Malaysia, explains that the village was 'created, recreated, sub-divided, rejoined and sometimes imagined to exist in the name of development' (1989: 20). Space, in short, was bureaucratized to put people in their place.[3]

This process is important in so far as it has laid the foundations for the impression that Southeast Asia remains a region of farmers. Ironically perhaps, the administrative reforms described above were so successful that a very real sense of village-ness was, over time, created. Vital statistics are collected by village heads, cooperatives operate on the basis of the village unit, schools are arranged with catchment areas centred on villages or groups of villages, health centres are located within villages, funds are dispersed to village development committees, and so on. Villages, whether they be Indonesian *desa*, Malaysian *kampung*, Thai or Lao *bân*, Philippine *barrios* or *barangays*, Cambodian *phum*, Vietnamese *thôn*, or Burmese *myo* define for most people their place in the world.

### Who is a farmer? What is a household?

The process described above has led to two important distortions in perception regarding the nature of the rural 'world'. First, it has meant that households tend

159

*Plate* 5.2 A Thai nuclear village in the Northeastern Thai province of Mahasarakham. Ban Noon Tae, not far from the main road between Khon Kaen and Mahasarakham towns, has electricity and was voted one of the best-kept villages in the area

to be recorded as living in villages even when individual household members may live a greater proportion of their time in urban areas. In Thailand, the intricacies of the household registration system means that there is a strong disincentive to register a change of residence. When migrants expect to work only temporarily in the city, few can be bothered to go through the rigmarole of changing their residency. And those few who do take the decision to try to re-register often find that it is impossible because the owners of the buildings where they happen to live, perhaps with hundreds of short-term tenants and wishing to avoid the bureaucratic hassle, refuse to enrol them – a requirement for re-registration (Fairclough 1995b). The second distortion in perception is that it has led to the assumption that because the countryside is largely rural, and because the word 'rural' is synonymous with 'agriculture', then these people are, virtually by definition, farmers. It should be added that this is not just a case of the state imposing its own notion of people's identities on rural inhabitants. Urban residents from rural areas themselves will often say they live in a village – even that they are farmers – when they live and work in the city. To some extent, village-ness has become a measure or badge of identity that people take with them to urban areas. They do not leave it behind with their paddy fields and ploughs – they wear it as a statement, as a distinguishing marker that sets them apart from other urban residents.

There may be a semantic problem here, compounded by complexities associated with different ways of 'knowing' and of categorization. To ask someone the apparently simple question 'Where do you live?' may be interpreted not so much as 'Where do you live your life?', but rather 'Where are your roots?' or 'What is home?'. Many migrants in Southeast Asian towns and cities do not expect to live their entire lives in those places. The intention, often, is to return 'home' to the countryside and they often view themselves as sojourners in the city. The fact that they left that countryside fifteen or twenty years previously may not alter this belief, and indeed many 'rural' people die in urban areas as old men or women, still anticipating, on their death beds, the day when they might return to their village roots. It is also worth asking whether Eurocentric conceptions of a single 'home' can be applied cross-culturally. Chapman, in his 'autobiographies' of Pacific islanders, quotes one Melanesian, John Waiko, who explains: 'We Melanesians are all engaged in a tapestry of life, where the threads of movement hold everything together' (1995: 257). Chapman goes on to hazard: 'Consequently the essence and meaning of a people's mobility becomes far more comprehensible when conceived as an active dialog between different places, some urban, some rural, some both and some neither, as incorporating a range of times simultaneously ancient and modern, and as the most visible manifestation of a dialectic between people, communities, and institutions' (Chapman 1995: 257). In other words, 'home' and 'place' are ambiguous and shifting notions, where multiple identities – both 'rural' and 'urban' – can be simultaneously embodied.

As if this did not make things complicated enough, there is the added problem of categorizing people and households with diverse and multiple occupations within a simple, binary classification. How is someone who works as a farmer for six months and a textile worker for the other half of the year to be classified? It seems that often researchers are only too ready to apply their own designations and determinants of membership, often based on overly-rigid classifications rooted in conceptual exercises that may make perfect sense in an arid academic context, but which in the context of the village are patently inapplicable. Alternatively, the researcher may adopt an emic approach, asking villagers themselves to decide the composition of their households. As such emic classifications are usually more nuanced and dynamic than researchers would like, it leaves the 'household' not as a unit with a clear-cut membership, but as a shifting entity whose membership is not just blurred at the edges but which may also change shape through the course of the year and between years. Not very helpful, in short, for scholars who wish to know the answer to apparently simple questions like, 'What proportion of household income is derived from non-farm sources?'.

There has been much discussion – at the methodological, theoretical/ conceptual and practical levels – as to where a 'household', as a unit of analysis and identification, begins and ends. Given evolving rural–urban and farm–non-farm interactions, a spatial determination of 'household' (i.e. the co-residential dwelling unit or domicile) has become increasingly difficult to sustain. At a functional level, there is a strong argument that all those people who actively participate in the reproduction of the household should be included as members. Part of the reason for the differences of opinion is that the household is, at one and the same time, a conceptual construct, a functioning system, and a methodological unit of analysis. The demands of each tend to lead scholars to arrive at different definitions of the 'household' (see Netting *et al.* 1984).

A broader, related, challenge lies in the question of whether the household is a useful or relevant unit of study at all. The study of household economy has tended to view all members as equal members, subsumed within and driven by one household-determined ethic, and bound together by necessity and by kinship in a largely, at least from the analytical point of view, undifferentiated mass. However, Diane Wolf, for one, disputes the view that young women are merely reacting to household-determined necessities. She was surprised to find in Java that 'Rini and other factory daughters sought [factory employment] without their parents' encouragement and even without their awareness or approval' (1992: 5, see also Wolf 1990 and Ong 1987). These daughters were not acquiescing to a collective household strategy; they were independently building their own lives. The household, like the village, is a much more fractious social unit than such assumptions would lead one to believe. It may be a unit riven by competing needs and conceptions, by antagonisms and conflicts. If one is to accept that not all is peace and traquility within this mythic household, then to talk of household 'strategy', as if each member sits down to map out the common future whether that be based on a strategy of economic survival, consolidation or accumulation, is patently inadequate.

Notwithstanding the above difficulties of identifying where the 'household' and the 'village' begin and end, recent studies of village economies consistently show that between 30 per cent and 50 per cent of total household income is derived from non-farm, or off-farm, sources (see, for example: Ritchie 1993 on Thailand; De Koninck 1992 on Malaysia; Morrison 1993 on Sarawak; and Effendi 1993 and Cederroth 1995 on Java).[4] For a significant number of households, non-farm activities represent their major source of income.[5] In the East Javanese village of Bantur, of 1,880 people, 792 (42 per cent) had jobs outside agriculture and in total there were over sixty different kinds of occupation (Cederroth 1995: 110).

So, in terms of identity, identification and activity, there has been a divergence. On the one hand people have increasingly been identified, and have identified themselves, in village terms. And yet structural change in the economy, coupled with rising expectations and a radically improved transportation system, is doing the reverse. It is likely that this divergence will resolve itself as people adapt and accept their new ways of life, and as governments alter the ways in which their populations are administered, recorded and measured. But for the present at least, statistics such as those presented in Table 5.1 should be treated with caution: 70 per cent of the population of Southeast Asia do not live in rural areas; and 58 per cent do not work in agriculture. Both figures are lower. Furthermore, the distinctions are becoming blurred as households increasingly occupy, or have representation, in both the rural and urban worlds and, more to the point, earn a livelihood in both agricultural and non-farming activities. This area of interconnectedness, and the evolving interactions between rural/agricultural and urban/industrial are addressed in more detail in Chapters 7 and 8. Suffice to say at this point that agencies charged with the task of administering or analysing development, like the ILO, tend to treat rural and urban-based enterprises separately, in the process disguising the degrees of interaction and dependence binding the two (Timberg 1995). Partly as a result, there is the linked tendency of continuing to characterize non-farm economic activity as supplementary and marginal to the 'key' agricultural endeavours (this is discussed at greater length in Chapters 7 and 8).

## 'In the shadow of agriculture': non-farm work in historical perspective[6]

It has been usual to describe rural areas of the pre-modern developing world as almost entirely agricultural in character. However, just as recent work has emphasized the degree to which even pre-modern villages were part of a wider money economy, so there is also increasing evidence that there was often much more to such village economies than agriculture. White, for example, has estimated that in 1905, 30 per cent of the rural population of Java were primarily engaged in activities outside agriculture (see Figure 5.1). He further argues that peasants were not being marginalized into such occupations, but were taking them up as a means of consolidating their incomes and accumulating wealth.

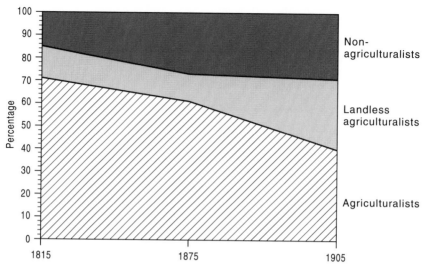

*Figure 5.1* Growth in non-agricultural employment, rural Java 1815–1905 (percentage of economically active population)
*Source*: White 1991

People were weaving, making hats, trading, running stores, and much else besides (White 1991; Boomgaard 1989: 109–35).

What happened to rural industry in the face of cheap imports from the West has been the subject of some debate. The decline of traditional industries has often been portrayed as a process of 'de-industrialization', although Peter Boomgaard writing on Java maintains that it is more accurately described as 'de-specialization' (Boomgaard 1991: 35–6). While professional artisans like metal workers and ship builders usually managed to incorporate new production methods and technologies, and remain competitive, peasants sometimes stopped making crafts and concentrated their energies in agricultural production.

It was not just on Java where the rural economy was surprisingly diversified. Katherine Bowie's study of textile production in nineteenth-century Northern Thailand illustrates the degree to which rural communities were engaged in commodity production and were integrated into the wider economy. She concludes that 'this examination of textile production reveals a society with a complex division of labour, serious class stratification, dire poverty, a wide-reaching trade network, and an unappreciated dynamism' (Bowie 1992: 819). Clearly, areas differed in the degree to which non-farm work played a role in the village economy, but the general point seems to be that we should not be surprised if pre-modern villages were supported by diversified economies, and that latter-day diversification has built upon a tradition which emphasized multi-stranded livelihood strategies as a means of reducing risk and boosting incomes.

## Aspirations and needs: from the next meal to the next motorbike

I hope the lives of my children will be different from my life. I want their futures to be the same as those of rich families' children.

(Luong Xuan Hoang, a Vietnamese farmer in Thanh Hoa
province, quoted in *Far Eastern Economic Review* [*FEER*] 1994)

In former times we were happy when we had enough to eat. We were happy when life was stable and peaceful. But today we need more. Now we know about the necessity of development, we know better what we really need and how we can get it. But to reach it we have to work hard, we need a higher output.

(Mr. A., a 'peasant' in North Sulawesi, Indonesia,
quoted in Weber 1994: 199)

Modernization and development have encouraged people in Southeast Asia to expect more from life; their 'pressure of needs' is intensifying and expanding year by year. The observation that 'basic needs' are in fact 'relative needs' has been recognized for many years. Adam Smith, for example, argued as long ago as 1776: 'By necessities I understand not only the commodities which are indispensably necessary for the support of life, but whatever the custom of the country renders it indecent for creditable people, even of the lowest order, to be without'. The challenge is for standards of living to keep pace with escalating expectations.[7] Alison Murray, in her detailed description of an urban *kampung* in Jakarta, writes that '[a] vinyl three-piece suite, coffee table and lacquered buffet are considered essential symbols of social standing, and it was considered shameful that one of the RT [*rukun tetangga* or neighbourhood] leaders in RW [*rukun warga* or community] 'B' had no lounge suite and had to entertain visitors at his father-in-law's house' (Murray 1991: 37).

Nor is the driving concern for consumer goods and prestige measured in terms of wealth, just a function of city life. Robert Hefner, in his study of the village of Tosari in the Tengger Highlands of East Java, quotes a farmer as he laments the loss of subsistence innocence:

It's not like before. In the old days people here [in the highlands] were different from those in the lowlands. They weren't interested in wearing fine clothes that drew attention to themselves, or in eating special foods like those you see today. Even though some people owned more and some less, people dressed and ate the same. . . . Now it's different. Those who are well off want to give orders and keep their hands and feet clean of earth. They keep track of everything they give and everything they get in return. It's just like the lowlands. Everything is counted up and owned.

(Hefner 1990: 1)

This may not be a case of gradually rising expectations that can be managed and met through careful and diligent economic planning. It is clear that the change is, in some cases, extremely rapid. In a study of the village of Klong Ban Pho in Chachoengsao province in Central Thailand, Chantana Banpasirichote notes that the pick-up truck made the transformation from 'status symbol' to 'necessity' in the space of just five years (Chantana Banpasirichote 1993: 58) (see Tables 5.2a and 5.2b). In other words, in less time than it takes many books on 'contemporary' rural life to make the transition from proposal to bookshelf, their contents may be rendered obsolete by the rapidity of change in those areas they profess to describe.

In 1991, the Indonesian *Biro Pusat Statistik* (Central Bureau of Statistics) asked a nationwide sample of Indonesian families what they thought about changes in their welfare by comparing conditions at the time of the survey with those three years earlier (Table 5.3). Although most thought their situations had generally improved, in some areas a significant proportion – between a fifth and a half – reported a deterioration in their welfare. But the responses in the categories 4–6 are unlikely to indicate an absolute decline in the provision of, or access to, the welfare criteria in question. Instead, they are likely to reflect a rising level of expectation, so that the formerly acceptable, becomes unacceptable. The areas of concern are enlightening. There is a clear dissatisfaction with the level of access to secondary schooling. This highlights parents' concern not so much for their own well-being and prospects, but for their children's (this is reflected in the quotes at the beginning of this section). Primary level education may now be virtually universal in Indonesia and across Southeast Asia, but this has created

*Table 5.2a* Consumer goods in Ban Non Tae and Ban Tha Song Korn, Northeast Thailand (1982 and 1994)

|  | *1982* | *1994* |
|---|---|---|
| Pickup | – | 11 |
| Motorcycle | 16 | 63 |
| Television | 20 | 72 |
| Refrigerator | – | 43 |

*Source*: author's field research, 1982 and 1994
*Notes*
*n* = 78 (households)
Between 1982 and 1994, the television in Ban Non Tae and Ban Tha Song Korn made the transformation from comparatively rare object of veneration to common household item.

*Table 5.2b* Consumer goods in Jantinom, Central Java (1980–89)

|  | *1980* | *1989* |
|---|---|---|
| Car/minibus | 56 | 25 |
| Motorcycle | 57 | 2,512 |
| Television | 150 | 650 |
| Radio | 628 | 1,540 |

*Source*: Effendi and Manning 1994: 224

*Table 5.3* Changes in Indonesian welfare criteria (1988–91)

| | Great improvement | | | Much worse | | | Percentage in response |
|---|---|---|---|---|---|---|---|
| | *1* | *2* | *3* | *4* | *5* | *6* | categories 4–6 |
| Formal jobs | 0.1 | 4.4 | 26.6 | 46.7 | 18.4 | 3.7 | 68.8 |
| Access to reading material | 1.4 | 21.2 | 44.0 | 29.4 | 3.0 | 1.1 | 33.5 |
| Access to senior high school | 1.5 | 22.9 | 51.2 | 18.9 | 4.6 | 0.9 | 24.4 |
| Sports | 0.5 | 15.8 | 63.2 | 18.2 | 1.9 | 0.4 | 20.5 |
| Utilities | 0.5 | 21.6 | 58.6 | 18.1 | 1.1 | 0.1 | 19.3 |
| Access to television | 2.9 | 38.7 | 41.0 | 14.8 | 1.7 | 0.8 | 17.3 |
| Housing quality | 0.7 | 22.4 | 61.1 | 14.5 | 1.3 | 0.0 | 15.8 |
| Household income | 0.7 | 31.9 | 53.5 | 10.4 | 3.4 | 0.1 | 13.9 |
| Access to junior high school | 1.4 | 25.5 | 59.5 | 11.2 | 2.0 | 0.4 | 13.6 |
| Access to medicines | 1.4 | 33.6 | 51.4 | 12.0 | 1.1 | 0.4 | 13.5 |
| Access to transport | 5.1 | 49.6 | 32.4 | 11.2 | 1.2 | 0.5 | 12.9 |
| Food expenditure | 0.3 | 22.3 | 66.7 | 9.5 | 1.2 | 0.0 | 10.7 |
| Security from crime | 2.1 | 30.9 | 57.0 | 7.7 | 2.1 | 0.2 | 10.0 |
| Access to radio | 2.5 | 35.3 | 52.6 | 8.3 | 1.0 | 0.3 | 9.6 |
| Clothing | 0.5 | 27.4 | 64.1 | 7.4 | 0.7 | 0.0 | 8.1 |
| Health services | 1.4 | 38.1 | 52.8 | 6.8 | 0.8 | 0.2 | 7.8 |
| Health | 0.8 | 29.0 | 64.3 | 3.8 | 2.1 | 0.1 | 6.0 |
| Family planning service | 2.1 | 42.6 | 50.9 | 3.6 | 0.6 | 0.2 | 4.4 |
| Religious holidays | 1.8 | 29.1 | 66.4 | 2.2 | 0.6 | 0.0 | 2.8 |
| Access to primary school | 2.8 | 36.9 | 57.8 | 1.8 | 0.6 | 0.1 | 2.5 |
| Religious services | 3.0 | 39.3 | 57.0 | 0.6 | 0.1 | 0.0 | 0.7 |

*Source:* Terry Hull, personal communication

a heightened demand and expectation for secondary level schooling. One Northern Thai informant observed that while '[i]n the past, the parents' prime responsibility and concern was to feed their children, today . . . [it] is to educate their children' (quoted in Bencha Yoddumnern-Attig 1992: 20). Table 5.3 also seems to reveal that there is a significant demand for better housing and utilities and – surprisingly perhaps – for access to reading material.[8]

But most striking in the table is the degree of dissatisfaction with the level of access to formal jobs. This shows that despite rapid economic growth in Indonesia, security of employment in formal sector jobs is seen to be highly deficient. Why formal employment should be perceived to be important links back to rising educational standards. People who have successfully completed secondary education are prone to expect a job concomitant with their educational achievements; those who have degrees have higher expectations still. In general, though, the expansion of formal sector jobs has not kept pace with the development of the education system, creating what might be termed an 'expectation gap'. This notion that one of the by-products of rapid change is a yawning gap between provision and expectation operates at many levels, and at the same time both frustrates people who experience it while also driving them to attempt to narrow the gap.

*Plate 5.3* School assembly in the town of Maumere on the island of Flores in East Nusa Tenggara, Indonesia

The increase in levels of expectation across the region, but more particularly among the countries of Asean, is creating the impetus for thorough-going economic change. In some instances, these needs are being met through increasing agricultural output – whether by expanding the area under production, increasing yields, or moving into new agricultural endeavours. However, more often than not, these innovations and advances within the agricultural sector are not sufficient. A shortage of land and declining terms of trade between agriculture and industry have made increased production difficult in the first instance, and unattractive in the second. Allied to a lack of room for manoeuvre within agriculture, it is also true that education and the media have encouraged younger men and women to regard agriculture as a low-status activity, and the life of the farmer as hard and burdensome (see for e.g. Preston 1989: 51). In short, an occupation to be avoided. Thus, growing expectations and an absence of sufficient opportunities within agriculture are creating the conditions in which rural households are looking further afield, both spatially – beyond the village – and in sectoral terms – beyond agriculture. The expectation gap is a critical factor that drives and explains the process and pattern of development and change. This impetus for change is reflected in the comments of a 41-year-old man in a village in Laos, newly exposed to commercial opportunities by the upgrading of route 13, west of Pakkading near the Mekong:

> My dream is to buy a car, a pick-up. I am trying very hard to save money in the bank, for keeping cattle as capital is like hanging money in the tree. People see and steal it. You have to guard the cattle both day and night. Last year I sold cattle to be able to buy a small tractor. Some villagers asked to borrow it. They pay me a small amount or buy benzoin or oil for me. Life has changed for the better during the last five years. Step-by-step we have bought cattle, a motorbike, tractor, shop and now we have [a vehicle] repair shop. People will soon need to have repairs done. I wait for them. Sooner or later they will come.
>
> (Hågankård 1992)

## The transport revolution

Aspirations are driven by changing needs, and needs by advancing vistas of knowledge and experience. Knowledge and experience, in their turn, are linked to expanding networks of contact and communication. A key element driving rising aspirations, and perhaps enabling those aspirations to be met, therefore, is the greater accessibility that has been afforded by road expansion and improvement, and the availability of cheap transport (Table 5.4). In Indonesia, Dick and Forbes have termed this the 'quiet revolution' or the *revolusi colt* (1992).[9] In most instances the transport *revolusi* has involved an interplay of government investment and private capital. Governments have built the roads, while private entrepreneurs have provided the means to galvanize a latent productive resource – roads – into an agent of rural development:

*Table 5.4* Multiplicity of modes of public road transport in Indonesia

---

*Local, short distance*

**bicycle** – for personal use and for load-carrying and human transport, sometimes for private hire

**becak** – pedal tricycle (trishaw), for short-distance load-carrying and human transport, for private hire

**dokar** – horse and cart for short-distance load-carrying and human transport, for private hire

**cikar** – ox-cart for short-distance load-carrying and human transport, for private hire

**bajaj** – small three-wheeled motorized vehicle, usually urban use, for private hire

**bis kota** – town bus, running routes in larger cities, fixed fare

**ojek** – motorcycle taxi, for human transpor (rarely used as a load carrier), for private hire, usually local but occasionally used for longer trips of up to around 20 km

**bemo** – small vehicle running local routes in larger cities and linking market towns, fixed fare

**oplet** – small vehicle running local routes in larger cities and linking market towns, fixed fare

*Regional/national, long distance*

**bis ekonomi/bis express/bis a.c./bis VIP** (**'vip'**) – different standards of buses linking larger cities and towns, usually from bus terminal (situated out of town in larger centres) to bus terminal. *Oplets* and *bemos* run routes into the town centre. *Bis ekonomi* are usually the oldest vehicles, and offer the cheapest fares, the slowest service and the most cramped conditions. But they are also more likely to take sacks of market produce, chickens, bicycles and other livestock, too.

**bis malam** – night bus linking larger cities and towns, usually from bus terminal (situated out of town in larger centres) to bus terminal. *Oplets* and *bemos* run routes into the town centre.

---

> At the outset of the New Order [in Indonesia in 1968], walking was still the normal means out of the village, at least as far as the main road, if not all the way to the market town. During the 1970s this began to change . . . [and by] . . . the 1980s, at least in most parts of Java, Bali, and Lombok, transport between village and market town and between market town and city was quick, frequent, and cheap.
>
> (Dick and Forbes 1992: 270–1)

It is difficult today, in the countries of Asean, fully to appreciate the transport revolution that has occurred. Partly the problem was a simple lack of roads and consequent inaccessibility. But in some areas the state also limited mobility through decree and control. As is still the case in Myanmar, and until quite recently in Vietnam, Laos and Cambodia, a mobile population was viewed as a potentially dangerous and destabilizing one. Villagers wishing to leave the local area needed to obtain a permission note from the authorities before they could board a bus. At the beginning of the 1960s, villages in Gunung Kidal in the Special Region of Yogyakarta had to obtain an official letter of permission before they could take one of the few minibuses that made the journey between the city of Yogyakarta and Wonosari, just 35 km to the south-east (Rotgé 1992: 35). The

*Plate 5.4* Local transport in Indonesia: a *bendi* and driver in Bima-Raba, the main twin towns of eastern Sumbawa, West Nusa Tenggara, Indonesia. In Bima-Raba *bendis* are locally known as *ben hurs* – the film made quite an impact!

authorities also often limited the number of licensed vehicles on the road, creating a situation where the demand for transport far exceeded supply. The motivation behind this lay in the desire to keep such licences rare and valuable commodities – a lucrative source of extra income for poorly-paid regional officials. Even today in Thailand, villagers are required to have an ID card to travel beyond their province of residence, and in border areas the army routinely scrutinize bus passengers' ID cards. Although this requirement is not a constraint on mobility for the majority of Thais, the hill minorities have sometimes found their mobility restricted by the regulation, and for most hill people the main advantage of having an official Thai ID card is the freedom to travel that goes with it (see page 119).[10] Improved transport and communications does not just permit more efficient economic activity though, it acts as a catalyst for increased and different economic activities (Dick and Forbes 1992: 258). Booth, in the context of Indonesia, even suggests that improving transport links helps spread the ideology of economic development by permitting officials from local areas to travel to the centre (Jakarta) for briefings, courses and workshops, returning to their home patches imbued and energized with new ideas (Booth 1995b: 110). Perhaps because it appears so self-evident, there has been comparatively little work undertaken on the role of transport in economic development. Yet few would dispute that without such improvements, existing levels of economic activity could not have been sustained.

The role of government action in this revolution is critical. In Northern Thailand state investment in the 'development of adequate infrastructure (roads) provided the means, economic development and growth in urban areas provided the opportunity (jobs), and state policy favouring economic growth over distribution, and land titling (limiting opportunities in the village) provided the impetus for rural villages to become dependent on urban areas for their continued economic reproduction' (Ritchie 1993: 16; see also Hirsch 1989 and 1990). The one element that Ritchie omits in his sequence of explanation is the role of the vehicle-owning entrepreneur who 'activates' the roads. For the most part, the expansion of public transport has been a spontaneous response to increasing demand facilitated by an improving road network. Dearden, with reference to Northern Thailand, goes as far as to suggest that the development of the road network 'has probably done more to change the landscape of the North and the mindscape of [its] inhabitants than any other single factor' (Dearden 1995: 118). In Java, poverty has a key spatial component and regional patterns of poverty are closely related to geographical marginality and the high transactions costs that are associated with remoteness from administrative centres (Mason 1996). The same has been shown to be the case in Vietnam where, using the 1992–93 Vietnam Living Standards Survey, van de Walle demonstrates an 'intimate' link between 'poor infrastructure and high poverty' (van de Walle 1996: 43).

Most studies emphasize the degree to which road improvements integrate isolated and marginal people into the space economy. Masri Singarimbun quotes

*Plate 5.5* Public transportation in Vietnam remains antiquated. Da Nang's local bus station

a respondent in the village of Sriharjo outside Yogyakarta, Java as saying that improvements in transportation and communication had made previously remote places like Jakarta, Bandung and Surabaya, 'almost like your neighbours' (Singarimbun 1993: 268). However, roads can also further marginalize some groups, while being of scant interest to others. The concerted effort to improve Laos' road network (see Box 5.1), for example, may be causing off-road communities to become even less accessible and more remote. Businesses congregate by the roadside, traders refrain from travelling off-road to buy and sell goods expecting producers to come to the road to trade, and towns and villages with easy access to new or improved roads grow at the expense of communites with off-road locations.[11] In other words, and paradoxically, the effort to improve Laos' road infrastructure, rather than making peripheral areas more accessible, may be serving to increase spatial inequalities within these peripheral areas. Nor, for that matter, are the people who live in the vicinity of newly-built or improved roads necessarily those who benefit. Again in Laos, Ing-Britt Trankell notes how it is former residents of the capital, Vientiane, who 'have invested in new land in the area next to the road [in Bolikhamxai province], in expectation of economic development following the improved communications' (1993: 77).

Even in Laos, though, studies have shown that roads bring new business opportunities to local people, increase consumer choice, may selectively assist women by promoting marketing opportunities (marketing is often undertaken by women), improve access to services like schools and health facilities, promote the cultivation of secondary cash crops, improve access to agricultural inputs like fertilizers, and ease marketing constraints for produce. They also promote economic and social differentiation, and lead to land value inflation (Hågankård 1992).

## RURAL LIFE: MORE THAN THE SOIL

Pre-modern Southeast Asian life has tended to be characterized as essentially subsistence-based. As already noted, this emphasis on self-reliance and economic independence disguises the extent to which farmers produced for the market, and were tied into the market system (although often this was not capitalistic in the modern sense), even before the wholesale colonization of the region. None the less, the modern period has seen a significant increase in the commercialization and commoditization of the agricultural economy. Seven threads of change can be identified. While some build upon traditional activities, others are entirely new:

1 *Subsistence crops:* the application of increasing quantities of cash inputs like chemical fertilizers, and the growing use of wage labour. The sale of an increasing proportion of production. Growing mechanization of production.
2 *Cash crops:* the increase in the relative importance of cash crops, both in their contribution to the farm economy and in the relative area under cultivation.

*Box 5.1* Laos: the environment of transport and communications

Although the transport revolution may have come to the countries of Asean, and is begnning to make an impact in Vietnam, Laos, particularly, suffers from serious infrastructural constraints. Of the country's already limited network of 13,300 kilometres of roads, just 20 per cent are paved. Even roads which have been relatively recently upgraded, like route no. 9 which was upgraded in 1988, are deteriorating because of lack of maintenance and the effects of the tropical climate. The Lao government considers the upgrading of transport and communications to be possibly its most important task. Between 1986 and 1990 government expenditure on transport averaged 40 per cent of total expenditure (Zasloff and Brown 1991: 149), while in the 1991/92–1995/96 Public Investment Programme 40 per cent of total planned outlays were allocated to the transport sector (Nielsen 1994: 184). This was the highest allocation for any sector. The problem facing the government as it attempts to upgrade the country's road network has three key facets:

- First, 80 per cent of the country is classified as mountainous, presenting a considerable engineering challenge. This, allied with a monsoon climate where torrential storms are common, means that roads are difficult and expensive to build and need to be constantly monitored and repaired.
- Second, Laos' population density, at 17 people/km$^2$, is one of the lowest in Asia. As a government report prepared for a meeting on Least Developed Countries (LDCs) in Geneva in 1989 stated, 'the cost per capita of providing the necessary infrastructure [in Laos] is in many cases prohibitive' (Lao PDR 1989: 49). The 1993–94 budget provided for total state investment of just 122 billion kip or US$168 million. Of this, 57.6 billion kip was allocated to 'communications, transportation, postal services and construction' – a mere US$79 million (Dommen 1994: 168).
- The third facet of the infrastructural challenge facing the country concerns the damage sustained during the Vietnam War. It has been estimated that 2.1 million tonnes of explosives were dropped on Laos during 580,000 bombing missions in the clandestine American attempt to stifle the flow of arms along the Ho Chi Minh Trail from North Vietnam, through Laos, into former South Vietnam. The Lao PDR has the dubious honour of being, in per capita terms, the most heavily bombed country on earth – two tonnes of ordnance per capita at a total cost of US$10 billion (Evans 1994). This damage is still being repaired, and continues to hamper attempts to upgrade the country's physical infrastructure. In 1993 bombs *and bombis* – mostly disturbed by farmers in their efforts to clear and prepare land for agriculture – killed 100 people in Xieng Khouang province alone (Evans 1994: 58). Not only does this increase the dependent population and represent a drain on household and national budgets, it also inhibits new land development. Inventive and adaptive as ever, villagers have found productive use for this legacy of the war – as war scrap for sale, water troughs, house pillars and storage bins.

*Source*: adapted from Rigg 1995c

*Plate 5.6* Roads in Laos are very poor and in the most part unsurfaced. A large proportion of public investment is being directed into upgrading the country's stock of roads. Here a converted Japanese truck waits at a roadside stop south of Paksé on route 13. Many of the villages along the country's main highways derive a significant proportion of their income from selling snacks and drinks to

*Plate 5.7* In some areas in Laos, especially in Xieng Khouang province which was very heavily bombed during the conflict in Indochina, unexploded ordnance still disrupts daily life and kills or maims hundreds of people a year. Photograph: Jonathan Miller

Their cultivation using increasing cash inputs and wage labour. Growing mechanization of production.

3 *Raising and sale of livestock*: the raising of livestock for the market and the specialization by some households in livestock production, including fisheries.
4 *Wage labouring, on-farm and local*: a decline in reciprocal labour exchange and a shift to agricultural and farm-linked wage labouring.
5 *Non-farming activities, on-farm*: the expansion of on-farm, income-generating activities not linked to farming, such as cloth and mat production, and pottery and furniture-making. Some authors prefer to use the term off-field (e.g. Cederroth 1995).
6 *Non-farming activities, off-farm but local*: the growing importance of part-time, seasonal and full-time non-farm employment in off-farm activities such as factory work, public sector jobs and activities within the informal sector. Workers usually return home to the village nightly.
7 *Migration for employment*: the increasing role of migration to employment beyond local areas. This may be to capital cities, regional urban centres, other rural areas or abroad. In each case, the migrant maintains a rural base.

Sometimes this process of economic change is presented not so much as one of economic differentiation (as here), but of proletarianization. In other words, from a position of self-employment on peasant farms, to wage labour in which individual control over resources is lost (see Eder 1993). Although proletarianization can, and does, occur within agriculture – for instance in the spread of estate agriculture and contract farming – more often, proletarianization is part and parcel of economic differentiation, and not distinct from it. The seven strategies listed above represent two approaches to increasing output. The first three represent intensification *within agriculture*; the latter three, intensification *out of agriculture*. The fourth strategy is a result of the first three. In the past there been a tendency to see the sixth and seventh strategies listed above as being both spatially and economically distinct. Thus, employment in export-oriented garment and footwear factories would invariably entail migration beyond the local area; while those villagers who commute to work would be employed in small, local factories and in local service provision. However, the location of large factory enterprises in rural areas, particularly on Java but also in the Central Plain and Northern regions of Thailand, Central Luzon in the Philippines, and in the Klang Valley around Kuala Lumpur in Malaysia, has opened up the possibility for factory employment while maintaining a home base. This expansion of the city into rural areas has been termed *kotadesasi* in the Asian context (see page 264), and some studies talk of 'industrialized villages' (e.g. Wolf 1992).

As the discussion below will show, diversification, and interpretations of the causes and effects of diversification, are themselves becoming more diverse. Generalizations and single-cause explanations for the changes affecting lives and livelihoods in rural areas of Southeast Asia cannot do justice to the multifarious processes underway. The very terms used imply that such activities are subsidiary to farming and a residual product of essentially agrarian processes of change: 'non-farm', 'off-farm', 'off-field' and 'non-agricultural', for instance. This narrow, agriculture-centred mode of explanation is simply inadequate in the context of modern Southeast Asia (see Koppel and James 1994). As the earlier discussion emphasized, village studies show that in increasing numbers of villages the majority of household income can be defined as non-farm.

Finally, it is necessary to emphasize the semantic problems connected with discussing 'non-farm' activities. Scholars like Evers (1991) and Cederroth (1995) have expended considerable scholastic energy trying to define their terms. Farm/non-farm and on/off-farm may refer to geographical criteria; they may highlight a sectoral distinction (agricultural/non-agricultural); or they may distinguish between different types of households (those engaged in farming, those who are not, and those who do both); or they may refer to one, two or all three of these at the same time. It is tempting to get around this semantic morass by claiming, rather like the unicorn, that 'you will know it when you see it'. The distinguishing categories used here (5 and 6 in the list above) are also ambiguous and there is a degree of blurring between them and the other categories.

### Case studies in rural change from Southeast Asia (Figure 5.2)

A striking example of fundamental rural change is Ban Lek in the Chiang Mai Valley, Northern Thailand (Ritchie 1993). This village has been studied over almost two decades from 1974, at which time it was a community of farmers (Table 5.5). The construction of roads brought two-way access with regional centres. As a result, villagers gained access to markets and urban employment, while outsiders like traders, labour recruitment agents, and land speculators gained access to the village. The granting of titles to land created a resource that could be bought and sold, and land began to become concentrated among an emerging group of more wealthy families. Young men and women from poorer households, unable to buy land now that it was scarce and expensive, moved into non-farm employment. In various material and non-material ways, and through direct and indirect policies, 'the Thai state facilitated the emergence of a hybrid rural class structure tightly articulated with urban areas' (Ritchie 1993: 1–2). Over twenty years, Ban Lek has become a mixed community of both agriculturalists and wage labourers. Ritchie observes that although the village may still be situated in a rural setting, 'the people and households which are involved in agriculture are now in a minority' (1993: 9). In 1974, only 7.5 per cent of all households in the village mixed farming and non-farm occupations; the figure in 1991 was over 30 per cent (Table 5.5). Wage labour is now the dominant occupation. Furthermore, while in 1974, 94 per cent of wage labour was agricultural labouring, by 1991 it was only 14.5 per cent. Daily commuting to the city of Chiang Mai had become common, the largest single form of wage labouring

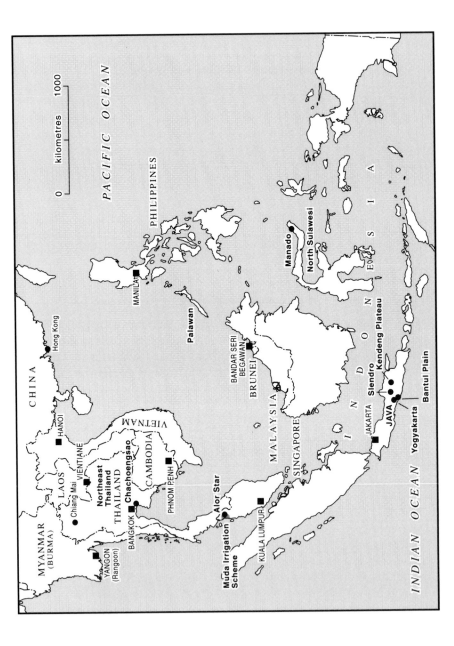

*Figure 5.2* Location of case studies

*Table 5.5* Occupational change in Ban Lek, Northern Thailand (1974–91)

| Occupation | 1974 (%) | 1985 (%) | 1991 (%) |
|---|---|---|---|
| Farming | 52.0 | 47.0 | 4.8 |
| Farming and wage labour | 2.2 | 17.8 | 18.3 |
| Farming and other | 5.3 | 2.6 | 12.5 |
| Wage labour | 32.0 | 26.7 | 51.0 |
| Self-employed | 5.7 | 4.4 | 9.6 |
| Government employment | 2.6 | 1.5 | 3.8 |

*Source*: Ritchie 1993: 10

being urban construction work (some 33 per cent of the total). What is striking, and challenging from both a policy and conceptual angle is that '[w]hile these transitional villages are physically placed within agrarian systems of production, the majority of individuals within the village are dependent on urban wage labour' (Ritchie 1993: 14). Ban Lek is not an example of urbanization, since the village is still rural. It represents the merging of the rural and urban worlds in a composite community (see the discussion of EMRs in Chapter 7).

In De Koninck's longitudinal study of two villages in the Muda Irrigation Scheme on Malaysia's Kedah Plain between 1975 and 1987, he identified the main cause of growth in non-farming employment as being the increase in women's participation. In turn, he interpreted this development as a 'consequence of the marginalization of their [women's] skills in the padi production process' stemming from the mechanization of rice production (1992: 177–8). There were interesting differences between the strategies employed by residents of the two villages in question. In the vicinity of Matang Pinang, located within the Muda irrigation scheme, there were comparatively few local opportunities for off-farm employment. As a result, a comparatively much larger proportion of those involved in non-farm work migrated and took up employment beyond the locality. In Paya Keladi, a far greater role was played by employment in the locale. In analysing the results of the study, De Koninck notes that mobility is becoming so pronounced that 'it is difficult to identify a "family development" cycle regulating access to land' (1992: 184) and postulates that 'an increasing number of the descendants of the peasants of Paya Keladi and Matang Pinang will need to break out of the community circle, if only to break out of agriculture' (1992: 186). Hart's longitudinal study of Sungai Gajah, also in the Muda irrigation area, revealed that in the decade between 1977 and 1987 the proportion of married men engaged primarily in non-agricultural activities rose from 5 per cent to 30 per cent (1994: 58–9).

In 1986 the official population of Slendro on Central Java's Kendeng Plateau was 2,782, in 620 households (Firman 1994). Of those men and women classified as economically active, 772 out of 811 (95 per cent) were defined as land-owning agriculturalists. Yet the frequency of temporary movement out of the village to search for work as construction labourers (*buruh bangunan*), pedicab drivers (*tukang becak*), domestic servants (*pembantu rumah tangga*), on plantations, or to trade, and the scale of remittances connected with that work, implies that non-farming endeavours were playing an important, and increasing role in the household economy. The author argues that for '*most* Slendro households, having family members working outside the village is a *matter of survival* [emphasis added]' (Firman 1994: 98). It is this area of production and employment which determines whether families have *cukup* (enough) to meet their needs, or do not. What is also of note in this case is that even local registers considerably understate the level of movement and rural–urban interaction. In theory, whenever a resident wishes to travel he or she must obtain an official letter of permission or *surat keterangan bepergian* from the village secretary or *carik*. In practice, this requirement was usually ignored.

Rotgé, working in two villages on the Bantul Plain near Yogyakarta on Java, also notes the growing role of non-farming employment in the household economy (Rotgé 1992). It seems that while comparatively few farmers were willing to give up agriculture altogether (just one in five households), there were many who strove to combine agriculture with non-farming employment. White and Wiradi in their study of nine villages in West, Central and East Java show a similar emerging pattern and record that for eight of the villages, the contribution of non-farm income to total income was more than 50 per cent – and that in a year when agriculture performed relatively well (1989: 294–5).

The poorest region of Thailand is the Northeast or Isan. A National Statistical Office study undertaken in 1986–88 found that 45 per cent of migrants in Bangkok were from the Northeast, and 89 per cent from rural areas (Parnwell 1993). Another study estimated that the dry season population of the capital was 9 per cent more than the wet season population – implying an in-flow of one million seasonal migrants (Apichat Chamratrithirong *et al.* 1995).[12] Klong Ban Pho, in the province of Chachoengsao in Thailand's Central Plains region, provides another example of thorough-going economic and social change in a formerly rice-based economy, dating from the late 1980s (Chantana Banpasirichote 1993). In this instance the intensity and speed of change appears particularly acute. The author argues that knowledge of farming is not being passed on to the younger generation, and that children are encouraged by their parents to look for opportunities outside agriculture. Whereas just a few years ago factory work was regarded as supplementary to farming, by the early 1990s such employment had become perceived as preferable to farming. Local people, apparently, 'foresee a time when land is left idle' (1993: 11) – as it already is over large areas of Peninsular Malaysia (see Box 7.2 page 253).

Not all Southeast Asian villages, of course, are experiencing internal structural change at a rate, or to the degree, of those described above. Many settlements can still be characterized, essentially, as farming communities. A study conducted among 230 families in the Minahasa area of North Sulawesi in 1985, for example, showed that only 6 per cent of income came from off-farm employment, and more than 93 per cent from labour in agriculture (Sondakh 1994: 171–2). But even in this instance the study did indicate that those households with smaller landholdings earned significantly more from off-farm employment indicating that such activities were becoming an important way for land-poor families to meet their needs (Sondakh 1994: 173).

Eder's longitudinal study of the village of San Jose on the Philippine island of Palawan in 1971 and 1988 shows that the proportion of households engaged in agriculture declined from 80 per cent to 55 per cent, during a period in which the population of San Jose increased from 112 to 278 households. He writes:

> Both self-employment and wage work are common. Frequent occupations of the self-employed include fishing, store-owning, market vending, charcoal making and tricycle driving. . . . Common forms of wage work included unskilled day labor, skilled labor (e.g., carpentry and masonry), salaried employment at private businesses in Puerto Princesa City, government employment (e.g., teaching, nursing, clerking), and contractual overseas employment.
>
> (Eder 1993: 657, see also McAndrew 1989 for another
> Philippine case study)

Two key issues comes through in Eder's case study which are of particular relevance to this discussion. First, there is considerable complementarity, in terms of labour availability, between agricultural and non-agricultural activities. In other words, households select activities so that they either reinforce or support established agriculture, or at least do not conflict with demands in established agriculture.[13] And second, agriculture in San Jose is not experiencing a process of absolute decline; it remains dynamic and productive.

The series of case studies mentioned above are drawn from the market economies of the region. Detailed local studies from Indochina and Myanmar are, by comparison, far less numerous and the 'characteristic' village would also continue to fulfill the designation 'agricultural'. None the less, rural economic differentiation, especially in Vietnam, is occurring and in some places very rapidly. Dang Phong writes, rather inelegantly, of a process of 'de-rice-ification' in the Red River Delta as farmers have moved out of rice and into non-rice crops and non-farm activities (1995: 169–70). He also reports that mechanization has allowed ploughing and threshing to be contracted out to entrepreneurs. In other areas of the region this is often an attractive proposition, even for households with comparatively small holdings, because a shift into non-farm endeavours has left them in labour deficit.[14]

# IMPLICATIONS OF RURAL CHANGE FOR GROUPS IN RURAL SOCIETY

## Age and gender implications

The shift within rural communities towards livelihood strategies that avoid agriculture, whether by necessity (a shortage of land) or design (a wish not to be a farmer), tends to be generational. Thus, while among older cohorts, agriculture remains the dominant occupation, in younger cohorts it is, often, wage labouring (see Ritchie 1993; Morrison 1993) (Figure 5.3). There are two possible explanations for this. The first is that it is a life-cycle shift, and when these young men and women enter middle age they not only marry and raise a family, but also settle back into farming. There is certainly some evidence to support the contention that off-farm wage labouring is a *rite de passage* in some rural areas of Southeast Asia. The process of *merantau* among the Minangkabau of West Sumatra, of *pai thiaw* in Thailand, and of *bejalai* among the Iban of East Malaysia, are partially founded on the belief that migration is bound up in the process of attaining maturity.[15] However, a more convincing argument in most rural areas is that the growing importance of non-farming pursuits among the young reflects a permanent change in the complexion of the rural economy brought on by a radical shift in resource availability, and specifically in land (see Ritchie 1993: 13). But this should not necessarily be interpreted as either displacement of young people into poorly-paid non-farm activities nor the marginalization of the old in stagnant agriculture. It seems, and this is discussed below, that diversity has attractions in itself. Older people retain access to land, providing a source of stability, while younger household members search out and gain employment in non-farm work (see Hart 1994: 67).

A second characteristic of this change in the rural household economy is the growing importance of female employment. In Thailand, for example, a study conducted by the National Statistical Office (NSO) in 1986–88 revealed that 65 per cent of migrants in Bangkok were female – a significant change from the 1970s when most migrants were male (Parnwell 1993). A slightly more recent NSO labour force survey broadly confirmed these findings, revealing that between 1985 and 1990, 53 per cent of migrants to the capital were female (Clover 1995). The rising female labour force participation rate is a general trend, but is most pronounced in export-oriented activities such as the garment industry, electronics, and the footwear sector where 90 per cent or more of workers may be female (see page 216).

There are a number of factors that lie behind this increase in female off-farm employment. Some may be viewed as 'enabling'; others as 'determining'. From the village perspective, the social unacceptability of women engaging in off-farm work, which formerly deterred many from leaving the villages, has diminished (see Parnwell 1993: 8–9). Coupled with such factors as declining fertility, delayed marriage, greater mobility and rising female education levels, this has

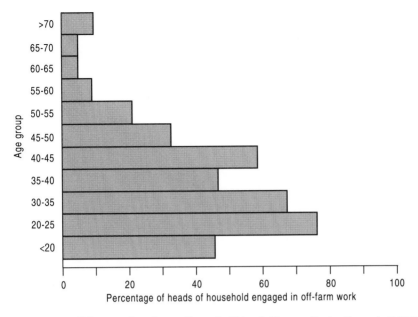

*Figure 5.3* Off-farm work and age of household head, Kamena Basin, Sarawak (1990)
*Source*: adapted from Morrison 1993: 63

created the conditions where women, for the first time, are in a position to work beyond the confines of the village.[16] But although this may be a general trend, there remain some areas where there are still strong cultural impediments to female participation. In West Java, for example, it seems that cultural constraints remain comparatively strong (Manning 1987: 63), and the same is true in the Muda area of Peninsular Malaysia, especially with regard to the newer labour markets (Wong 1987: 166). In Central Java, by contrast, the evidence indicates that such social sanctions are far weaker (although see Preston 1989: 53).

As well as this easing of previously powerful social barriers to female migration and off-farm employment (an enabling factor), developments in the agricultural economy may also be forcing – or perhaps permitting – women to take up off-farm employment. Some studies have argued that mechanization has selectively displaced women, thereby freeing them for, or displacing them into, other work (see page 242). At the same time, a series of changes in the structure and nature of the industrial economy have encouraged and/or permitted an increase in female labour force participation (see page 216 for a fuller discussion). The lesson which underlies such studies is that rural people no longer see their future solely, or even mainly, in farming. Mike Parnwell writes of the Northeast of Thailand, for example: 'Thus we are faced with a situation whereby livelihoods for . . . a large and increasing number of people in the North-East region are becoming dependent upon the continued economic dynamism, and associated

*Plate 5.8* Young men wait for a fare in the Northeastern Thai town of Khon Kaen. Most *saam lor* drivers are from local villages, and all are male. They usually hire their machines, sleep in them, and return home periodically to work on the family farm. Most are also young, although some *saam lor* drivers work through to old age.

income-earning opportunities, which exist elsewhere [i.e. in urban areas, and in particular in Bangkok]' (Parnwell 1993: 9).

## Income and class permutations

In addition to age and gender, it has long been assumed that variations in wealth are also a key factor determining the pattern of non-farm employment. Although the picture is still sketchy, it seems that two groups within rural society are more likely to be involved in non-farm employment: the comparatively wealthy, and the poor. This distinction is often expressed in terms of land, defining the wealthy as the land rich, and the poor as the landless or the land-poor. The problem with such a view is that it equates wealth in terms of agriculture, by using land ownership as a surrogate for income. It thus falls into the trap, noted above, of assuming that we can understand the rural world in agricultural terms.[17] Admittedly, in some case studies it seems that such an assumption about the close relationship between wealth, livelihood strategies and farm size does seem to hold true (e.g. Preston 1989). However, elsewhere, and increasingly, it does not. For example, Rotgé notes in his village study on the Bantul Plain in Yogyakarta, Java that of the one in five households who gain no income from agriculture, most are relatively wealthy (Rotgé 1992). Similarly, Eder writes of the village of San Jose in Palawan, the Philippines that 'relatively few of the land sales by core households between 1971 and 1988 reflected economic distress, and at least some reflected "strategies to prosper"' (Eder 1993: 660).[18]

A key question is whether prior wealth enables villagers to engage in non-farm employment, or whether non-farm employment makes them wealthy. The difficulty here is that longitudinal studies are comparatively rare, and so most accounts are forced to 'create' the past. None the less, the work on migration and occupational diversification does appear to indicate that there are two, perhaps three, strategies underlying the process: migration as a strategy of further accumulation, for the already comparatively rich; migration as a strategy of survival, for the poor; and migration as a strategy of consolidation for middle income households (see Rigg 1989a; Hart 1994; Effendi and Manning 1994: 216). This division is also supported by White and Wiradi's study of nine villages in Java, mentioned above. They note that while there is an inverse relationship between agricultural and non-farm incomes among the landless and very small landowning groups, for middle and large farmers the relationship is positive. In other words, they hazard, while for the members of the former group it is the 'inadequacy of agricultural sector incomes which propels [them] into nonfarm activities as a survival strategy', for the latter group there is a 'dynamic strategy of accumulation, in which surpluses derived from one activity are used to gain access to the other' (1989: 296). They conclude that such non-farm activities not only allow middle and large landowners to accumulate wealth but at the same time allow smaller landowners, with 'sub-livelihood' holdings 'to achieve subsistence incomes without the distress sale of their . . . plots' (1989:

*Plate 5.9* Sack-sewers in Manado, the capital of North Sulawesi, Indonesia. Piece-work like this is some of the lowest-paid employment and is a characteristic non-farm occupation of the poor

299). Indeed, it is striking the extent to which small-scale landholdings are persisting in the face of modernization and commoditization. The anticipation that small-scale farmers would be elbowed out by land concentration driven by the forces of technological and economic change has not materialized (Hart 1994 and see Morrison 1993). Cederroth, who anticipated that Green Revolution-induced change would marginalize large numbers of poorer farmers, explains the fact that it did not in terms of the 'unprecedented upswing in off-field employment' which not only compensated for the predicted decrease but allowed some of the poor to increase their standards of living (1995: 42–3).[19] Maurer's expectations were much the same when he studied four villages in Bantul Regency, Yogyakarta, between 1971 and 1987. Yet he found there was 'no real evidence of any strong land concentration or absolute improverishment process' in the face of Green Revolution-driven agricultural modernization (Maurer 1991: 97). As Maurer admits, he made the mistake of focusing on the agricultural economy, when it was the non-farm economy which was under-going the greatest change.[20] In large part, the emergence of occupational diversity has put off the displacement of the rural poor from their land and has, perversely, helped to sustain 'traditional' rural economies by allowing farm production to maintain an image of centrality. This view of there being a division of non-farm employment between petty trading, such as door-to-door hawking, for the poor and more capital intensive activities like operating taxi services or small-scale manufacturing enterprises for the rich is also supported by Effendi's work on Jatinom in Central Java (Effendi 1993) (see Table 5.6).

The difficulty, however, with assigning particular responses to particular groups of people – usually labelled as representing different 'classes' and 'class interests' – is that these groups, though perhaps clear and distinguishable enough in conceptual terms, are often remarkably hard to identify on the ground. The fluidity and complexity of rural change, the overlapping and conflicting nature of class 'interests', and the lack of a class consciousness among the very people who are supposed to be embroiled in class 'conflict', makes it questionable whether such notions have any utility whatsoever. James Scott, for example, in his classic study of agrarian change in the Muda irrigation area of Kedah, West Malaysia, writes:

> It is all very well to identify a collection of individuals who all occupy a comparable position in relation to the means of production – a class-in-itself. But what if such objective, structural determinations find little echo in the consciousness and meaningful activity of those who are thus identified?
>
> (1985: 43; see also De Koninck 1992: 191–2)

A further difficulty is that most local studies take the 'household' as the unit of study, while it is individuals who make up classes – not households. To place the members of entire households in one class or other makes little sense when those members may, individually, be best allotted to different classes (see: De Koninck

*Table 5.6* Non-farm activities and income

| Household | Poor | Middle | Rich |
|---|---|---|---|
| | Hawking | Trading | Taxi and truck services |
| | Petty trading | Local shops | Vehicle repair shops |
| | Home-based shops | Hairdressing | Food manufacturing |
| | Wage labouring | Tailoring | Money lending |
| | Construction work | Foodstall owning | Teaching |
| | Trishaw driving | | Government employment |
| **Strategy** | *Survival* | *Consolidation* | *Accumulation* |

*Note*: the categories and the activities listed here are only indicative.

1992: 192 and the discussion earlier in this chapter, 'Who is a farmer? What is a household?' ). Laslett argues that a family represents a 'knot of individual interests', and observes that 'to make a clear and satisfactory analysis of the relationship between household production and membership in a wider collectivity it would be necessary to examine in detail the interest, roles, and above all the power of every class of member of households of every type and in every region or economy' (Laslett 1984: 370).[21] As he admits, such an exercise is plainly impossible. None the less, it emphasizes the difficulties and dangers involved in teasing out generalizations from a system which is so complex and nuanced.

So although we can, with relative certainty, argue that people engaged in non-farm pursuits do so for very different reasons, to allot discrete groups into neat categories is probably impossible. We lack the longitudinal information required to undertake such a task at a cross-country scale; groups blur one into the next; rural residents themselves may have little conception of belonging to such groups or classes; and the usual unit of analysis – the household – does, itself, represent an amalgam of competing interests. Finally, and this is the bane of so much 'empirical' work based on case study material, it is uncertain how far such studies can be used to build global propositions about the nature of rural change (Eder 1993: 648). Are the poor displaced into low return non-farm work? Is technological change marginalizing the poor? Are richer classes able to capitalize on their position and create yet deeper intra-village inequalities? In all these, and many other, key areas of debate, the evidence from village studies can be used to support any one of a range of positions, ideological or otherwise.

### Strategies of non-farm employment

The increasing diversity of the household economy described above is not always reflected in the local and regional economies.[22] This may, on first sight, appear contradictory. However, while the household economy is, in essence, socially constructed, being based on a family, or extended-family unit, the local/regional economy is usually spatially designated. As a result, while the former may include income from activities that occur far removed spatially (even remittances from abroad), the latter implies those activities and industries which are located within the spatial boundaries of a district, province/state or region.

Rotgé notes this tendency in his study of villages on the Bantul Plain in Yogyakarta, Java. For while a substantial share of the population work in non-farming activities, generating a considerable share of household income, 'the development of locally-based off-farm activities which are not retail trading, remains conspicuously low' (1992: 19). Instead villagers commute to urban areas for work. It is the movement to work outside the local area which is the basis, in many instances, for the diversification of the household economy.

### Rural industrialization (and de-industrialization)

It is perhaps because of this disjuncture between the general unavailability of local non-farming employment on the one hand (although see below), and the need to satisfy a growing requirement for income on the other, that governments in the region are attempting to promote what has been termed 'rural industri-alization':[23]

> Rural industrialisation can be defined as a process involving the growth, development and modernisation of various forms of industrial production *within the rural sector generally and rural villages specifically.*
> (Parnwell 1990: 2 [emphasis added])[24]

There appear to be three interrelated and overlapping reasons for this policy drive. First, governments and some scholars perceive rural industrialization as a means by which spatial inequalities can be narrowed, whether they be rural–urban or regional, or – most likely – both. Greater spatial equality is assumed to be a desirable aim for a variety of economic, political and social reasons, and rural industrialization can help in achieving this end. Second, rural industrialization may help to relieve congestion and the pressure on services in rapidly-growing urban areas, reducing the environmental costs of growth and easing demand for housing and jobs. In this way, keeping farmers on the farm helps in urban management.[25] The third attraction of rural industrialization is that it provides employment for rural people, generating income and raising living standards. In arguing for rural industrialization in Thailand, Parnwell states that 'as part of the strategy for alleviating the problems of rural areas and peripheral regions . . . [the case] is not a difficult one to make' (1990: 5), and he proceeds to list no less than thirteen positive attributes of rural industrialization (see Table 5.7).[26]

But rural industrialization would not be a realistic policy option if there were not also some sound economic reasons for companies to base themselves in the countryside. Among the attractions of a rural location are generally lower wage rates, cheaper land and utilities, and fewer and less stringent environmental controls (notwithstanding some local opposition by environmental groups).

*Table 5.7* Positive attributes of rural industrialization

| |
|---|
| 1 Employment and income generation |
| 2 Diversification of the rural economy |
| 3 Mobilization and fuller utilization of local resources |
| 4 Utilizing comparative advantage |
| 5 Building upon an existing framework of traditional rural industries |
| 6 Flexibility |
| 7 Increased female labour participation rate |
| 8 An alternative to land reform |
| 9 Reduction of out-migration |
| 10 Slowing the pace of urban concentration and associated problems |
| 11 More balanced pattern of industrial activity |
| 12 Better balance in the urban–rural division of labour |
| 13 Strengthening of the linkages between towns and rural hinterlands |

*Source*: Parnwell 1990: 5–8

Some companies also find that planning controls in urban areas inhibit their expansion plans. There may even be local sources of some raw materials. To set against this, investors usually need to take into account, and contend with, higher transportation costs, the absence of a pool of skilled workers, difficulties in enticing managers to locate in comparatively remote and undesirable areas, and the inevitable dislocation from the key centre(s) of financial and bureaucratic decision-making (i.e. the capital city).[27]

Broadly-speaking there is a distinction between industries which are entirely new and alien to the rural setting, and those that build upon existing skills and activities. In the poor Northeastern region of Thailand for example, the Canadian shoe company Bata, in collaboration with T-Bird (Thai Business Initiative in Rural Development), an off-shoot of the Population and Community Development Association (PDA), have established four small shoe factories in village settings. These factories employ 200 villagers and produce 9,000 pairs of uppers a day. In 1994 the workers were paid 110–120 baht a day, as against the prevailing agricultural wage of about 50 baht for a day's labour in the fields. Other companies working in collaboration with T-Bird include Bangkok Glass Industry, Swedish Motors, 3M and American Express. But rather than relocating their activities to rural areas like Bata, the latter three companies are providing money and other support to adapt and develop indigenous industries for a modern market. American Express, for example, has lent its support to silk production and weaving – a traditional village 'industry' in the Northeast (Fairclough and Tasker 1994; Tasker 1994a; Fairclough 1994).

But, it should not be assumed that because industries are located in rural areas they necessarily employ, and benefit, rural people. This seems to be particularly the case in the vicinity of medium and large-sized towns. Small companies may create workshops or factories in rural areas so that they can benefit from the advantages noted above, but then employ non-local settlers – so as to avoid some of the associated disadvantages. Therefore the emerging and growing

*Plate 5.10* Boiling silk cocoons in a Northeastern Thai village. Silk weaving is a traditional rural industry in this poor part of Thailand which has received considerable outside assistance in an effort to transform it into a significant local income earner

interactions between rural and urban areas may not only bring rural people to the city, but also city dwellers to the countryside. If this occurs – as Rotgé, for example, records in his study of employment in the vicinity of the Javanese city of Yogyakarta (1992: 31–34; see also Wolf 1992: 105–8) – then rural industrialization may not assist local rural people. Indeed, in both Thailand and Indonesia (Rotgé 1992: 32), researchers have noted the extent to which local people are 'displaced' from accessible rural areas to more remote places. Enticed by the rising value of their land, especially if it is located near a transit corridor, but unable to benefit from the process of rural industrialization that is occuring around them, they sell their land. The proceeds are then used either to buy a larger parcel of land in a more remote location (Rotgé 1992: 32), or 'squandered' on luxuries (Ross and Anuchat Poungsomlee 1995).[28] If this process is widespread, then 'rural industrialization' should perhaps be reformulated as 'industrial extension and rural displacement'.

Many studies of rural industrialization stress the degree to which external agencies must support and intervene in the process (e.g. Parnwell 1993: 20–1). The apparent inability of rural people spontaeneously to establish their own businesses is because they lack the management and technical skills, financial resources, collateral, contacts, and marketing knowledge to make the leap from farmer to small businessman or woman. It is these gaps which NGOs and some government programmes aim to fill. But although there may be an attractive logic in such efforts, successful examples of self-sustaining and profitable rural enterprises are hard to find. As Udom Buasri of Khon Kaen University in the Northeast has been quoted as saying: 'Many NGOs have tried to teach people skills like traditional handicrafts so they'll stay at home. Mostly they fail. They're not able to keep people in the village' (Fairclough and Tasker 1994: 23). This, perhaps, is because rural industries are being seen, in the policy context, as *alternatives* to migration and urban work, whereas for rural households they are viewed as an *addition* to existing strategies, helping to fill the 'expectation gap' noted earlier.

Although much recent policy prescription has emphasized the promotion of rural industrialization, there has always been an element of 'industry' in rural areas (see the short discussion above on the historical basis of non-farm activities). And while governments may be pursuing one policy objective, the forces of economic restructuring and technological change may be encouraging just the reverse. This is especially true in those areas with a tradition of rural manufacturing. Hans Antlöv and Thommy Svensson's study of the traditional textile producing area in the vicinity of the town of Majalaya, 35 km south-east of the West Java city of Bandung, shows just such a shift as the locus of production moved, beginning in the early 1970s, from small workshops in the rural areas surrounding Majalaya to large textile factories in Majalaya town itself:

> Today's urban-based textile industry was born in the countryside. . . . Rural villages provide the reservoir of labour for the large-scale industrialization

which is now taking place. But the villages also benefit economically from the textile workers who bring home and spend their factory incomes in the local setting. It is striking that most textile workers continue to live in the villages and commute daily to the town. On all the roads in Majalaya, at 6.30 a.m. and 5.00 p.m., innumerable mini-buses bring workers to and from the factories.

(Antlöv and Svensson 1991: 124)

As Antlöv and Svensson describe, in some areas of Southeast Asia, for example in the Central Plains and Lower North of Thailand, the western seaboard of Peninsular Malaysia, and especially in Java, the close proximity of rural villages to centres of urban-industrial employment has allowed villagers to commute daily, maintaining their rural base but taking up formal and informal non-farm employment.[29] At the same time, large export-oriented manufacturing enterprises have also found it desirable to locate in rural areas. In Java, Wolf describes 'ten large-scale "modern" factories, driven by Western machinery and technology . . . [squatting] in the middle of the agricultural land of two villages that still have neither running water nor electricity . . . Some . . . [nesting] in rice fields, disrupting neat rows of rice shoots with metal fences and guards' (1992: 109). To some extent this evolution of highly integrated rural–urban economies links closely with the concept of the Extended Metropolitan Region (EMR) discussed in Chapter 7 (see page 264). However, commuting to work in local towns is more widespread than the distribution of Southeast Asia's EMRs might suggest, and is becoming more so as transport links improve and the tendrils of public transport bring more people within commuting distance of an urban centre. The agricultural and industrial elements of rural livelihoods are not in competition, but often fill complementary roles. Income from urban employment is fed into agricultural investments, and – as discussed above – allows families working small, marginal and sub-marginal farms to continue to maintain a presence in the countryside (see page 267 for a discussion of the 'virtuous cycle' of rural–urban relations). With cities spreading outwards, factories relocating in surrounding areas, and people moving faster and more frequently across a vital landscape of economic interaction, it is becoming increasingly hard to establish where urban ends and rural begins – certainly in functional terms, and often in physical terms too. Brookfied *et al.* talk of '*in-situ* urbanization' to describe this process in the Klang Valley area around Kuala Lumpur (1991).[30]

### Rural industrialization, exploitation and accumulation

In the literature, rural industrialization is usually presented as a favourable alternative to urban employment. It keeps families together, reduces rural–urban migration, results in higher incomes in rural areas, and means that rural inhabitants can continue to enjoy the (assumed) better quality of life in the countryside. It may also be construed as a means of improving poor people's livelihoods without having to resort to politically sensitive and disruptive agrarian reforms such as land redistribution, making it highly attractive to embattled 'weak' states. However, there is also an argument that rural industrialization permits the hyper-exploitation of rural people (this is also discussed with reference to the urban employment of rural people in Chapter 6). As Wolpe famously argued with reference to the employment of rural migrants in South African industry:

> since in determining the level of wages necessary for the subsistence of the migrant worker and his family, account is taken of the fact that the family is supported, to some extent, from the product of agriculture in the Reserves, it becomes possible to fix wages at the level of subsistence of the individual worker.

(Wolpe 1972: 434)

Although Wolpe was focusing on the movement of people to work, the same argument could be applied to rural industrialization where the work moves to the people. Pleyte, in 1911, was highly dubious about the supposed advantages of small-scale rural industry in Java:

> small-industry propogandists, you know only the outward appearances and judge native industries by them. Go into the kampungs [villages], visit the slums and hovels where the pieceworkers live and work, and see with your own eyes how the dispossessed in native society do their work and eke out their existence!

(Pleyte 1911: 38, quoted in and translated from the
Dutch by White 1991: 49)

The contention that rural industrialization involves just the relocation of urban sweat shops to rural areas does receive some support in the contemporary literature. For example, in Ban Thung, a village in Thailand's central region, some households subsist entirely on the home-based piece-work production of *ngop*, or farmers' hats. The income this work generates though is so meagre that it does not allow any room for accumulation. As one villager explained, 'you can't do it quick enough to eat' (Arghiros and Wathana Wongsekiarttirat 1996: 131; see also Tambunan 1995 on Java). In those instances where rural-based non-farm work does generate an adequate return it does not necessarily follow that it will narrow inequalities in the face of highly skewed patterns of land ownership. White notes that despite the fact that well over 50 per cent of rural incomes in Java in 1987 came from non-farm sources, those generating these additional returns tended also to be the larger landowners. In other words, and unlike such East Asian countries as Taiwan and South Korea, there was a positive relationship between levels of farm and non-farm income (White 1991: 59–61).

## The new rural world

Relatively few studies investigate, in detail, the non-farm (or non-agricultural) elements of the 'agrarian' economy. Yet, at the same time, almost all studies – and the literature on agrarian change is truly voluminous – emphasize that there is more to rural life than agriculture – and that this 'more' is getting larger and more significant. This strange coincidence of recognition and then neglect is linked to the conceptual difficulties of dealing with farmers who might be industrial workers, and industrial workers who might be involved in farming. As this chapter has attempted to show, understanding how and why these inter-actions and linkages occur cuts across spatial, gender, class, age and temporal divisions. For example, at different times during the year a household may have its members involved in different activities. The seasonality of agricultural production – highlighted in the reference to the *musim mati* ('dead season') in Peninsular Malaysia, or *musim lapar* ('hungry season') in eastern Indonesia – illustrates the pronounced temporal fluctuations that occur in village produc-tion. The activities, often built in and around these fluctuations, may involve daily movements of just a few kilometres, seasonal movements of hundreds of kilometres, or periodic movements of thousands of kilometres. Further, as each individual works through his or her life cycle their economic roles may well change. Nor will households act – whether by accident or design – in the same fashion. Different 'classes' within the rural population will become involved in the economy in different ways. Conceptual bases for understanding rural change have, without exception, failed to deal effectively with these myriad and nuanced processes and interactions.

Although this chapter can hardly fill such a conceptual void, there are some tentative observations that can be offered. Most important, the processes of social and economic change sweeping through rural areas of Southeast Asia can rarely be understood in terms of just agricultural modernization. The forces of change have their roots, more likely than not, in the non-farm economy. Driven by rising needs, and often constrained in agriculture by a contracting resource base and unfavourable terms of trade, rural inhabitants are turning to non-farm activities. Researchers in different countries tend to use different terms – Grandstaff (1988) and Thomas (1988) in Thailand talk of a 'diverse portfolio of activities', White in Java (1976), of 'occupational multiplicity', Muijzenberg (1991) in Luzon, the Philippines of 'diversification for survival' – but the underlying meaning of the terminology is the same. Households usually wish to maintain and develop both agricultural and non-farm activities: giving up agriculture altogether is often perceived to be a risky strategy, and one to be avoided wherever, and whenever, possible. Hence the emphasis in so many studies on the attractions of diversity to rural households. However, it is difficult to escape from the tyranny of the forces of economic structural transformation: that ultimately, most farmers will give up agriculture altogether (see Leinbach and Bowen 1992: 349). But for the present at least, most rural households have not reached this stage. Instead, the demands of these non-farm activities rebound on agriculture, which adapts to the new production environment. At the same time, the presence of an army of rural workers willing and needing to take up industrial employment fuels the growth of manufacturing. These linkages and interactions are discussed in detail in Chapter 7. It is perhaps sufficient to end just by stating the obvious: that the Southeast Asian rural world(s) is (are) an unmistakably new rural world(s). The direction of change appears relatively clear – towards greater diversity, integration and commercialization. The effects of change, though, remain hazy.

## Notes

1 Moerman writes: 'From December to May, when the ground is dry, it takes rice trucks eight hours to travel the 45 miles from Phayao to Chiengkham Town. A few nights of rain may increase the traveling time to twelve or fourteen hours, or prevent truck travel entirely. For five months the road is impassable even to four-wheel-drive vehicles equipped with winches. It then takes two full days on foot or horseback to travel the 35 miles between Chiengkham Town and Dork Kam Taj, where there is a stone-filled roadbed leading to Phayao' (1968: 6).

2 David Wyatt describes Damrong in the following terms: 'An indefatigable worker, incisive and direct, with an administrative genius unmatched in his generation, Damrong was given a free hand by the king' (Wyatt 1982: 209).

3 The historical evidence appears to indicate that rulers were involved in very similar practices in the pre-modern period in Southeast Asia, they just lacked the resources to enforce their wishes at a wider scale. After King Rama III of Siam overwhelmed and sacked Vientiane, and captured King Anou of Laos in 1828–29, he set about resettling the population of Laos in over 40 newly created *müang* (districts) in present day Northeastern Thailand (Wyatt 1982: 172). Grabowsky claims that '[f]orced resettlement campaigns . . . [were] an important aspect or even the main rationale of wars in traditional Thailand and Laos . . . ' (1993: 2). He quotes the old Northern Thai (Yuan) proverb *kep phak sai sa kep kha sai müang* – 'put vegetables into baskets, put people into towns'. In addition, the motivation for these acts of human control were similar: wealth was measured in people, and to increase the wealth of a kingdom it was necessary for a ruler to increase the population under his sway.

4 For a more general study see Anderson and Leiserson (1980) who quote a range of 30–40 per cent of total household income from non-farm sources.

5 Non-farming is used here to describe activities, both on-farm or off-farm, which are not agricultural (whether cropping or livestock raising). They include both traditional non-farming activities like pottery-making or weaving, and new activities like factory work.

6 This title is adapted from Alexander *et al.*'s book (1991) *In the shadow of agriculture: non-farm activities in the Javanese economy, past and present.*

7 Post-developmentalists see the discourse of 'needs' as central to that of 'development'. Ilich writes of the 'historicity of needs' and argues that development/modernization required an identification of needs (i.e. unfulfilled wants) in order to justify progress which, in turn, underpinned development (1992).

8 This latter point possibly hints at a difficult and nebulous area – that of political freedoms. The Indonesian government now accepts, in broad terms, that economic development also creates demands for political development, specifically greater freedom of expression and opposition. General Sayidiman Suryohadiproyo, formerly head of the National Defence Council, observed in 1994: 'It's obvious that at the

current pace of economic development, the people will become more critical and demanding. The State simply must meet their aspirations.'

9  Dick and Forbes use 'quiet' to mean low profile and often ignored, rather than in the aural sense! 'Colt' is the term used in many parts of Indonesia to describe the ubiquitous minivan, a combined people-mover and load-carrier that seems to link every settlement, no matter how small, with local towns, at a fare of just a few hundred rupiah. 'Colt' was the name of the very first of these minivans, and has now become generic in Java, although some regions have their own local terms (for example, Daihatsu in the hill town of Bogor, '*bemo*' in Surabaya and elsewhere).

10  Thomas Enters who worked in the Chiang Mai valley found that at least 90 per cent of the hill people he questioned claimed increased mobility to be the main attraction of having an ID card (personal communication).

11  Gina Porter writes with reference to off-road communities in Nigeria: 'It has been suggested in this paper that off-road communities may become increasingly disadvantaged, increasingly unequal, following road construction, particularly if bush track maintenance is abandoned and off-road transport services decline: market systems are reshaped and many off-road markets suffer, agricultural production is affected in terms of profit per crop tonne and possibly yield per hectare, personal welfare suffers as access to health, education and other facilities is reduced' (Porter G. 1995: 10).

12  The study also estimated that one fourth or 14.5 million Thais were migrants in the five years before the survey, and that the balance between male and female migration was roughly equal but that women outnumbered men in the younger age group (<25 years old).

13  Manning makes the same point with reference to urban *becak* (trishaw) drivers from villages in West Java (Manning 1987: 69).

14  The emergence of these entrepreneurs follows a trend that is evident in other countries of the region. Dang Phong reports that the cost (in 1993) of ploughing one hectare of paddy was 200,000 dong or US$19 at the prevailing exchange rate (1995: 177). In Thailand the rate ranged between US$25 and US$37.50 (author's research, 1994).

15  Traditionally this would have applied to males only. For females other achievements were markers of maturity. For example, in many societies, skill and dexterity in weaving was, and remains in some areas, important. In Northeast Thailand, a woman was only considered *suk*, or 'ripe' for marriage, when she had mastered the loom; until that time she was regarded as *dip*, or 'raw'. Many girls were expected to weave gifts for presentation at the wedding ceremony and for display in the home. One Lao text from the Northeast states: 'A good wife is like a ploughshare. If she is skilled at weaving, then her husband can wear fine clothes. A wife who talks harshly and is unskilled at her loom makes a family poor and shabby in dress' (quoted in Conway 1992). This quote, in addition, links weaving skills not just with maturity but also with wealth. Also see the discussion on page 207.

16  Such has been the rate of fertility decline in Thailand that the Ministry of Education is already planning the closure of primary schools in rural areas. In Phrao District, Northern Thailand, the 59 primary schools operating in 1994 are to be reduced to 56 in the first round of closures, to be further reduced to 25 schools over the next decade, and finally reaching just 11 schools in 20 years' time (Clover 1995).

17  It is also true that using size of land holding as a proxy for income makes the assumption that land is all of the same quality and worth. When one hectare of good irrigated riceland can be as productive as 20 hectares of marginal, rain-fed upland this makes little sense. There are also signficant differences in productivity between different classes of, say, paddy land (see Rigg 1988a).

18 By 'core household', Eder means those households present in both 1971 and 1988, the two periods of fieldwork.

19 The 1995 study can be usefully contrasted with Cederroth and Gerdin's 1986 paper. White, in his 1979 paper, expresses surprise that census data at that time were not indicating a concentration of holdings. He questions the accuracy and reliability of the information on which the census data are based, and wonders whether the apparent stability of farm size distribution is not disguising a significant – and for the poor, deleterious – change in the structure of ownership and tenure. He writes that the census data 'surprises those who have done field research in Java and have returned with the strong conviction (even if they were unable to demonstrate it by accurate measurement) that a fairly rapid concentration of holdings has been occurring, including rapid increases in absentee ownership by urban élites' (1979: 102).

20 He writes: 'Though I kept stubbornly enquiring about the social differentiation effect of agricultural modernization, I was surprised to find no real evidence of any strong land concentration or absolute impoverishment process. . . . Could it be explained differently by looking "beyond the *sawah*", on which I had probably focused too exclusively?' (Maurer 1991: 97).

21 In their study of off-farm employment on transmigration settlements in South Sumatra, Leinbach *et al.* write that 'decisions are made in accordance with a family survival strategy, which is based on family labor resources, an evaluation of time commitments to family and farm work, and a cognizance of physical capabilities and existing skills to participate in specific types of off-farm jobs' (1992: 44).

22 Although see the discussion of the Extended Metropolitan Region (EMR) in Chapter 7.

23 It is perhaps unfortunate that the world's greatest experiment in rural industrialization was such a developmental calamity: China's 'Great Leap Forward' (1959–61).

24 In a later paper, Parnwell notes the difficulty of using the term 'rural' in this context because of the tendency for the Thai word for rural/countryside – namely *chonnabot* – to be interpreted as 'up country', or simply 'not Bangkok'. Parnwell makes it clear that rural industrialization is not just another form of regional development, but *village-based* industrial development (1994: 30).

25 With reference to Thailand, Parnwell writes that '[o]ne consequence [of migration] . . . is the phenomenon of "over-urbanisation", most manifest in the proliferation of slum dwellings and petty commodity production which may be inadequately served by urban infrastructure and public utilities, thereby contributing further to a deterioration in the quality of the urban living environment . . . ' (1994: 29).

26 Chantana Banpasirichote, working in a formerly agriculture-based community in the Central Plains of Thailand, refers to an elderly person in the village who observed 'that family life is becoming more like urban life where young parents work [in rural-based factories] late into the evening, leaving their children alone, or with grandparents' (1993: 40).

27 The most significant attempt by the Thai government to disperse industrial activity away from Bangkok has been the Eastern Seaboard development project, focused on the provinces of Chonburi and Rayong. A complaint that was frequently heard from investors was that this comparatively accessible and developed part of the Kingdom, within two or three hours' drive of the capital, lacked the necessary social infrastructure to make it attractive to middle management.

28 'Squandered' is used here with a degree of reluctance. There is no doubt that some people do indeed 'waste' their money, and are viewed as doing so by their neighbours. But in other instances, what might be classified as 'waste' in Western eyes may, in fact, be productive investment in local terms. The building of a new

house, the purchase of a car or motorbike, or the donation of a sum for the construction of a new *wat* (Buddhist monstery) building or *surau* (Muslim prayer hall) may raise the status of a family in local terms allowing, for example, a couple to marry their son or daughter more advantageously. Religious donations may also, of course, offer the prospect of a better life in the next world.

29 This applies particularly to Java which, though a single island, has a population of over 110 million, representing around one quarter of Southeast Asia's total population of 461 million (1992).

30 Although the EMR concept is usually presented as uniquely Asian, there are many aspects of rural–urban development that mirror changes in the West. The shift from circulation to commuting, the location of industries in city environs, and the evolving dominance of non-farm employment in rural communities.

# 6

# THE FACTORY WORLD

## Introduction

The World Bank's publication *The East Asian Miracle* (1993a), makes a brave attempt to distill the 'Essence of the miracle'. The rudiments include such things as a high savings rate, the creation of human capital, and a market-friendly approach to development (see Chapter 1). However, the key tangible product of these – often nebulous – ingredients has been the emergence of a vital manufacturing sector geared to export. In the eyes of many commentators, the visible signs of success are to be found in the factories of the region, and are reflected in statistics detailing the growth of foreign direct investment (FDI) and the parallel expansion of manufacturing exports (Table 6.1 and Figure 6.1). In its *World development report 1995: workers in an integrating world,* the World Bank argued that the 'problems of low incomes, poor working conditions, and insecurity . . . *can* be effectively tackled in ways that reduce poverty and regional equality [emphasis added]', suggesting that to do so governments must 'pursue market-based growth paths', and 'take advantage of new opportunities at the international level, by opening up to trade and attracting capital' (World Bank 1995: 2). These, as the authors of the report put it, 'are revolutionary times in the global economy', adding as a catch-all caveat that 'rapid change is never easy' (1995: 1). To critics of the World Bank's preoccupation with this market-based, globally-structured and internationalized recipe for successful industrialization, the risks and faults are all too evident at the human level in the factories of the developing world. Although, they argue, foreign investment flows may be large and profits huge, simply graphing such trends says little about whether the impacts of industrialization are beneficial to individuals, families, communities and the environment (see Hill 1993: 8–10) (Figure 6.2).

This chapter will not present the macroeconomic picture of this industrial transformation, which has been well-covered elsewhere (see Hill 1993 for a overview). Rather, and following the previous chapter, it will adopt a grassroots or, more accurately, a factory-floor perspective. The emphasis here will be on factory life and the factory world. How do people come to find themselves making trainers for Reebok and Nike or disk drives and semi-conductors for

Table 6.1 Structure of merchandise exports (%)

| | Fuels, minerals, metals | | Other primary products | | Machinery & transport equipment | | Other manufactures | | Textile fibres, garments and textiles | |
|---|---|---|---|---|---|---|---|---|---|---|
| | 1970 | 1993 | 1970 | 1993 | 1970 | 1993 | 1970 | 1993 | 1970 | 1993 |
| Brunei | n.a. | n.a. | n.a. | n.a. | n.a. | n.a. | n.a. | n.a. | n.a. | n.a. |
| Cambodia | n.a. | n.a. | n.a. | 15 | n.a. | 5 | n.a. | 48 | n.a. | 17 |
| Indonesia | 44 | 32 | 54 | n.a. | 0 | n.a. | 1 | n.a. | 0 | n.a. |
| Laos | 36 | n.a. | 33 | 21 | 30 | 41 | 1 | 24 | 3 | 6 |
| Malaysia | 30 | 14 | 63 | 82 | 2 | 2 | 6 | 9 | 1 | n.a. |
| Myanmar/Burma | 7 | 7 | 92 | 17 | 0 | 19 | 2 | 58 | 1 | 9 |
| Philippines | 23 | 7 | 70 | 6 | 0 | 55 | 8 | 25 | 2 | 4 |
| Singapore | 25 | 14 | 45 | 26 | 11 | 28 | 20 | 45 | 6 | 15 |
| Thailand | 15 | 2 | 77 | n.a. | 0 | n.a. | 8 | n.a. | 8 | n.a. |
| Vietnam | n.a. | n.a. | n.a. | n.a. | n.a. | n.a. | n.a. | n.a. | n.a. | n.a. |

Source: World Bank 1995
Note: Textile fibres are part of 'other primary commodities'; textiles and clothing are part of 'other manufactures'

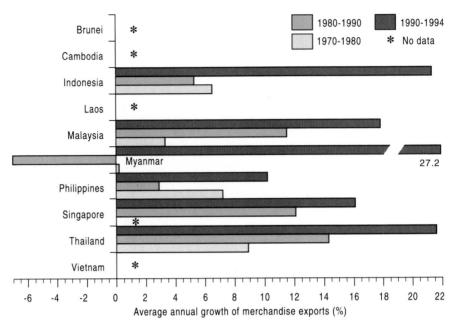

*Figure 6.1* Annual growth of merchandise exports (1970–94)
*Sources*: World Bank 1995 and 1996

Texas Instruments and Seagate Technology? Who throngs the factories of the region? How do they get there? Are conditions poor and dangerous and how do workers resist and protest? Are women marginalized and exploited? Such questions tend to result in scatter shot answers, ending with the disclaimer, 'it depends'. Reflecting this, the discussion does not arrive at any hard-and-fast conclusions about the nature of the industrialization process in the region and its impacts on the people at the bottom of the development 'heap'. Although the various interpretations presented here do partially reflect important differences in ideology, far more important is the simple truth that there is a range of experiences and a coincidence of gains and losses.[1] As almost all studies seem to agree, whether they be radical (e.g. Kemp 1993; Mather 1983) or conservative (e.g. World Bank 1995), industrialization and the growing availability of factory work has been a 'mixed blessing'. For Kemp, the acquisition of new skills and additional income has to be set against the price in terms of 'health, stress and an uncertain future' (1994), while for the World Bank the 'great gains' to labour are counterbalanced for some by their 'increased vulnerability to volatile international conditions' (1995: 4).

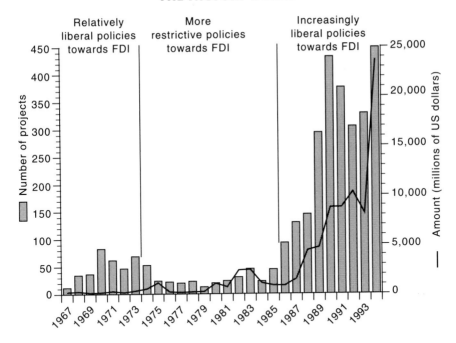

*Figure 6.2* Approved foreign investment in Indonesia (1967–94)

*Notes:*
1  These figures exclude investments in the petroleum, banking and finance, and leasing sectors.
2  Indonesia's Foreign Investment Law was enacted in 1967.
*Source:* drawn from data in Thee Kian Wie 1995.

## COUNTING THE BENEFITS OR REAPING THE COSTS? FACTORY WORK AND HUMAN LIVES

### Pushed out of agriculture or pulled into factories?

She . . . earned about 46 to 77 [U.S.] cents daily, considerably less than the already low minimum wage of 625 rupiahs, or 96 cents. Rini's family was poor by any standards.
(Rini, a worker in an export-oriented garment factory in Java.
Quoted in Wolf 1992: 2)

A clear divergence in the tone of the literature on industrial employment and factory work lies in the question of whether workers are displaced from agriculture and therefore 'pushed' into industry, or whether rural people are attracted to, and therefore 'pulled' into factory work (see White 1979: 97, Tambunan 1995).[2] Is, in other words, the increasing role of factory and other non-farm work in rural livelihoods a sign of distress or progress? In the first scenario, which is an essentially pessimistic perspective on the development process, problems in the

205

agrarian sector – lack of land and work, low and stagnant incomes, and widening inequalities – force villagers to look for non-farm work. This process creates the army of displaced men and women willing to work in unskilled, poorly-paid and sometimes dangerous and degrading jobs. The second scenario, or 'trajectory' as Koppel and Hawkins refer to it, is one where a vibrant agricultural sector, coupled with rising expectations and better education, free rural people for work in often better paid non-farm occupations (see Koppel and Hawkins 1994). There may also be complementarities between farm and non-farm work. Where scholars stand on this issue appears to relate to three issues. First, the status of agriculture and human–land resource conditions in the countryside. A shortage of land and rising landlessness, coupled with the effects on labour of agricultural mechanization, would seem to indicate displacement. Second, it relates to the balance between real wage rates in agriculture as opposed to urban employment. And third, to the ideological prism through which the forces of capitalist transition are viewed.

The first two of the above issues may, however, seriously overlook the integration of the urban and rural labour markets. As is discussed in some detail in Chapters 5, 7 and 8, workers in agriculture and industry are often the same individuals who exploit opportunities in both sectors. In many cases they, or more usually their households, are not giving up/being forced out of agriculture and into industry, but are maintaining a presence in each. Thus questions of displacement and relative wage rates become somewhat redundant for this group of people. During the off-season when no agricultural work is possible or available, farmers are often willing to tolerate very poor conditions and low wages in order to secure work. There is no agricultural alternative with which they can balance and gauge such a decision. Thus, Duong, a Vietnamese farmer, earns the equivalent of US$10 for 38 hours of work in the rice field, but he can only work for six months of the year; for the remaining six months he earns almost nothing from agriculture. As a result, his 13-year-old son has left home to work as a street trader in town (World Bank 1995: 1). Masri Singarimbun acknowledges this shift from farm to non-farm solutions to poverty, when he admits – like Maurer in the last chapter (see page 200) – that he did not anticipate the increasingly important role of off-farm employment in raising the incomes of households in the village of Sriharjo, Yogyakarta. He explains that when he began undertaking field work in 1969 he 'wrote about rural poverty mainly in the context of too many people for too little land'. By the late 1980s, access to land was no longer a major factor in identifying and characterizing the poor. 'Now', he writes, 'the problem for the majority of villagers is that too many people are competing for jobs' (Singarimbun 1993: 272).

None the less, there are more than a few studies which indicate that factory wages are lower than those in agriculture. White (1993), for example, in his study of garment and footwear factories in West Java found monthly wage rates of Rp45–55,000 in footwear factories (1990), and Rp70,000 in garment enterprises (1989). He notes that such rates are equal to or lower than those available in agriculture, especially when higher urban living costs are taken into account.[3] In an earlier paper, White concluded that the evidence on employment in farm and non-farm occupations indicated that the 'shifting employment patterns . . . represent not so much a healthy competition between agriculture and non-agricultural production, but rather the effects of a process in which low-income households are being progressively excluded from involvement in agricultural production, and at the same time lack the non-labour resources [i.e. skills, etc.] which might provide them with higher labour incomes in other occupations' (1979: 100). Joan Hardjono's work in the village of Sukahaji near Majalaya in West Java shows that wages in 1993 for various agricultural tasks ranged between Rp1,000–2,500 for a five-hour day. By comparison, 10 hours of work in a weaving factory paid Rp2,000–3,000, depending on the level of skills (Hardjono 1993; see also Cederroth 1995: 113–39). Studies like these which argue that wages in non-farm work are equal to or lower than returns in agriculture are counterbalanced by those which show the reverse to be the case (see Table 6.2). Sjahrir, for example, quotes the daily wage for a construction *tukang* (worker) in 1992 in Jakarta as being Rp7,500–10,000, depending on the worker's level of skills. Even an apprentice received Rp5,000/day while *becak* (pedicab) drivers could expect to earn Rp5,000–7,000/day. By comparison, the most a wage labourer in rice cultivation could earn was Rp1,500–2,000/day (Sjahrir 1993: 250) (see Table 6.2). It would be sensible to assume that rural people are being *both* pushed out of agriculture and pulled in to industry. Tables 6.2 and 6.3 indicate that while the lowest wages paid for unskilled factory and other work are similar to those in agriculture, for skilled and semi-skilled work the wages appear to be considerably higher. Admittedly what the figures in the table do not reflect are the additional costs of living and working in urban areas.[4] None the less, it would seem fair from the evidence to state that those with skills to sell can often better themselves, at least financially, by taking up factory employment.[5]

The emphasis on relative wage rates in agriculture and industry has tended to obscure the fact that even when factory wages are lower, and agricultural work is available, many younger villagers opt to work in industry. White, for example, in the study noted in the previous paragraph, goes on to write that '[e]arnings levels are certainly not themselves the main source of the attraction of factory employment for rural youth, with a larger role being played by considerations of relative security of earnings and the relative "cleanness" and status of the job' (White 1993: 134). Given that wages sometimes did not even cover basic living expenses for a single, young worker, it is tempting to see factory work as a transitionary stage in the life cycle; one where young men and women are given the opportunity to work in urban areas, enjoying the camaraderie of such work and escaping the over-weening clutches of the household, only to return in time to the village to take up the responsibilities of farm work and family life.[6] Wolf, for example, found that 30 per cent of her sample of commuters from villages in Java contributed either nothing to the household budget or actually took money from

Table 6.2 Wage rates and the minimum wage in Southeast Asia

| | Unskilled industrial wage (per day) | | Informal sector income (per day) | Illegal migrants' wage (per day, unskilled) | Agriculture (per day) | | Minimum wage (per day unless stated) | |
|---|---|---|---|---|---|---|---|---|
| **Indonesia** (Rp) | 2,000–3,000 | (1992)[1] | 5–7,000 (1992)[7] | | 1,500–2,000 | (1992)[4] | 3,000 | (1993)[5] |
| | 2,500–5,300 | (1993)[2] | | | 1,000–2,500 | (1993)[8] | 2,600 | (1993)[6] |
| | 2,600–4,000 | (1993)[3] | | | | | 3,800 | (1994)[31] |
| | 2,000 | (1993)[9] | | | | | | |
| | 4,500 | (1995)[28] | | | 3,000 | (1995)[26] | | |
| | 15,000 | (1995)[29] | | | 1,200 | (1995)[27] | | |
| **Thailand** (baht) | 164 | (1992)[10] | | 50 (1996)[24] | 100–120 | (1992)[11] | 125/145 | (1993–95)[16] |
| | 118 | (1995)[30] | | | 50–60 | (1994)[12] | 110/126 | (1993–95)[17] |
| | | | | | 40–50 | (1987)[13] | 102/111 | (1993–95)[18] |
| | | | | | 90–112 | (1989–90)[14] | | |
| | | | | | 135–165 | (1989–90)[15] | | |
| | | | | | 115 | (1996)[23] | | |
| **Malaysia** (M$) | | | | 20 (1996)[25] | | | | |
| **Philippines** (Peso) | 105 | (1993)[19] | | | | | | |
| | 149 | (1993)[20] | | | | | | |
| **Vietnam** (US$) | | | | | | | 30–35/month | (1995)[21] |
| | | | | | | | 11/month | (1995)[22] |

*Table 6.2* continued

*Notes*

1 Average wages of unskilled labour (Manning 1993b: 81)
2 Wages in the small-scale shoe industry (Thamrin 1993: 147)
3 Wages for work in a large, export-oriented garment factory; bonuses add another Rp600–1,500 to this daily wage (Lok 1993: 164)
4 Average wage in rice agriculture, Central Java (Sjahrir 1993: 250)
5 Minimum wage, Jakarta (Hadiz 1993: 197)
6 Minimum wage, West Java (Hadiz 1993: 197)
7 Pedicab (*becak*) driver's return for 8–10 hour day (Sjahrir 1993: 250)
8 Wage in agriculture, West Java (Hardjono 1993: 287)
9 Unskilled wage (Hardjono 1993: 287)
10 Factory work in Chachoengsao, Central region of Thailand, including overtime (Chantana Banpasirichote 1993: 13)
11 Agricultural work, Central region, Thailand (Chantana Banpasirichote 1993)
12 Agricultural work, Northeastern region, Thailand (author's research, 1994)
13 Average agricultural wages across the Kingdom (Nipon Poapongsakorn 1994)
14 Transplanting wage, Central region (Nipon Poapongsakorn 1994)
15 Harvesting wage, Central region (Nipon Poapongsakorn 1994)
16 Minimum wage, Bangkok and surrounding provinces (Alpha 1994 and 1996)
17 Minimum wage, Chiang Mai, Chonburi, Korat, Phangnga, Ranong, Saraburi (Alpha 1994 and 1996)
18 Minimum wage in provinces other than those listed under [16] and [17] (Alpha 1994 and 1996)
19 Basic wage in the Mactan EPZ (Chant and McIlwaine 1995b)
20 Average wage in the Mactan EPZ including cost of living allowance and average overtime (Chant and McIlwaine 1995b)
21 Minimum wage in foreign-invested factories (Schwarz 1996)
22 Average pay in state-owned factories (Schwarz 1996)
23 Daily agricultural wage in Central Plain (author's own research, 1996)
24 Unskilled factory wage (Fairclough 1996)
25 Wage as a rubber tapper in Peninsular Malaysia (Silverman 1996)
26 Harvesting and planting in a plantation in Lampung, Sumatra (Elmhirst 1995a: 10)
27 Daily tasks like weeding carried out mainly by women and children in a plantation in Lampung, Sumatra (Elmhirst 1995a: 10)
28 Daily wage for a trainee at a knitware factory in Tangerang. Rp1,000 was deducted to pay for the canteen (Elmhirst 1995a: 19)
29 Daily wage for a trained worker at a knitware factory in Tangerang. Rp1,000 was deducted to pay for the canteen; a shared room in a hostel cost Rp30,000/month and the worker in question was remitting Rp100,000/month to her mother in Lampung, south Sumatra (Elmhirst 1995a: 19)
30 Minimum wage at a frozen seafood factory in Songkhla, southern Thailand (1995) (TDN 1995a: 70).
31 Minimum wage for Jakarta, Bogor, Bekasi and Tangerang, 1994 (Fane 1994: 37).

*Table 6.3* Bantul, East Java: the wages of work (1987)

|  | *Rp/day* |
| --- | --- |
| Agricultural wage | 1,000 |
| Basket plaiting (traditional handicraft) | 450 |
| *Krupuk* production (new enterprise) | |
|     *krupuk* fryers | 800 |
|     baker | 3,000 |
| Roadside petrol salesman | 2,000 |
| *ojek* rider | 1,500 (per night) |
| Minibus operation | |
|     driver | 4,000 |
|     conductor | 2,000 |
| Factory work | 1,500–2,000 |
| Skilled carpenter | 3,000 |

*Source*: adapted from Cederroth 1995
*Note*: a *krupuk* is a deep-fried cracker; an *ojek*, a motorcycle taxi

it.[7] And in those instances when something was returned, it tended to be in the form of gifts, rather than cash. Workers, including young female workers, tended to control their own money and in many cases it was used for pleasure and entertainment, not to support the wider household, even should the household be poor (Wolf 1992: 180–8). Yet interestingly, and this lends credence to the belief that life cycle changes are important in understanding off-farm work and its links with the sending household, while during the main period of Wolf's fieldwork parents complained of their daughters that '[s]he spends her salary as she pleases, for shopping; she doesn't even tell me what she earns', on the author's return some four years later she found most parents claiming that their daughters did contribute regularly and significantly to the household budget. Wolf puts this important shift down to the fact that they had become, by then, wives and mothers (1992: 235).

Chantana Banpasirichote writes of the long hours (0630–1800 [including travel]), with compulsory overtime for many, in the factories of Chachoengsao, south of Bangkok. Many of the workers complained of fatigue yet 'when asked if they wanted to return to farm work, none of them said yes' (1993: 36). He suggests that '[t]he definition of hard work for this generation of factory workers has changed – no matter how hard the work is in the factory, it is still better than outdoor work like farming' (1993: 36). The social and cultural forces that underlie the shift to factory work, and which factory work itself encourages, are equally as important in explaining the shift as relative wage rates and the availability of land. Murray, on the slum-dwellers of Jakarta writes that 'Girls, especially new girls, who arrive in particularly unflattering clothes (calf-length flares, socks and sandals etc.), are generally put down by [other girls] as cheap, common, or, the ultimate insult, too "*kampung*" [village, country bumpkin]' (Murray 1991: 115). The young are not just escaping from low wages and lack of opportunity in agriculture. They are also escaping hard and dirty work, simple clothes and a quiet life, trading these values – and by implication the lives of their parents – for modernity and excitement. 'In Jakarta', Murray writes, 'the kampung youth are increasingly disinterested in their parents' way of life, but the old appear to be too fatalistic to want to change' (Murray 1991: 131).

Taking all the perspectives noted above, it may well be that the 'pushed or pulled' question is, in fact, a non-question, because not only do both forces operate together, but they also influence one another. Wolf's study shows that most factory workers in her sample in Java came from the poorest land-owning/landless 'class'. Yet in many cases, the factory daughters, the subject of her work, tended to leave home against their parents' wishes. But then, having taken the plunge, these daughters' decisions were ultimately usually accepted by their parents. One female worker, Yukamah, is quoted by Wolf as explaining: 'They accepted the money [her first monthly wage packet] and cried, maybe because they felt sorry for me', before adding, 'Then I told my younger sister she could also get a job in the factory, and now she likes it here [too]' (1992: 173). This problem partly lies in the fact that while wealth ranking is determined at the household level (rich, poor, middle), the decision whether to take up factory work is increasingly being mediated at the level of the individual. The dangers of assuming that we can talk of 'household strategies' at all, in this instance, are all to clear (see page 162).[8] The second reason why the push/pulled question may not mean very much is because rising industrial wages feed into prevailing wage rates in agriculture. Thus, in Malaysia, the differential between the average wage for a plantation labourer (the agricultural workers generally paid least) and a worker in manufacturing remained roughly constant between 1973 and 1989. Both groups saw approximately the same average annual increase in real wages over the period – by 3.0 per cent and 3.5 per cent respectively (World Bank 1995: 18). In 1989 workers in manufacturing were earning double the wage of those on plantations. For Java, between January 1990 and October 1993, nominal wages for agricultural work (in this case, hoeing) rose at an annualized rate of 13 per cent, or in real terms by 8 per cent per year, tracking although not keeping up with rises in the industrial minimum wage (Fane 1994: 37).[9]

## Has factory work benefited labour?

To some extent where a study falls on the issue addressed above – 'Pushed out of agriculture or pulled into factories?' – pre-empts the answer to the rather wider question of whether industrialization has benefited labour. Again there are sharply divergent views. The positive position, often favoured by institutions such as the World Bank, is that growing industrial employment has been to the advantage of workers, companies, countries and agriculture alike. The negative perspective is that the development process, although it has benefited economic and political élites and foreign investors, has been fundamentally inimicable to the interests of labour. They have suffered a real drop in earnings, poor and

*Box 6.1* It's who you know

There exists the persistent view that the urban workplace is like a roulette wheel. Potential workers mill around and, given similar attributes of age, gender and education, are selected randomly. However, the literature on Southeast Asia clearly shows that discrete networks of contacts operate at all levels, from the village meeting room to the factory gate, channelling certain individuals into particular jobs. Interpersonal networks founded on geography, ethnicity, kinship, friendship, membership, patronage and clientage mould the human landscape.[a] Why one village should send ten young women to work in a garment factory outside the national capital while another community one kilometre away sends none, or why one young man obtains work while another remains jobless, are inextricably bound up in these networks. To have a *lung* ('uncle', Thailand) or *babak* ('father', Indonesia) makes all the difference in the world. Lee goes so far as to suggest that 'an analysis of social networks at the micro-level provides the basis for generalizations about macro social phenomena' (Lee 1986: 110). The difficulty for social scientists is that, because these events are rooted in unique relationships, logical explanation may be impossible. These social networks can be long and convoluted. Hugo quotes an example of labour migrants from East Java passing through the hands of no fewer than seven intermediaries before reaching their prospective employers in Malaysia (Figure B6.1).

Both parties, employer and employee, tend to see benefits in exploiting such networks. Workers feel more confident about conditions and wages if they can secure a job in this way, rather than through a faceless agency; while companies transfer some of the costs and risks of recruitment (and, to some extent, training) to intermediaries – *mandors* or *calo* in Indonesia, and *nai naa* in Thailand. As Hugo writes, such social networks tend to be trusted by poor migrants because, although they may be long and consist of numerous links, they start in the home village 'with a *calo* who has to bear the results of a failure in the system or some form of exploitation' (1993: 58). The importance of such social networks for applied and academic research is that they mean that even after the most nuanced assessment of economic and social factors, bald statistics may still fail to provide an explanation, let alone an understanding, of employment patterns. Culture, so to speak, gains hegemony over economics.

*Northeast Thailand*

A great amount of information transference occurs in Isan [Northeast Thailand], particularly with the onset of the dry season and the holding of religious festivities by each village. At this time kin come to visit and spread the news of opportunities in new areas. Once a family from Baan Dong Phong moves, a continuing migration stream occurs for some years, consisting especially of other close relative.

(Lefferts 1975: 177)

continued . . .

*Figure B6.1* The social route to riches: social networks and labour recruitment, East Java to Malaysia

*Bandung, West Java, Indonesia*
   *mandors* [labour subcontractors] prefer to recruit workers from their own villages or from persons known through previous construction projects. . . . Consequently, it is extremely hard for villagers to obtain jobs on a large constructon site without being part of the *mandor's* regular gang.

(Firman 1991: 99)

*Minahasa, North Sulawesi, Indonesia*
   as jobs are not offered through newspaper advertisements or through a labour office, social relations often have great relevance for income generation. . . . it appears that the popularity of village clubs and

continued . . .

associations as well as other social networks in Minhasa stems, at least in part, from the pooling of information about potential jobs available, preferably to men of merit from a club.

(Sondakh 1994: 174)

*Hanoi, Vietnam*
[The rural district of] Xuan Thuy's connection to waste recycling . . . dates from the 1930s, when French colonial rulers hired a man named Nam Dien from the district to collect garbage in Hanoi. Dien brought many of his neighbours with him, and, in the decades since, many of their ancestors have settled in O Cho Dua, a southern suburb of Hanoi which specialises in recycling waste.

(Hiebert 1993: 36)

*Roiet, Northeast Thailand*
Baan Don Samran, like other Northeastern villages, sends its migrant workers to the same occupations in Bangkok, like a sort of transplanted village guild. Theirs is a village of poultry slaughterers and *tuk-tuk* drivers.

(Sanitsuda Ekachai 1988)

Note

a These are highlighted, for example, by: Ulack (1983) for migrants in squatter communities in Mindanao, the Philippines; Paritta Chalermpow Koanantakool and Askew (1993) for slum residents in Bangkok, Thailand; Hiebert for waste scavengers in Hanoi, Vietnam; Firman (1991) for workers on construction sites in Bandung, Indonesia; Ong for factory workers in West Malaysia (1987: 153–5); Wolf (1992) and Mather (1983) for factory workers in Java; Nipon Poapongsakorn for labour recruitment in agriculture in Thailand (1994: 184); Lee (1986) with reference to ethnicity in urban Malaysia; Cederroth (1995) in finding a job in an East Javanese village; Hugo (1993) for international labour migration between Indonesia and Malaysia; Aphichat Chamratrithirong *et al.* (1995) for migration in Thailand in general; and Rigg (1989a) for international migrants from Northeast Thailand.

dangerous working conditions, and the abuse afforded by authoritarian and heavy-handed working regimes (see Manning 1993a, and see Table 6.4). Kemp reports factory workers in Java poetically saying: *kaya makin kaya, miskin makan sinkong* ('the rich get richer and the poor eat cassava') (Kemp 1994).[10]

As Manning remarks with respect to labour welfare in Indonesia, where different authors stand 'is determined by the relative importance one attaches to economic (and especially pecuniary) benefits accruing to labour as against those which involve less tangible labour rights and freedoms' (1993b: 85). Scholars taking the former perspective tend to argue that, for the poor, the need to increase income and meet basic subsistence needs overrides all other concerns; while the latter see this emphasis on wage rates as merely a disguise and sop for exploitative labour practices and degrading conditions (see page 26). Thus Hill,

*Table 6.4* Both sides of the Indonesian ledger

**Gains**

*More jobs* – to absorb the increase in the urban labour force in Indonesia which tripled between 1971 and 1990 to 20 million

*Rising female labour force participation rate*

*Better jobs* – training and the expansion of education has helped to promote better jobs

*Higer wage rates* – wage rates have risen by the order of 30 per cent to 60 per cent between the late 1970s and 1990s

*Comprehensive labour protection* – Indonesian statutes on labour protection are fairly comprehensive

*Declining poverty* – the expansion of employment in industry has been instrumental in poverty decline (see Table 3.1 and Figure 3.1)

**Losses**

*Employment growth has been in marginal work* – most workers have been displaced into marginal jobs, many in the informal sector

*Low real wage rates* – real wage rates in manufacturing have been stagnant or have increased only very slowly

*Labour laws are widely flouted*

*Trades unions are institutionalized* – the government-created All-Indonesia Workers Union (SPSI) is not concerned with protecting workers' rights, representing their interests, or confronting government and industry

*Source:* compiled from Manning 1993b

an economist, with reference to the critical textile and garment sector, while admitting that workers may endure '"sweatshop" employment conditions, cramped living and working premises, and limited career possibilities', argues that this misses the point for 'the industries provide badly needed employment opportunities, real earnings appear to be rising over time, and employment conditions are probably superior to many informal sector and agricultural activities' (1991: 95). For Hill, wages and conditions in industry must be viewed within the context of the labour market in general. In contrast, Ozay Mehmet makes the case for a 'pro-labour' development strategy that 'enhances human beings as creative agents, while employment creation dignifies the individual as a worker' (1995: 147). He makes a plea for the protection of workers through 'clear and enforceable legal codes' designed to ensure 'adequate wages and healthy working conditions', supported and enhanced by effective implementation to promote 'economic justice'. Nowhere does he mention such trifles as competitive wage rates and profitability.

Clearly, factory employment cannot be painted all with the same brush. Often, foreign firms are more likely to adhere to statutory regulations over health and safety than local firms (see below), pay above the minimum wage, and offer their employees additional fringe benefits such as medical insurance (see Chant and McIlwaine 1995b: 164). Despite addressing a litany of grievances from health problems to lack of vertical mobility, long hours, tough working environments, and strict work routines, Chant and McIlwaine state that:

> Women in the [Mactan Export Processing Zone in the Philippines] . . . earn at least the legal minimum wage (usually more), receive substantial benefits over and above those required by law, and may be in a better position than many of their male (and female) counterparts in other industries in the locality, not to mention of other sectors of employment.
>
> (Chant and McIlwaine 1995b: 166)

## Women and work

It has often been observed that young female labour dominates many export-oriented industries. Lok's study of a large garment factory on the outskirts of Bandung, West Java revealed that 92 per cent of shop floor employees were women, nearly all single and with an average age of just 19. Wolf – also in Java – found that many of the firms she studied insisted that female workers bring a letter from the *lurah* (village head) confirming their unmarried status (1992: 115). Of the 3,935 female workers in her sample, 90 per cent were under 25 years old. Akin Rabibhadana notes much the same in the canning factories of Samut Sakhon in Central Thailand, where factory managers employ single women because they are more flexible, allowing them to work shifts unhindered by the demands of housework and childcare (1993: 29). The same pattern of employment – with a preponderance of young women – was revealed by Chant and McIlwaine in their survey of export manufacturing industries in the Mactan EPZ, the Philippines. Some 80 per cent of employees were female, and most employers would only consider taking on women under 25 years of age (1995b: 155).

Although research conducted in the region, and in other parts of the world, clearly shows that firms engaged in export-oriented manufacturing employ a disproportionate share of female workers, such feminized labour recruitment is neither set in stone, nor is it universal. Chant and McIlwaine in the Philippine case study mentioned above, note that most women workers expected to be employed for a long period and the rapid turnover of young women reported in studies from elsewhere did not apply in Mactan. As they hazard, this may 'relate to the recognition by employers of the high profile of economic activity within Filipino women's roles throughout the life cycle' (1995b: 162). The implication seems to be that in the Philippines at least there may be occurring a shift towards the recruitment and retention of older, married women. Differences in gender relations between countries seem to be highly important in determining the gender characteristics of the labour force as well as the conditions in which they work.

There are also important differences between multinationals of different nationality. In the Philippines it has been noted that gender-typing is more pronounced in Japanese firms, where working conditions are also more patriarchal, than it is in US (and Filipino) companies (Chant and McIlwaine 1995b). Ong found much the same in her work in foreign-owned factories in Penang and

Selangor, West Malaysia. In factories operated by East Asian MNCs, a Confucian, patriarchal ethic was used as a basis for company discipline and organization; while in North American and European MNCs, Western notions of individualism were more in evidence. It is this, Ong argues, which is the 'source of [Western] factory women's *bebas* ('loose') reputation, and the secret envy of their sisters in staid Japanese factories' (1990: 396).

Despite this evidence of limited change in the Philippines – which it is tempting to see as exceptional and not indicative of a similar shift in other countries – the broad picture remains one where there is a preponderance of young female employment in export-oriented manufacturing. A range of explanations have been proffered to account for this pattern of employment (see Table 6.5). First, the changing structure of the urban economy has led to the growth of industries such as textiles and electronics where the work is 'light' but requires some dexterity. Women – and especially 'Oriental' women – are perceived to be better suited to such work than men with their legendary 'nimble fingers' (see Ong 1990: 396; 1987: 151; Chant and McIlwaine 1995b).[11] Allied to this, and second, the preference for female labour among many factory managers is due to the belief that they are harder-working, more controllable, less troublesome, easier to sack, and less inclined to unionization and industrial unrest. In the Indonesian context they are expected to be *takut dan malu* (fearful and shy) – attributes deemed appropriate of women in public (Mather 1983; Kemp 1993: 20).[12] Ong suggests that the counterpoint to 'nimble fingers' is 'slow wit' (1987: 151).[13] Men, by constrast, were characterized by Wolf's informants as too *berani* (aggressive, plucky, assertive) (Wolf 1992: 117).[14] There are also a set of explanatory social and economic factors that reside, so to speak, in rural areas (see page 184).

What is interesting in the literature on Southeast Asia, as distinct from that on Africa and Latin America, is that there is almost no reference to the femi-nization of the labour force being associated with economic recession. In other parts of the developing world, writers are only too ready to see the process being a response to poverty, unemployment and declining incomes, and especially the loss of male formal employment opportunities. In Southeast Asia, although there may be a range of explanations offered, rising female participation rates are only rarely associated with cyclical perturbations in the national economy (see Gilbert 1994).

A second characteristic of the literature on the region is that the 'margin-alization of women' thesis has gained fewer converts than in other parts of the developing world. Where it is most pronounced is in literature on the Philippines, and it is perhaps notable that many scholars working on such issues in the Philippines have also undertaken work in Latin America, rather than in other countries of Southeast Asia. Proponents of the female marginalization thesis argue that women are being displaced into factory work. Gilbert (1994: 618), however, sees confusion regarding what different authors mean by the term. Some use it to mean 'housewifization' – banishment to the home. Others,

*Table 6.5* Building the stereotype: factors highlighted to rationalize the preference for (young) women workers in manufacturing industry

**Physiological factors**
- 'Nimble fingers'
- Dexterity
- Light touch

**Psychological factors**
- Patience
- Concentration
- Harder working
- Passive and docile
- Controllable
- 'Slow wit'
- Higher boredom threshold

**Socio-cultural factors**
- Non-political
- Less prone to unionization than men
- Easier to sack than men
- 'Fearful and shy'

**Situational factors**
- No dependents
- No parenting responsibilities

*Note:* for most of the above, women are regarded as having these qualities to a greater degree than men.

displacement into poorly-paid, informal or casual work. And yet still more, as displacement from traditional occupations which are then co-opted by men. It can also be argued that different women experience 'marginalization' – or the absence of it – differently. Work in export-oriented factories tends to be restricted to young, unmarried women and it is they who selectively benefit from the growth of the export sector. As this sector upgrades, and employment becomes more skill-intensive, the field is being further narrowed to those young women with a secondary school education. Working mothers and older women, especially those with just primary level education, are finding themselves either left out altogether or restricted to marginal work – and in these terms, 'marginalized'.

There can be little doubt that working women in the industrial sector are often paid less than men, but it is rarely possible to add the crucial clause 'for the same work', because men and women usually do different work. The difficulty for women is in gaining access to (better-paid) men's work. For example, in garment factories sewing, cutting, button-holing, sorting and packing are all dominated by women, while men are employed as technicians, warehouse staff, drivers and security guards. Therefore scholars working on the region are generally reluctant to talk, in an all-embracing way, of industrialization being inimical to the interests of women. While women have tended to be assigned the more menial, lower-paying jobs, the industrialization process has also allowed them to gain access to a far wider range of employment opportunities (see Price 1983). Nor does it seem to be necessarily true that women are concentrated in such activities because they are female. Increasingly, education is becoming a key factor in determining job opportunity and placement. None the less, it might reasonably be argued that as females have fewer opportunities to pursue education beyond primary school, then this is still of function of gender, merely one step removed (see O'Brien 1983). The debate over women and factory work will doubtless remain highly complex and prone to stereotyping. But despite the abuses, the poor pay, and the unhealthy and dangerous conditions (see below), very few micro-studies unearth workers willing to agree with the statement that they would have been better off staying at home. This simple fact is too often overlooked.

In Chapter 5 it was suggested that the shift of young women and men out of agricultural work and into non-farm employment was indicative not of a generational shift (i.e. that when they reached middle age they would return to farming), but of a permanent change in the complexion of the rural economy (see page 184). However, there is some evidence that generational changes remain important within the industrial sector, with unmarried women without responsibilities working at first in the export-oriented sector, and then shifting to more flexible employment in subcontracting work or in smaller firms as they get married and become mothers (Table 6.6).

*Table 6.6* Generational shifts in female industrial employment: the Indonesian perspective

| Approximate age | Working status |
| --- | --- |
| 0–7 | Dependent |
| 7–15 | Family helpers |
| 16–30 | Large-scale manufacturing enterprises, often geared to export. Unmarried women with no responsibilities and willing to work flexibly and for long hours are preferred by employers |
| 31–40 | Smaller-scale manufacturing enterprises, home- or locally-based and often geared to the home market. Hours are flexible or regular, and young mothers find they are able to mould the demands of work around the demands of childcare |
| 41–50 | Home work (sub-contracting). Highly flexible from the worker's perspective and attractive to mothers with two or more children |
| 51+ | Home work, self-employment, family labour. Older women find it hardest to secure employment and often have to accept the lowest paying option |

*Source:* adapted from Grijns and van Velzen 1993

*Box 6.2* Strategies underpinning non-farm work in two villages in Lampung, south Sumatra

Becky Elmhirst's research in two communities in the *kecamatan* (sub-district) of Pakuon Ratu, north Lampung, Sumatra provides a particularly detailed analysis of the complex mosaic of cultural and economic reasons that lie behind particular patterns of non-farm work. Tiuh Baru with a population of 1,360 in 255 households is a Lampungese (Way Kanan) village; Negara Jaya with 3,845 inhabitants in 910 households is a Javanese transmigrant settlement. In both instances, non-farm work was highly important to livelihoods.

In Negara Jaya and Tiuh Baru, Elmhirst suggests, households were seeking alternatives to agriculture and were endeavouring to diversify their livelihood strategies in order to seek greater security. Yet the difference between the two communities in the means by which they achieved these similar ends was striking. Local planatation work and the collection and sale of wood to a sawmill were the two main forms of non-farm work in Negara Jaya. The work generated daily incomes ranging from Rp1,000–3,000. Indeed, in Negara Jaya no household was able to meet its needs from agriculture alone and remuneration from plantation work accounted for 30 per cent of total income. In Tiuh Baru, while the population would rarely engage in local labouring of the type common among the residents of Negara Jaya, large numbers of young women engaged in factory work in Tangerang, Java. This contrast between non-farm work in the two communities was most clearly reflected in the economic roles of unmarried daughters.

In Negara Jaya, unmarried girls were permitted to take part in almost all agricultural activities, both on the farm and on local plantations. Economic conditions were such that girls had to work, but their wages – just Rp1,500 per day – were controlled by their mothers. Elmhirst hazards that there were strong social pressures on daughters not just to work but to contribute all their income to the collective good of the household. In addition, because daughters were also required to work at home – fulfilling both agricultural and domestic tasks – their economic value was maximized only if they remained local.[a]

In the Lampungese village of Tiuh Baru, it was not desirable for women to work in the fields. Indeed, to do so was regarded as shameful. However, the exigencies of poverty did force some women to break these normative constraints on their mobility. One woman explained: 'If you are poor here, if you don't work, you don't eat. I was ashamed, but I had to go to the field to work'. But 'married women', Elmhirst writes 'are not allowed to go out of the village [to the fields] unaccompanied . . . widows will even withdraw a son from school for a day to chaperone them'. For unmarried girls the additional risk of abduction for marriage[b] made working away from the safety of the house even more hazardous and, potentially, even more shameful. But

continued . . .

while women found their freedom of movement sharply curtailed within the village and the surrounding area, every household with unmarried women aged between 15 and 20 years – over 100 – had a household member engaged in factory work in Tangerang, Java. Initially parents were reluctant to let their daughters leave the village, but they were eventually persuaded that they would live with other Way Kanan girls and be under the supervision of older and respectable fellow villagers. What is surprising though, at least in the light of conditions at home, is that girls enjoyed comparative freedom to do as they pleased in Tangerang. They made trips to Jakarta, wore what they liked, could date and go to the cinema – all actions that would have been strictly taboo at home. The question is why was this permitted. Elmhirst believes that the answer lies in the adage that 'there is no shame in what cannot be seen'. Through working in the local plantation, as the Javanese of Negara Jaya did, '*gensi* [prestige] is lost [because] you are seen by [local] people and it brings shame'. By contrast, 'my parents do not mind what I do in Tangerang, they just care what I do in the village'. In addition, working in Tangerang does not hold the risk of abduction for marriage. Elmhirst summarizes the strikingly different views and approach to non-farm work in Negara Jaya and Tiuh Baru in these terms:

I have tentatively suggested that attitudes to the relative autonomy of Javanese daughters [in Negara Jaya] have hardened in the face of adversity, as their labour is crucial to family survival; while among Lampungese, spatial restriction on daughters has loosened, with *adat* reworked to accommodate work in Tangerang – at a distance from the village.

(Elmhirst 1995a: 22)

*Sources*: Elmhirst 1995a, 1995b and 1996

Notes

a In the case of Negara Jaya it seems that talking of 'household' livelihood strategies does have some veracity.
b *Caripaksa*, the abduction of girls for marriage, is sanctioned by local *adat*. Girls – willing or not – can be abducted by a man and his accomplices and then virtually forced into marriage. Although girls who do not consent to marriage can return to their families, this brings great shame on the girl's family. The girl would be highly unlikely to find a marriage partner in the future.

Finally, and underlying all this discussion of the effects of industrialization on women, is the equally contentious issue of the general status of women in Southeast Asian society. There has been a tendency to view gender relations in the region, traditionally, as relatively egalitarian when compared with neighbouring East and South Asia. The role of women in trade and other money-making activities, the preponderance of matrilocal/uxorilocal and even matriarchal (most famously, the Minangkabau of West Sumatra) societies, the general lack of gendered pronouns in many Southeast Asian languages, the important position that women hold in household decision-making, the greater emphasis on the conjugal unit, and the prominent roles that women have played in history, for example, have all lent credence to the argument that women have traditionally enjoyed high status (see Reid 1988: 146–50; Mason, K. 1995). Women may not have been equal to men, but they did have power and influence and it is notable that 'the value of daughters was never questioned in Southeast Asia as it was in China, India and the Middle East' (Reid 1988: 146). As a well-known Thai proverb has it, 'Man is padi, Woman is rice'. Combined with high labour force participation rates, relative equality of educational opportunity, and comparative freedom when it comes to making reproductive decisions, this has shored-up the belief that the Southeast Asian experience is substantially different from that of Africa, the Middle East or Latin America (see Mason 1995 for a general discussion, and Richter and Bencha Yoddumnern-Attig 1992 for a discussion with reference to Thailand). There is a wide range of possible indicators of the situation of women in Southeast Asia. Table 6.7 records the preference ratio for the sex of the next child, and it shows that although male children may be preferred in the region (except, notably, in the Philippines), the differential is significantly lower than in the countries of neighbouring regions. It should be added, though, that this view – that the cultural traditions of Southeast Asia are comparatively benign to women – has not gone undisputed. Critics point to,

*Table 6.7* Preference ratios for the sex of the next child among currently fecund married women awaiting a next child

|  | Preference ratio* |
| --- | --- |
| **Southeast Asia** | |
| Indonesia | 1.1 |
| Malaysia | 1.2 |
| Philippines | 0.9 |
| Thailand | 1.4 |
| | |
| **East Asia** | |
| South Korea | 3.3 |
| | |
| **South Asia** | |
| Bangladesh | 3.3 |
| Nepal | 4.0 |
| Pakistan | 4.9 |
| Sri Lanka | 1.5 |

*Source*: Mason, K. 1995: 17
* Ratio of women preferring a son to those preferring a daughter, with undecided allocated equally.
*Note*: it seems that preference ratios in Vietnam would favour male offspring to a greater degree than in the other Southeast Asian countries listed above. In a recent survey, 720 respondents were asked what combination of children they would like if they could have only two: 88.4 per cent replied that they would like a boy and a girl, 10.1 per cent said they would prefer two boys, and 1.5 per cent, two girls. While these figures do not indicate a strong preference for sons, significantly almost half did say that their ancestors would be displeased if they did not have any sons (Goodkind 1995: 351).

for example, the tendency (noted above) for women to be paid less than men, for Buddhism to assign to women a lower place in the hierarchy of birth and rebirth, for Confucianism to sustain patriarchy, for women to be denied access to influential positions in the civil service and big business, and for political power to remain largely in male hands.[15] Nevertheless, the countries of Southeast Asia do appear to come out well in the women's situation stakes, in significant part due to the social and cultural traditions of the region. In addition, although her conclusions will be contentious, Mason hazards in a wide-ranging review paper that the 'majority of the changes in women's status [in Asia] associated with development . . . have been positive' (Mason K. 1995: 18).

## Health, safety and the wages of work

All countries in the region set minimum wage rates (which often vary between regions and industries) and health and safety standards (Table 6.8 and see Table 6.2). However work by scholars, labour activists, NGOs and others has shown that conditions in factories in Southeast Asia regularly fail to meet stated government levels *vis-à-vis* safety, and labour laws aimed at ensuring workers' health, safety and minimum wages are commonly flouted. There are three, not mutually exclusive, ways in which minimum labour standards can be enforced. First it is possible to allow vibrant, independent unions and other workers' organizations to fight and bargain on behalf of labour. Second, governments can take upon themselves the task of representing the interests of workers. This can be through the work and action of government agencies and ministries, or through institutionalized, quasi-independent unions. And third, various informal arrangements which help to regulate labour standards can be encouraged. These may range from the work of NGOs and other pressure groups, to the operation of cultural norms regarding acceptable practice.

In Indonesia, in annual consultation with the All-Indonesia Workers Union (SPSI), the government sets minimum daily wage and overtime rates. The law allows for a 40-hour week, usually spread over six days. Officially, female workers receive paid maternity leave if they have worked at the same factory for more than two years, and leave without pay if they have worked for more than three months. They are also entitled to two days' menstruation leave a month. However, Wolf's work on female workers in factories in Java, for example, found that the entitlement to menstruation leave was almost never honoured, and in one factory any worker asking for such leave was punished by the loss of her monthly bonus (Wolf 1992: 119–20). In other enterprises, workers had to prove to the company nurse that they were menstruating, a degrading excercise that most, understandably, wished to avoid. The required three months' maternity leave was also rarely respected by the factories Wolf studied.

Even more worrying is the widespread non-observance and infringement of health and safety laws. Labour laws requiring that workers be provided with protective clothing (masks, saftey glasses, gloves, hard hats, etc.) and that

*Table 6.8* Enforcing and regulating labour standards in Indonesia

| Type of intervention | Guarantees and policies | Examples |
|---|---|---|
| Establishment and protection of workers' rights | Right to associate, organize, protest, strike and bargain | All factory employees belong to the government-controlled SPSI (All-Indonesia Workers Union). Independent unions are not encouraged, and labour activists have been persecuted by the state. Genuine collective bargaining has been discouraged since 1965 and labour controls are tight |
| Protection for the vulnerable | Minimum working age | 14 years, although the Dutch-era Measures Limiting Child Labour and Night Work of Women (1925) is still effective, and sets the minimum working age at 12. A 1987 regulation allows children under 14 to work up to 4 hrs/day when it constitutes an essential element of household income. |
| | Equality of wages and employment opportunities | |
| Minimum wage legislation | Minimum wages, setting of overtime pay rates | Varies according to industry and region (see Table 6.2). For child labour, which is allowed in some instances, there is no minimum wage. |
| | Non-wage benefits | Maternity (three months) and menstruation leave (2 days/ month) |
| Setting of health and safety standards | Maximum hours of work Minimum health and safety standards | 40 hours/week |
| Income and employment security | Social security | |

*Sources*: adapted from World Bank 1995: 71, using US Department of Labor 1995, Manning 1993c

employers ensure a safe working environment are on the statute books of all Southeast Asian countries. However, the majority of studies which examine working conditions in any depth admit that such laws are widely transgressed. Wolf illustrates this with respect to garment and glass factories in Java (1992: 120–2), while a survey conducted in the Philippines revealed that 81 per cent of inspected firms violated at least one health and safety standard (World Bank 1995: 77).

Child labour also appears to be prevalent in some areas and sectors, although most studies admit that documentation is very limited and tend to rely instead on anecdotal evidence supported by individual life histories (see Lee-Wright 1990 on Thailand). The lack of data relates in part to the illegality of child labour. But equally important are definitional difficulties associated with the questions 'who is a child?', and 'what is work?' It is widely acknowledged that child labour in rural areas is more of a problem than in urban areas, that those who work in the domestic, household context are often overlooked, and that definitions of 'child labour' that take as their age range 10 to 14 years will inevitably overlook a large number of children aged less than 10 (Myers 1991). As a result the emphasis tends to be on older, usually male children working in factories.[16] Various estimates have put the number of children (i.e. between the ages of 10 and 14) in the work force in Indonesia at between 2.2 million (1990 Indonesian population census) and 3.3 million (unofficial NGO estimate, *FEER* 1996). The UN Commission on Human Rights put the figure in 1994 at 2.7 million while in 1993 Azwar Anas, the Coordinating Minister for People's Welfare, reported that 2.2 million children were working in rural areas, and 250,000 in urban areas (US Department of Labor 1995) (Table 6.9). An extreme estimate of 10 million has been quoted in the campaigning magazine *Inside Indonesia* (Bessell 1996: 18).

A study undertaken by the Committee for the Creative Education of Indonesian Children (KOMPAK) in 1991 claimed that children in export-oriented factories in Tangerang, outside Jakarta, were working long days (7–13 hours) in dangerous conditions for just Rp8,000 (US$4) per week (US Department of Labor 1995). Child *pembantu* – domestic servants – are reported to earn as little as Rp20,000–50,000 (US$10–25) a month (Bessell 1996). Enforcement of laws regarding child labour, and penalties for employers who contravene those laws, are comparatively meagre. In 1994 the maximum fine was just Rp100,000 (US$50) or three months' imprisonment, and the US Department of Labor claimed in 1994 that no Indonesian employer had been charged with violating child labour laws (US Department of Labor 1995). There is some divergence of opinion concerning whether child labour *per se* should be censured, or just child labour without adequate regulation and supervision. Some scholars fear that outlawing child labour without first building an understanding of its role in household survival strategies and the conditions and sectors in which children work will merely drive it underground (e.g. Myers 1991). It is noteworthy that the Indonesian authorites prefer to talk not of child labour, with its exploitative undertones, but rather of *anak yang terpaksa bekerja*: children who are compelled to work (Bessell 1996).

*Table 6.9* Estimates of child labourers in the workforce (millions)

|  | *Official estimate* | *Unofficial estimate* | *Population <16 years* |
|---|---|---|---|
| Indonesia | 2.2 | 3.3 | 69 |
| Philippines | 0.8 | 5.0 | 29 |
| Thailand | 1.6 | 4.0 | 20 |

*Source*: *FEER* 1996: 55

Three reasons can be highlighted to explain the non-observance of government-set minimum labour standards. First, enforcement agencies are generally weak and underfunded. In Indonesia, the Ministry of Manpower is charged with responsibility for enforcing minimum wages and conditions at work. However, at the beginning of the 1990s, less than a quarter of factories were registered with the Ministry and it lacked the human and physical resources to oversee even this small share of the total (Kemp 1994). In 1993 it was estimated that the Ministry employed just 1,320 inspectors, of whom 700–800 were operational, the remainder being support staff. This translates into one inspector for every 3,895 companies (US Department of Labor 1995; see also Manning 1993c: 76).[17] It seems that pressure on firms to meet government-set minimum wage rates is coming more from a combination of market forces, media attention, and pressure from NGOs than it is from government action and intervention. One report in the Indonesian news magazine *Kompas* estimated that 40 per cent of infringements in West Java were brought to the attention of the Ministry of Manpower by media reports (Manning 1993c: 76). With such a weak regulatory environment, it is not surprising that there is only limited incentive for companies to comply with the regulations. Wolf's work suggests that although most larger urban and peri-urban enterprises in Java do pay the minimum wage, among small and medium-sized operations, especially those in rural areas, non-payment is common (1992: 117). This pattern seems to relate to the greater awareness of workers in and around urban areas, the greater demand and shortage of labour in these areas, and the greater visibility of such firms.

A second reason why enforcement is poor is because the bribing of factory inspectors to overlook trangressions of minimum wage rates and safety standards is commonplace (see Wolf 1992: 145–6; Manning 1993c). And third and finally, there is a case that the governments of the region have been pressured into placing such laws on the statute books by external agencies (foreign governments, NGOs and unions, for example) who have threatened trade reprisals if internationally agreed minimum standards are not set. The US Department of Labor and the US Trade Representatives Office have in the past linked labour practices in Indonesia with eligibility for trade privileges under the US Generalized System of Preferences (GSP) (see Manning 1993c: 77–9). Under US law, a country is only eligible for inclusion under the GSP if that country is 'taking steps to afford internationally recognized worker rights', including freedom of association, acceptable conditions of work, restrictions on child labour, the elimination of forced labour, and the right to collective bargaining (Awanohara 1993). When laws are placed on the statute books to appease such external bodies, it should not be surprising if the administrative apparatus necessary for their policing is lacking. Indeed, by their actions, there is ample evidence that governments are not fully committed, certainly to the concept of the minimum wage and to some extent to such assumed 'core' standards as the outlawing of child labour and acceptable safety measures.

In poorer countries, it can also be argued that setting a minimum wage does not protect or benefit the very poorest who tend to be concentrated in the agricultural and informal sectors, and indeed there are economists who suggest that setting a minimum wage discourages the growth of formal employment by raising costs in the process (see World Bank 1995: 75). Those who might be characterized as employment 'realists' suggest that the market should, in most cases, be used to determine wages and conditions. Thus Manning writes that: 'it is quite unrealistic to attempt to enforce the same set of labour standards in Indonesia in the early 1990s as now apply in many developed countries, or even those that are increasingly applied in middle income countries such as Singapore, Malaysia, Korea and Taiwan' (1993c: 90).

It tends to be only when tragedies bring the issue of working conditions and safety to the fore that governments feel pressurized into taking action. In Thailand, for example, a series of human disasters during the 1990s raised the question of whether the Kingdom's enviable economic growth rate had been achieved at the expense of workers' safety, not to mention the environment (see Table 1.7, page 28). In particular, a fire at the Buddha Monthon (Kader) toy factory outside Bangkok on 10 May 1993 caused the death of 188 workers. The toys were being produced for export and most of those who died were young women. An investigation showed that many of the factory's fire escapes were blocked or locked and that safety laws had been widely breached. But although attention tends to focus, understandably, on the high profile 'tragedies' with loss of life, even when health and safety standards are being adhered to, the type of work, the hours worked, and the conditions in which work occurs may have significant health implications. Chant and McIlwaine in their study of factories in the Mactan EPZ, for example, write that 'occupational hazards such as back-ache, eye strain, blurred vision and headaches are also encountered by female assembly workers as a result of sitting and concentrating for long periods, not to mention complaints such as chest infections and skin irritations which arise from the handling and inhalation of chemical substances' (1995b: 164).

## ORGANIZATION, RESISTANCE AND PROTEST

Although workers may be willing to accept (or at least to endure) poor working conditions and low wages, or perhaps not be in a position to confront them, the evidence from Southeast Asia shows that organization, resistance and protest are becoming more common, and in many cases more effective.[18] In Indonesia, all factory employees belong to the *Serikat Pekerja Seluruh Indonesia* (SPSI) or All-Indonesia Workers Union, formed in 1973 following the emergence of a consensus 'between the military and their economic advisers that a unified trade union movement was in the best interests of political control, economic stability and growth' (Manning 1993c: 68).[19] Not surprisingly given its origins, the SPSI has so far failed to gain recognition as an independent and legitimate workers' body, despite fervant attempts by the Indonesian government to promote it as such. Most workers view the SPSI as little more than 'a tool of state control' (Hadiz 1993: 190) – a critical component in the state's labour strategy. Mulya Lubis, during a spate of labour disputes in the early 1980s, labelled the SPSI, because of its relations with government and business, *Gerakan Buruh Salon* – the Salon Workers' Movement or, as Manning translates it, Armchair Workers' Movement (Manning 1993c: 73). Activists at the time talked of 'envelope workers' representatives' (*wakil buruk emplop*), referring to the bribes paid to such officials (Manning 1993c: 73).

Partly because of this institutionalization of workers' interests, a number of independent – and therefore illegitimate (in statist terms) – organizations emerged during the 1990s to represent the interests of workers. For example, the SBM Setiakawan (the Solidarity Independent Workers Union) was formed in 1990, and the SBSI (Indonesian Prosperous Workers Union) in 1992 (Hadiz 1993: 190). There are also many more NGOs involved in protecting and representing the interests of labour. These organizations tend to maintain a low profile and try to avoid provoking the authorities. But, and notwithstanding this evidence for the emergence of a more vibrant labour movement in Indonesia, it has been argued that the form of capitalist industrialization in Southeast Asia has tended to inhibit the process of unionization. The preponderance of small firms, the importance of seasonal workers who maintain a rural base, and the fragmentation of the workforce by location, age and occupation, all tend to act as obstacles to labour organization. Interestingly, it appears that it is in the larger, multinational-owned factories where the barriers are least imposing and where, consequently, unionization is most marked (see Hewison and Brown 1994: 502–3 on Thailand).

There has been a marked increase in industrial unrest in Indonesia – at least if strikes can be used as measure of such things – coinciding with the rapid expansion of export-oriented manufacturing and concentrated in enterprises geared to export production. In 1988 there were 39 strikes; in 1990, 61; and in 1992, 177 (Hadiz 1993: 187; see also Manning 1993c) (Figure 6.3).[20] In 1993 the number of strikes declined slightly to 130 (Fane 1994). However, it should not necessarily be assumed that rising levels of unrest are necessarily indicative of poor and declining wages and working conditions. Indeed, conditions and wages were probably marginally improving, on a broad front, just at the time when unrest was increasing. Data on earnings should be treated with caution, but it seems that real wages increased by 10 per cent and 50 per cent between

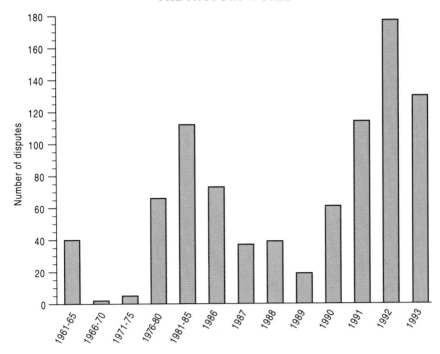

*Figure 6.3* Industrial action in Indonesia, strikes per year (1961–93)
*Sources*: Hadiz 1993: 187, Manning 1993c, Fane 1994

the early 1970s and early 1990s (Jones and Manning 1992: 382). What Indonesia has not experienced though – despite high and sustained economic growth – is an equally sustained increase in real wages. Most commentators put this down to the fact that Indonesia remains in labour surplus (e.g. Jones and Manning 1992: 382). The literature highlights four principal explanations for the increase in labour unrest, and although they are here linked to the Indonesian experience, they can be seen to have wider application.

First, in the Indonesian context, there is some evidence to support the contention that workers are freer now to demonstrate their grievances. While, formerly, troublemakers were simply sacked and put themselves at some considerable physical risk, companies, the police and government appear to have become more circumspect in their actions during the 1990s. Likewise, while the period of industrial unrest in the early 1980s led to a heavy-handed response from government, the later spell of strikes saw a much more balanced response, with ministers blaming employers as well as employees (Manning 1993c: 83). That said, this is not to imply that taking industrial action is not without its risks. In May 1993 a female labour activist in East Java, Marsinah, was found murdered after she had helped to lead a strike at a watch factory 40 km south of Surabaya. She was tortured, raped, sexually violated with a sharp instrument, and subsequently bled to death. To begin with, the owner of the factory where Marsinah worked, seven other employees and the local police chief were charged with her murder. Though the regional court sentenced the nine to prison terms of between seven months and seventeen years, an appeal to the Supreme Court quashed the convictions as unsound and extracted through torture. As the convoluted trial and investigation continued, so it became clear that much more senior officials were implicated, including members of the military (see McBeth 1995). The case of Marsinah became something of a *cause célèbre*, a exemplar for all that is bad in Java's factories and in the government's approach to labour standards and regulation. 'The tragedy of Marsinah', Waters argued shortly after her murder, 'represents the dark side of the growing success of economic development which foreign and domestic capital are fuelling' (Waters 1993: 13).

Second, Hadiz believes that the rise in industrial action in Indonesia is intimately tied to the emergence of a new industrial working class which is challenging the 'viability of the existing framework of state–labour–capital relations' which kept such industrial struggles at bay during the 1970s and for much of the 1980s. In 1961 there were only 1.9 million workers employed in the manufacturing sector in Indonesia; by 1990 this figure had reached 8.2 million (Figure 6.4) (Hadiz 1993: 191; Manning 1993c: 67). It is through being part of the process of development, of which the growth of manufacturing is a critical component, that workers are creating a sense of class identity which in turn has created the conditions in which the potential for industrial action has increased. It is, Hadiz argues, 'through participation in the social relations embedded in these factories, [that workers] are developing an understanding of what it means to take part in modernising and industrialising Indonesia' (1993: 193). The fact that such industries tend to be spatially concentrated has, arguably, also helped to create the 'critical mass' necessary to forge a 'class' consciousness and therefore a basis for industrial action. In 1991, 70 per cent of strikes occurred in factories in the JABOTABEK region surrounding Jakarta, and nearly half in large enterprises employing 500 or more workers (Manning 1993c: 83). There is no doubt that workers, made more aware of their rights by government legislation and by the criticisms of international and domestic groups, have rising expectations regarding acceptable labour standards.

The third possible reason why industrial unrest has increased is because although wages may have risen, those in export-oriented industries have, it is argued, been suppressed through government labour controls and macro-economic policies (Manning 1993c). This has led to accusations that foreign investors are exploiting the Indonesian people, adding nationalist overtones to an already incendiary issue. Mochtar Lubis, the influential author and journalist, for example, wrote in 1991 that low wages in foreign-controlled firms were an 'insult' (*penghinaan*) to the Indonesian people (quoted in Manning 1993c: 86). When the profits and costs of export goods are set against wages, the differentials are striking. The labour cost in a garment sold ex-factory for US$10 has

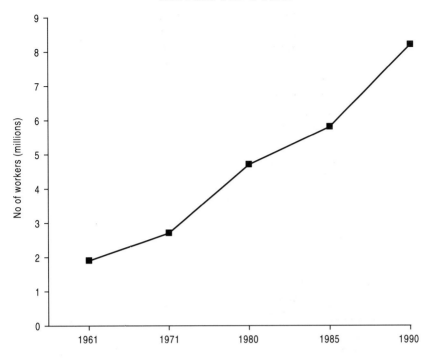

*Figure 6.4* Workers employed in manufacturing in Indonesia (1961–90)
Source: Hadiz: 1993: 191; Manning 1993c: 67

been estimated to be just US$0.70–0.80 (Lok 1993: 162). Materials cost US$6 and the garment will then retail for US$40 or more in the developed world. When it is calculated that labour costs in such a garment therefore account for just 1.75–2.0 per cent of its retail value, it is easy to see why Lubis can use the term 'insult' (Figure 6.5). The government appears sensitive to these accusations, and yet the reality of the global market place gives it little room for manoeuvre (see page 24). Having embraced foreign investment-driven export-oriented industrialization, the government is aware that should wages in low-skilled industries rise excessively investors will simply move their operations and their orders to lower wage locations like Vietnam, southern China, and Bangladesh (see the discussion of this issue in Chapter 1, page 26). Although few economists would endorse Myanmar-style economic autarky (circa 1960s– 1970s) as a viable alternative, Indonesia's very success in enticing FDI has brought to the fore, just as it did in Singapore, Thailand and Malaysia, the costs and risks of such a strategy. Manning writes that 'the fundamental reason for "labour repression" is poverty and near poverty, and associated absence of human capital' (Manning 1993c: 89). The dilemma for the government is that such justifications, though they may make excellent economic sense, are an ineffective sop to workers with very real grievances.

231

Factory labour
US$0.75

Profit and
other costs

Materials
US$6

Labour
US$0.75

Ex-factory price
of a garment US$10

Retail price in United States
US$40

*Figure 6.5* Breaking down the price of a shirt
*Source:* Lok 1993

A fourth explanation concerns the identification of the factors lying behind industrial unrest. Macro-studies tend to emphasize such 'fundamental' concerns as wage levels, safety standards, and contractual issues. However, micro-level research seems to indicate that worker dissatisfaction and protest is often triggered not by such fundamental complaints over pay and safety, but by penalties for minor infringements such as broken machine needles, late arrival, or the refusal to allow menstrual leave. Kemp, for example, found the following notices posted at an East Java shrimp paste factory (Kemp 1994):

> To all workers. A charge of Rp100 [US$0.05] will be deducted from pay every Saturday for the cost of washing the dishes used to eat meals at work.

> Workers wanting to wash before going home will be required to pay Rp50 [US$0.025] and another Rp50 will be charged to workers parking their bicycles in the factory grounds.

232

Such fines and penalties may be small, but they are indicative of an environment where workers are poorly treated and may act as a trigger for industrial action.

The experience of Thailand represents an interesting counterpoint to Indonesia. Although Thailand's unionized labour force remains comparatively small at less than 5 per cent of the total, the number of unions has increased significantly since 1980 (see Table 6.10). What has not occurred, though, is any systematic increase in the number of days lost through strikes (although lockouts appear to have increased). Partly this can be attributed to the relatively undeveloped nature of the labour movement in Thailand. However, in addition, the government has been successful in its use of industrial relations machinery to limit industrial unrest and defuse potential disputes (Hewison and Brown 1994: 505). During the 1980s, a tripartite system reminiscent of Singapore's National Wage Council (NWC) brought together the government, employers and employees to negotiate over labour relations. In 1993 a new Ministry of Labour and Social Welfare was created to manage labour issues, taking responsibility over from the Ministry of the Interior. This, as Hewison and Brown point out, represents an important shift in outlook. While labour relations were under the aegis of the Ministry of the Interior (which controls the police), its management was perceived to be as much a political (and security) issue as an industrial one. Now that the Ministry of the Interior has relinquished control it seems that 'trade union struggles are seen to be restricted to wages and conditions, while broader social and political reforms are to be effected through the political system and political parties' (Hewison and Brown 1994: 510–11). In Indonesia the independent labour movement remains as much about politics as about wages and conditions – a fact which partially explains why the government has been so reluctant to allow it free rein.

Strikes and protests are the most visible ways in which workers vent and demonstrate their grievances. But low-intensity methods of 'resistance' (the factory equivalent of James Scott's 'weapons of the weak') may be even more important because they are everyday events, far more commonplace, and probably more significant in reducing industrial output. It seems that these less overt forms of protest are particularly attractive among female workers who, for various cultural reasons, may be less inclined to confronation than men (see Wolf 1992: 128). Among the strategies employed are 'go slows', claiming that there are ghosts or spirits, stayouts or mass-absenteeism, feigning ignorance of orders or technical details, citing religious reasons or 'female problems', and spirit possession (see for example, Wolf 1992: 128; Ong 1990: 417–22). These attempts at resistance, for Ong, represent a struggle not just against the dehumanizing effects of capital, but also of capital manipulated in the interests of men. Thus the patterns of resistance reflect a dual struggle, in her view, of workers against industry, and female workers against male domination (Ong 1987).

Some studies have highlighted a tension arising from the attempt to implement modern measures of efficiency and discipline, whether they be Western (US/European) or Asian (Korean/Taiwanese/Japanese), in the Southeast Asian

233

*Table 6.10* Unionization and industrial disputes in Thailand (1976–91)

| | Number of unions | Number of disputes | Number of workers involved in strikes | Days lost in strikes | Days lost in lock-outs |
|---|---|---|---|---|---|
| 1976 | 184 | 340 | 194,469 | 495,619 | n.a. |
| 1980 | 254 | 174 | 58,461 | 5,356 | n.a. |
| 1982 | 377 | 376 | 100,959 | 116,795 | n.a. |
| 1984 | 430 | 86 | 32,752 | 183,698 | n.a. |
| 1986 | 470 | 168 | 27,982 | 157,858 | 4,875 |
| 1988 | 562 | 120 | 29,178 | 32,386 | 7,464 |
| 1990 | 713 | 118 | 35,792 | 22,775 | 48,792 |
| 1991 | 657 | 128 | n.a. | n.a. | n.a. |

*Source*: Hewison and Brown 1994: 505

context. Managers complain of laziness and a lack of discipline; workers of hard-hearted employers and a lack of understanding. Workers respond by seeing ghosts, spirits and demons requiring, in the case of some factories in West Malaysia, that a *bomoh* or spirit healer be engaged to ritually cleanse the factory by making animal sacrifices (Ong 1990; see also Wolf 1992: 135).

There is evidence that even in supposedly authoritarian states, governments do respond to workers' unrest and protest. In Indonesia it has been suggested that the rise in the minimum wage in Java has been a response to industrial unrest, while so too has the emergence of a harder line against employers who flout the minimum wage and other labour regulations (Hadiz 1993: 198). It would be pleasing to think that these responses reflect a growing awareness of the need to legislate for, and enforce, minimum labour standards. However, it is likely that the impetus is not driven purely by ethics and morality. The wave of industrial unrest in Indonesia during the early 1990s was concentrated in the country's export-oriented manufacturing sector, and particularly in textile, garment and footwear factories. This led to fears that foreign investors, worried at the scale of industrial unrest, would invest their money elsewhere (Hadiz 1993: 188). The motivation, in these terms, was largely one rooted in economic rationality.

The pattern of strikes and resistance which is well documented in countries like the Philippines, Thailand and Indonesia is also becoming evident in Vietnam. There, the Vietnam General Confederation of Labour, with 4.2 million members, is the only legally recognized union and comes under the control of the Ministry of Labour and, ultimately, the Vietnamese Communist Party. It is reported to be highly unpopular among the workers it was established to represent and is regarded as little more than a means to control labour in the interests of foreign investors. A labour code introduced in early 1995 is widely flouted, and minimum wage regulations ignored. The institutional structure of the union places it uncomfortably between labour, capital and the state and it lacks the strength to enforce the code which it helped to draft. As a result, 'wild cat' strikes have mushroomed, often at foreign-owned factories geared to export (Schwarz 1996). In 1993 there were seventeen strikes, and in the first two months of 1994, eleven in Ho Chi Minh City alone (Heng Hiang Khng 1995: 370). The debate over wage levels and factory conditions in Vietnam mirrors that in Indonesia. So while the Ministry of Labour recommended in 1996 that the minimum wage be increased from US$30–35 to US$40–45 per month in foreign companies, other sections of government were worried that this might scare off foreign investors. Tran Dinh Hoan, the Minister of Labour, explained at the time: 'Vietnam cannot set its minimum wage higher than other regional countries . . . Otherwise foreign investment will not come to Vietnam but will go elsewhere' (quoted in Schwarz 1996: 22). The situation in Cambodia is similar. Average monthly salaries are US$35, conditions are poor and dangerous and often contravene the country's labour laws[21] and yet Haing Sitha, director of the Ministry of Social Welfare, Labour and Veterans' Affairs labour inspectorate, warned against prosecution. 'Before we take action, we must think carefully about the possible consequences. If we punish them too severely we risk losing these investors' (quoted in Vittachi 1996: 15).

## Conclusion

If conditions are so poor, wages low, and government labour laws commonly flouted, the obvious question is why do workers continue to fill – with alacrity – jobs in the burgeoning industrial sector. There seem to be four main arguments employed when it comes to answering this question. First, and most obviously, there are those like Helen Lok and Hal Hill who maintain that although wages may be low and conditions sometimes poor, they are usually better than in agriculture (Hill 1991; Lok 1993). This question of remuneration in agriculture versus industry has been discussed at some length. Second, and linked to the first issue, manufacturing continues to enjoy conditions of labour surplus: for every thousand workers who are dismissed for labour unrest there are several thousand more willing to sign up for the same pay and with the same conditions. Third, there are those who take a more 'cultural' view of the issue. This line of argument tends to fall into one of two schools of thought. On the one hand there are those scholars who maintain that modernization and educa-tion are creating a population of young men and women who wish to avoid the drudgery of farming, even if wages in farming are the same or higher than those in industrial work (see above). While on the other, there are those scholars who suggest that workers, females especially, are prone to 'self-exploitation'. In this view, a history of subservience and fatalism has created a potential industrial army that is willing to endure poor conditions, long hours and low wages with little protest (see Wolf 1992: 134). The fourth explanatory thread challenges the very premise on which the question is built, arguing that it is essentially misconstrued. Work in industry may be hard, hours long, and the working environment autocratic and domineering, but work in rural areas is also arduous, and workers are not usually free and self-regulating.[22] Thus, Wolf, writing of female factory workers in Java, concludes:

> Although it is undeniable that factory work is exploitative, it is equally undeniable that young women prefer it to their other meager choices. Although factory work organization and discipline are strict and often brutal, female workers perceive factory employment as a progressive change in their lives, not as a gaping, un-healed wound.
>
> (Wolf 1992: 135)

The general tenor of this chapter has been to support Wolf's assessment. There can be little doubt that conditions in many factories are poor, that avenues for resistance and protest are circumscribed, and that remuneration is, at times, so low as to be reprehensible. Yet in the detail of people's lives it is clear that their choices are limited and in this context factory work offers a critical alternative or supplementary livelihood. Singapore and Malaysia's transition from low-wage locations to (comparatively) high-wage and high-skill production centres marks out the route that governments in the region hope to follow. The fact that Malaysians and Singaporeans have virtually priced themselves out of low-wage employment, sucking in cheaper labour from neighbouring countries to meet the shortfall (see page 126), reinforces the view that foreign investment-led, export-oriented industrialization does not forever relegate countries to a low-wage status.

## Notes

1　This chapter will disproportionately use material based on the Indonesian, and especially the Javanese, experience. This is partly because more work has been undertaken on this issue in Indonesia than in any other country of the region, but more particularly because the quality of the work is generally higher.

2　This question does not just apply to modern factory work. It is equally applicable to traditional (cottage) industries in rural areas (see the previous chapter).

3　Note, though, that factories are increasingly locating in rural areas (see Chapter 5).

4　It does not always follow that workers in factories have to be paid more if they are to match the value of agricultural wages. Chantana Banpasirichote, for example, notes that workers in textile, electronics, plastic products, and automotive part factories in Chachoengsao province south of Bangkok receive a wide range of non-pecuniary and pecuniary benefits: transportation, social security, some household utilities (such as electricity), health and education, for example (1993: 36).

5　Elmhirst's study of factory work in a Hong Kong/Malaysian-invested knitware enterprise in Tangerang, Java provides the following data on wages and costs: Daily wage: Rp15,000. Canteen costs: Rp1,000/day. Rental of a room in a hostel: Rp30,000/month. Daily wage after subtracting board and lodging: Rp13,000. The wages in this factory, where women receive three months of training, seem to be comparatively high by Javanese standards (Elmhirst 1995a: 19).

6　White notes a paradox in the findings of his research team: children working for even lower wages than the young adults 'all . . . contributed at least part of their wages to their parents' (1993: 135). Generally these children were from poorer families, indicating perhaps that in their case there was a necessity to remit money; while in the case of the young adults, the source households were effectively sub-sidizing their factory work.

7  This mirrors Mather's earlier study in Java where she noted that wages were barely sufficient to support one person and therefore should be viewed as supplementary to the household budget. The lowest daily wage she recorded – admittedly in the late 1970s – was only Rp150, sufficient to buy just a litre of rice and two bananas (Mather 1983: 8–9).

8  The World Bank for example talks of households as a decision-making unit, as if it is of one mind: 'Households must decide how to allocate their collective labor. . . . Households must also decide who will work and in what activities. . . . House-holds must also decide where to work' (1995: 23 and 25). Increasingly, it is not households that make such decisions, but individuals.

9  If rice is removed from the cost of living index – it has a very heavy weighting in the rural Java index – then the annualized real increase in wages declines from 8 per cent to 4 per cent (Fane 1994).

10  Cassava is widely regarded as 'poor man's food'.

11  Wolf writes that the management of some of the export-oriented factories she studied in rural Java insisted that new workers undergo a dexterity test. In a spinning and textile factory this consisted of 'quickly distributing spools in bottles and then pulling them out again'; in a garment factory, of tying knots in single sewing threads (1992: 114).

12  Akin Rabibhadana notes that managers in canning factories in Samut Sakhon, Central Thailand, 'refuse to take young men from neighbouring areas saying that they are rowdy and therefore undesirable' (1993: 29).

13  Women may also have 'lower expectations about working conditions and of impacts on their health . . . [and] lower "aspiration wages" – that is, they believe that their intrinsic worth as workers is lower than that of men' (Kemp 1993: 20). Lok, pre-senting the employers' case notes that management complain of a lack of discipline and poor work attitude, chatting, slowness, a lack of cleanliness and responsibility, and avoidance of tasks that might require greater effort (1993: 163).

14  Men, though, are employed for those tasks which require physical strength (see Chant and McIlwaine 1995b).

15  It is debatable whether, as an East Asian cultural import, Confucianism should not be used to sustain the argument that Southeast Asian culture is patriarchal.

16  In Cambodia the ILO notes that the number of children working in factories is low and is not a major problem. Child labour tends to be concentrated in rural areas where children work on salt farms, and in brick factories and small gem diggings, or in the urban informal sector where they work as garbage pickers and prostitutes (Fitzgerald 1995).

17  In 1984, there were just 50 inspectors covering the whole of Metro Manila, with responsibility for 30,000 enterprises (World Bank 1995: 77).

18  Although this discussion has focused squarely on the concerns, fears and complaints of workers, dissatisfaction is not purely the preserve of employees. Management also complain. For example, of a lack of discipline among workers (*kurang disiplin* in Java), inefficiency and laziness, a tendency to shy off work and manufacture illnesses, to extend breaks, and go for excessive numbers of bathroom visits (see Wolf 1992: 126–7).

19  This was formerly named the FBSI or All-Indonesia Workers Federation. The FBSI was established in 1973 and was renamed in 1985 (Hadiz 1993: 190). Broadly speaking, the Sukarno period (1950–65) was one of comparative freedom as far as workers' organization and action are concerned; increasingly tight controls on labour

date from the beginning of the New Order. For a review of industrial relations and the labour movement in Indonesia since Independence see Manning (1993c).

20 There was also a spate of strikes during the early 1980s (Manning 1993c: 70).

21 Buddhist Liberal Democratic MP Son Chhay alleged that '[T]he young women who work in these [garment] factories are no different from prisoners', adding that they 'have compared themselves to the many people who were forced to work under Pol Pot' (quoted in Vittachi 1996).

22 Ong, though, in her work in West Malaysia does maintain that young women go from a village environment where they enjoy self-determination in work to a factory environment where male supervisors take on the role of foster fathers or *bapa angkat*, and refer to their employees as their 'children' (*budak-budak*) (1990).

# 7

# RURAL–URBAN INTERACTIONS

## Introduction

A theme running through the last two chapters has been that the rural/agricultural and urban/industrial space economies are increasingly linked by flows of wealth, people and knowledge. However, what has not been discussed in any detail is how these interactions impact on production in rural and urban areas. If rural households are sending members to work in industry, what effect does the loss of labour have on agricultural methods and production? If, at the same time, urban areas are receiving influxes of labour from rural areas, how has this contributed to the character of the city and its activities? Nor is it just a case of labour movements contributing to change; the remittance flows which result from such movements, and which in many instances represent their very *raison d'être*, open up new avenues and possibilities for change.[1] It is the more tangible impacts of such interactions which are the concern of this chapter.[2]

There can be no doubt that both 'rural' and 'urban', and the bits between the two (of which, more below), have been influenced by these evolving and strengthening ties. The respective futures of agriculture and industry are inextricably linked. At a conceptual level much of the discussion has focused on whether such relationships should be viewed as parasitic, symbiotic (generative) – or simply commensalist. Here, although such issues are addressed, it is also questioned whether there is any point in highlighting discrete 'winners' and 'losers' at all.

Much of the emphasis in the literature, such as that on urban bias in development (see Lipton 1977 and 1984), has been on how urban areas and industry impact – often portrayed in a negative light – on rural areas and agriculture. This concentration on the rural makes some sense when it is remembered that the countryside is being transformed from what is usually understood as a formerly 'traditional' state, into some hybrid modern-cum-traditional entity, while urban areas, by contrast, are a creation, in large measure, of the operation of the forces of modernization.[3] In other words, change in urban areas is seen as self-imposed or self-generating, while in rural areas it is imposed and externally-driven. None the less, there is a case that urban areas are no less dependent on rural areas for their vitality and future, than rural areas on urban.

# INTERACTIONS IN RURAL AREAS AND AGRICULTURE

It is common for rural areas and agriculture, or more accurately the people who live in rural areas and work in agriculture, to be portrayed as victims of growth. A great deal of development theory is based on the premise that farming has been neglected and agricultural value extracted for use in urban areas (for an application of such a view to Thailand, see Bell 1996).[4] Looking through the literature on agrarian change, a distinction becomes evident between those studies which regard rural households as unwitting victims of processes beyond their control – pawns on a macroeconomic chess board – and those which view farmers as independent actors in their own right. James Scott's *Weapons of the weak*, based on a study of a village in Kedah, Peninsular Malaysia vividly illustrates that even the poor in rural areas have certain powers. The history of resistance, he suggests, is marked by the:

> persistent efforts of relatively autonomous petty commodity producers to defend their fundamental materal and physical interests. . . . At different times and places they have defended themselves against the corvée, taxes, and conscription of the traditional agararian state, against the colonial state, against the inroads of capitalism, against the modern capitalist state, and, it must be added, against many purportedly socialist states as well.

> (Scott 1985: 302)

Wolf, dealing specifically with 'factory daughters' also criticizes the tendency to accord more personality and animation to capital than to the women that it exploits. They are not shown 'as social agents who think about, struggle against, and react to their own conditions and who can also interpret their own situations', but as part of a puppet-like reserve army (Wolf 1992: 9). McVey expresses a similar view when she remarks on the degree to which Southeast Asia's economic transformation is described in terms of abstractions – capital, world systems, labour, the state, and so on. 'In the presence of these titans', she writes, 'the endeavours of mere humans seem the dithering of ants . . . ' (1992: 8). Partly this victim–actor distinction can be explained in terms of the differing viewpoints, or starting points, of each author: the work of those who see people as actors is often firmly based on, and informed by, village-level research; those who subscribe to the 'victims' perspective tend to adopt a more structuralist stance, beginning from a macro or global view, using this to explain local-level processes. Although this difference in view is, conceptually, quite fundamental, in practical terms the distinction is often rather less marked. Both sets of authors will tend to agree that the wider economy and state act as 'facilitators' or 'enablers' in rural change; and they will both also tend to accept that farmers do have choices and are not constrained to a single strategy dictated by outside events, forces and actors. The difference, then, is one of perspective, inclination and, to some extent, ideology.

## Making and saving labour

The clearest impact on agriculture of rural–urban interactions stems from the loss of household members to non-farm work. How this affects agricultural production, not to mention social and cultural processes and forms, is a key issue in understanding agrarian change in the context of rapidly evolving economies such as those of Southeast Asia. In a sense, almost all the major transformations in agriculture can be viewed as adaptations brought about – at least in part – by labour loss: land abandonment, mechanization, changes in land use, intensification, disintensification, changing tenancy arrangements, environmental degradation, and so on. The following discussion is rooted in a single, comparatively simple, question: how do farmers respond to the loss of an agricultural worker?

In Chapter 5, it was noted that most of the rural population moving into alternative, non-farm activities are young. In other words, households are losing some of their most productive members to alternative work. However, a not insignficant number of studies have shown that in terms of agricultural production and methods, such a loss of labour is insignificant (e.g. Kamphol Adulavidhaya and Tongroj Onchan 1985; Smart 1986). Generally, this is explained in three ways. First, that population growth in the context of limited land is more than making up for the loss of labour to non-farm work. Second, that alternative work is timed to coincide with the agricultural slack period when farm work is unavailable. And third, that because labour is inefficiently used (farm workers are said to be 'under-employed'), a more efficient allocation of labour and a shift to 'full' employment can easily make up for the loss of a farm worker (see Tongroj Onchan 1985: 472).

In many instances it is suggested that the effects of seasonal off-farm work are more pronounced in social and cultural, than in economic terms. The slack season is a period for courting, festivals and fairs, religious devotions, and the telling of tales. It is the season when households and villages collectively renew their bonds and identities. If young men and women are absent from such festivities, this creates the conditions in which the activities either become meaningless, or their meanings and roles change. As Tomosugi writes of Ban Tonyang in Thailand's Central Plains region, with only the elderly remaining, the village is 'bereft . . . of the vitality of youth' (1995: 103). The return of these young men and women may bring further tensions and difficulties. Female migrants may, in particular, find themselves culturally '[s]tranded between the city and the country' (Paritta Chalermpow Koanantakool and Askew 1993: 67). They may regard their period in urban employment as a temporary sojourn, but at the same time their experience of urban life may make them reluctant to marry men who have stayed in the village and whose horizons are so much narrower than their own. 'Whatever option these girls are going to take' Paritta and Askew suggest, 'it is going to be an awkward one' (Paritta Chalermpow Koanantakool and Askew 1993: 67). In addition, and despite the claim that if it is seasonally-timed, non-farm labouring has little effect on agricultural production, there is a strong case that the medium and long-term effects are potentially severe. For the slack season is not just a time for festivities – it is also a time when productive assets are repaired and developed (see 'To lose and regain the land', below).

Although some studies may show that the demands of non-farm work are not eroding the bases of agricultural production, there are many more which indicate that labour shortages, brought on by a shift in rural labour to non-agricultural activities, are pronounced in many areas – and becoming increasingly so. Households have tended to respond to such shortages either by selective mechanization, or by adopting new cultivation methods and crop types, or both. As the discussion in the last two chapters has shown, the impacts of these adaptations are felt unequally and differently according to age, gender and 'class'. In addition, different areas of Southeast Asia have undergone these changes at varying times: rice farmers in Kedah in Peninsular Malaysia, over large parts of the Central Plains region of Thailand, and in areas of Luzon in the Philippines adopted mechanical land preparation (tractors or rotavators) in the late 1960s and 1970s; tractorization spread to Java in the 1980s; it began to make a substantial impact in Northeast Thailand from the mid-1980s; and it was evident along the Mekong Valley in Laos from the early 1990s.

## Mechanization

Rosamond Naylor's study of 80 villages in East, Central and West Java undertaken in 1990, examined trends in technology adoption and labour use in rice production (Naylor 1992). Naylor argues that economic conditions in Java fundamentally altered during the 1980s, making mechanization attractive to many farmers for the first time. High labour costs and a particular shortage of male labour for land preparation, driven by the growth in non-farm activities, created a demand for mechanical land preparation and therefore the conditions for the emergence of a rental market in rotavators. In 1981 there were 0.82 rotavators per 1,000 hectares of harvested wet rice in Java; in 1988, there were three times this number (2.44). In Otsuka *et al.*'s study of farms in Luzon, the Philippines, the proportion of farmers using tractors rose from 11 per cent in the 1966 wet season to 95 per cent by 1990. The percentage using threshers rose from 77 per cent to 100 per cent over the same period (Otsuka *et al.* 1994: 89). Wong, in Peninsular Malaysia, saw the proportion of the rice crop mechanically harvested in her research area increase from 0 per cent to 70 per cent over just two years in the late 1970s (Wong 1987: 214). Similar stories of labour-deficit-induced mechanization have been reported from across the region.

The differential effects of mechanization on different classes of farmer, and on men and women, are considerable. However, there are a number of difficulties in interpreting the causes and effects of mechanization. One difficulty lies in understanding what those effects might mean for the poor as opposed to the

*Box 7.1* Klong Ban Pho: a case study in rural interactions and agriculture from Thailand

Chantana Banpasirichote's study of the effects of urbanization and indus-trialization on Klong Ban Pho in Chachoengsao province, east of Bangkok in the Central region, reveals a community that has witnessed no less than four economic transformations in little more than a decade. Before 1980, Klong Ban Pho was a traditional cash-crop and subsistence economy based on rice; from 1980 through to 1988, incomes were increasingly supplemented by new forms of farm production and non-farm work; in 1988, shrimp farming and factory work began to expand rapidly; and from 1991, while shrimp farming contracted, factory work continued to grow in significance (Chantana Banpasirichote 1993: 6).

By the early 1990s, labour shortages in the village had become severe as a result of the expansion of non-farm work and the drift of younger men and women into off-farm occupations. Rice farming, once constrained by a shortage of credit and land, became constrained by a lack of labour. Unlike some other parts of Thailand, like the Northeast, where off-farm work is often timed to coincide with the agricultural slack season, in Klong Ban Pho only a handful of migrants returned home to help during the rice growing season. In response to the shortage of labour, households were forced to adapt their cultivation methods. Buffaloes were replaced by mechanical rotavators (known across Thailand as 'iron buffalo' or *kwai lek*), and when the cost of hiring labour reached 100 baht/day, households switched from transplanting their rice, to broadcasting – a far less labour-intensive method of cultivation, but one which also produces lower yields.[a] Rice harvesting machines and mechanical threshers were also introduced to save labour. Chantana argues that these changes have been embraced independently of officials' efforts; 'Klong Ban Pho people respond cooperatively, but with indifference [to government policies]' (page 50).

Despite the declining importance of agriculture in the village economy, households engaged in non-farm work still endeavour to keep and farm their rice fields. This, though, does not apply to the younger generation who saw their futures lying in factory work. 'The local people themselves', Chantana writes, 'foresee a time when land is left idle' (1993: 11). In what the author describes as the 'worst cases', farming knowledge is not being passed on to the younger generation and they do not know how to manage a farm.[b] But this is not just a case of children being at odds with the wishes of their parents; parents, too, are supportive of this change in livelihood. In many areas, mothers and fathers are willing to sacrifice agri-cultural production to enable their children to continue with their education past primary level.[c]

*Source*: Chantana Banpasirichote (1993) *Community integration into regional industrial development: a case study of Klong Ban Pho, Chachoengsao*, Bangkok: TDRI

continued . . .

---

Notes

a In 1993, the minimum wage in Bangkok was 125 baht/day, while the going rate for a day's agricultural labouring in the Northeast was 50–60 baht/day.

b The use of 'worst' by the author implies that the development is to be decried. This says more about the desire of an outsider to see the traditional economy maintained than it does about the logic of such a wish from the villagers' perspective.

c This was true, for example, of the hill families in Thomas Enters' study community in a comparatively remote area of Mae Hong Son province in Northern Thailand (personal communication, 1995).

---

wealthy, and men as opposed to women. A second difficulty lies in knowing whether labour shortages, brought on by the increase in non-farm employment, have induced mechanization, or whether mechanization has displaced labour thereby forcing agricultural workers to look for employment in the non-farm sector. And a third difficulty concerns whether the 'blame' should be apportioned to the technology *per se*, or whether attention should instead be directed at the social and economic environment within which the technology is being adopted. Much of the evidence on these key questions is at odds.

Some studies show that the gender implications of technological change in agriculture have been to marginalize women. For example, the use of combine harvesters and mechanical hullers in rice cultivation and processing in Malaysia has tended to displace women from activities that were traditionally their preserve. Such a process either forces women to look beyond the village for employment, or to retreat into housework (see De Koninck 1992: 109–21, 177–8).[5] Parnwell and Arghiros, likewise, but with reference to the Thai experience, argue that mechanization has seen a decline in the importance of female labour in agriculture as mechanical innovations have become the sole preserve of men. As a result, a process of 'masculinization' of agriculture is occurring with 'a concomitant decrease in women's knowledge about rice production and technology', a decline in their status in rural communities, and an increase in women's dependency on men (Parnwell and Arghiros 1996: 21–22).

However, there are also studies which show that the emerging sexual division of labour under the effects of mechanization indicates quite the reverse. For example, in Laos, the mechanization of some aspects of rice production has selectively lessened the work load of men. Tractors and cultivators, particularly, have reduced the time involved in land preparation, a traditionally male-oriented task in Laos, as it is elsewhere in the region.[6] This has given men the chance to look for work outside agriculture, but not women, leading to a 'feminization of agriculture' and the subsidization of the reproduction of the non-farm workforce by women's subsistence production (Trankell 1993: 82; see also Schenk-Sandbergen and Outhaki Choulamany-Khamphoui 1995). Taking a structural stance, it is possible to argue that in this way agriculture is providing a subsidy to industry. From the farmers' point of view, however, the attraction of

*Plate 7.1* Ploughing a field the traditional way. This sight is rarely seen in Malaysia and increasingly rarely in Thailand too (where this picture was taken). Instead land is prepared using machinery, either a rotavator or a four-wheeled tractor

*Plate 7.2* The ubiquitous rotavator – or 'iron buffalo' as it is known in Thailand. Mechanization in rice production is transforming patterns of work in the countryside. Mechanization of land preparation saves labour when workers are scarce and expensive, and when many people have non-farm obligations to meet

such non-farm employment is that it diversifies and therefore stabilizes livelihoods (see the discussion of this on page 197) in the face of fluctuating economic
conditions (Hart 1994: 60).

It is possible to make some sense of this discrepancy in terms of the temporal
pattern of adoption of *different* mechanical innovations.[7] Generally, the first
forms of machinery to be adopted are cultivators, tractors, rice mills, and hullers.
These innovations have the greatest impact on male work loads (see for example,
Trankell's study from Laos) – although mechanical innovations connected with
post-harvest operations reduce female work loads. Only later are machines like
harvesters and transplanters adopted, which tend selectively to reduce female
work loads in agriculture (viz. De Koninck's study from Malaysia) (Figure 7.1). It
is usual to observe, none the less, that a feature of most mechanical innovations
– whether they be hullers, tractors or combines – is that they are controlled by
men. However, even this general principle is being challanged by the forces of
social and cultural change and the exigencies of cultivation in a context where
male labour may be scarce. In the Central Plains of Thailand, for example, not
only has the buffalo disappeared but it is not uncommon to find women using
rotavators – and effectively doing what has always been regarded as 'men's work'
(author's own research, 1996).[8]

*Combine harvesters: what do they do, and to whom?*

More than any other form of mechanization, combine harvesters are seen to
have the most pronounced effect on equity – in Kedah they have been referred
to as *mesin makan kerja*, 'machines that eat work' (Scott 1985).[9]

De Koninck and Wong, both working in Kedah, West Malaysia, observe that
the mechanization of harvesting has effectively disqualified women from much
of their agricultural work (De Koninck 1992: 116; Wong 1987: 123–5). The
introduction of the combine harvester displaced the need for women's work in
harvesting the crop – a largely female-dominated activity. Although other jobs
did multiply, in particular the need for people to bag and then transport the
harvested grain from the fields, along narrow tracks, to the village, this required
people with muscles and motorcycles. So although mechanization did create
new labouring opportunities, these were male-dominated. De Koninck, undertaking his fieldwork a little later than Wong, notes that in time even these
new jobs disappeared. The harvested crop began to be dumped by the roadside
rather than in the village and then, shortly after this, villagers worked out a
way to transfer the harvest direct from the combine harvester to the lorry. In
the competition between labour and capital, this pattern of change might
lead one to believe that, over time, labour lost out to capital. However,
farms have not been capitalized in the strict sense, for much of the machinery
is rented on a piece-work basis, not owned. This, indeed, is true of most studies
of mechanization in the region and includes not just expensive machines like
combine harvesters, but also comparatively cheap ones like the two-wheeled

*Plate 7.3* Post-harvest mechanization in the province of Nan, Northern Thailand

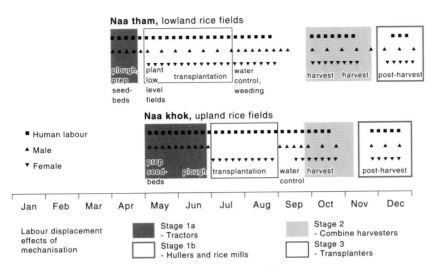

*Figure 7.1* Agriculture, the sexual division of labour and mechanization in Laos
*Source*: adapted from Trankell 1993.

tractor (the kubota).[10] Of the process of mechanization and its effects on women, De Koninck writes:

> Within the padi sector, where profit-making relies increasingly on the ownership and use of machinery, women are marginalised. If there is any redistribution [of benefits], it is among men; women are being virtually ousted from the fields.
>
> (De Koninck 1992: 120–1)

A key point that De Koninck makes, and one that James Scott's work also endorses, is that the agricultural transformations in Malaysia, and beyond, not so much exploit the poor, as simply remove them from the production process (De Koninck 1992: 189).

Does mechanization displace agricultural workers, or does a lack of agricultural workers – due to the loss of available labour through migration and the growth of non-farm work – stimulate mechanization? Here, again, the evidence is at odds. In Wong's village, there was a real shortage of men to carry out the task of threshing, and she believe that this shortage probably pre-dated the introduction of the combine harvester (Wong 1987: 124). Arghiros and Wathana Wongsekiarttirat's study of agrarian change in the Central Thai province of Ayutthaya also indicates that it was pressures resulting from changes in the non-farm sector that stimulated mechanization in the farm sector. Small and large land-owners alike found it difficult to procure labour at the right time, and farm labourers, with alternative non-farm employment beckoning, were demanding ever higher wages (Arghiros and Wathana Wongsekiarttirat 1996: 129). This is not to say that such a scenario can be applied to all groups. Wong suggests that the women in her Malaysian case study, unable to take advantage of non-farm work in the same way as men for cultural reasons, found that the introduction of the combine harvester did displace them from work. The poor were marginalized, but particularly the female poor. Pauperization, in these terms, took on a distinctly female hue. For De Koninck, though, the same story, in the same area of West Malaysia had – for the women of wealthy and middle households – slightly different implications. For these women, the combine harvester allowed them to retreat into the family unit, to be freed from the drudgery and stigma of farm work (De Koninck 1992).

Tractors, combine harvesters and threshers individually all tend to perform similar productive functions whether they are operated in Muda or Mindanao. But the evidence clearly shows that such are the variations between rural areas of Southeast Asia that to assume that they will also have similar effects on individuals, household and communities would be excessively reductionist.

### Changes in cultivation practices

Farmers can also save labour by adapting their cultivation methods. Perhaps the most striking such adaptation is the gradual disappearance over large areas of transplant rice culture in favour of broadcasting (direct seeding). This is often regarded as a retrograde step. Transplant rice culture is perceived as representing the pinnacle of rice agriculture; at once highly productive and sophisticated, the example *par excellence* of nature harnessed and manipulated in the interests of humans. Broadcasting has always been popular, especially in more marginal rainfed areas of Southeast Asia where water availability and yields are highly variable and farmers wish to minimize the labour they invest in cultivation. However, the recent spread of broadcasting to lowland areas, traditionally regions where transplant rice culture has predominated has, like mechanization, been driven largely by the desire to save labour and is being adopted even on the best land and the smallest plots (Table 7.1) (see for example, Tomosugi 1995: 63; Otsuka *et al.* 1994; Wong 1987: 214; Rigg 1995b; Hart 1994: 54). The importance of labour shortages in this process of change is illustrated by the emergence of 'weekend broadcasting' in some areas of the Central Plains of Thailand and West Malaysia. Farmers with full time commitments in non-farm work and no other available family members to fill the deficit find that the only way they can maintain cultivation without hiring labour and becoming absentee landlords in all but name is through saving labour at any opportunity and becoming weekend farmers.

Rigg's study of two villages in Mahasarakham, Northeastern Thailand revealed individual farmers both intensifying and disintensifying production – simultaneously (Rigg 1995b). Labour supply in the area was becoming increasingly stretched as men and women moved into alternative occupations. But rising incomes, largely generated from non-farm sources, allowed farmers to invest in yield enhancing technologies such as new high-yielding seeds and fertilizers. The strategies adopted reflected factor resource scarcities at the household level.

*Table 7.1* Labour use in rice agriculture (hours/hectare)

|  | Non-mechanized | | | Mechanized | |
|  | Transplant rice culture | | Broadcast rice culture | Transplant rice culture | Broadcast rice culture |
|  | Java | Thailand North | Thailand Central | Java | Luzon |
|---|---|---|---|---|---|
| Land preparartion | 300 | 179 | 58 | 125 | 56 |
| Sowing/nursery culture | – | 137 | 22 | – | – |
| Planting | 250 | 190 | – | 250[a] | 208[a] |
| Weeding/water management | 475 | } 126 | } 15 | 475[a] | } 40[a] |
| Fertilizing | 30 |  |  | 30[a] |  |
| Spraying | 30 |  |  | 30[a] |  |
| Harvesting/post-harvest | 375 | 451 | 376 | 200[b] | 200[b] |
| Total | 1,460 | 1,083 | 471 | 1,110 | 504 |

*Sources*: Java: Naylor 1992: 73; Thailand: Tanabe 1994: 216; Luzon, Philippines: Otsuka *et al.* 1994: 91

*Notes*
a Non-mechanized
b Sickle harvest, mechanical threshing

Farmers were saving labour by broadcasting their seed; saving land by raising yields; and increasing output by raising inputs.

The attraction of pre-emergence herbicides to farmers in many areas is not because they may enhance yields, but because they cut the need for weeding by 50 per cent (Naylor 1992: 87). Likewise, the availability of modern varieties (MVs) that have high seedling vigour, allowing them to compete with weeds, and the development of selective herbicides that can be used to retard weed growth on broadcast fields, have also allowed farmers to save labour. A cogent and convincing case exists that technologies and cultivation practices from rotavators to herbicides are attractive to farmers not so much because they might increase yields, but because they allow farmers to save labour in the face of non-farm induced labour shortages. In other words, the pressures and incentives driving agricultural change in Southeast Asia are as much, if not more, to do with extra- than intra-village factors.

## Too busy to farm?[11]

In villages where farming is only part of complex household livelihood strategies and where a range of opportunities exist for a more adequate living to be made other than from the land, it becomes increasingly necessary to use some land less than before.

(Preston 1989: 43)

Java is one of the most densely populated and intensively farmed areas of the world. 'Agricultural' population densities exceed 2,000 people per square kilometre in some areas and the average for the island is over 800.[12] In 1992 the yield of wetland paddy was more than 5 tonnes per hectare (BPS 1993). The experience of the Green Revolution in Indonesia, and the achievement of rice self-sufficiency (see Booth A. 1985 and 1988), tends to support the view that Java is a case of Boserup vindicated; of population pressure stimulating and supporting agricultural innovation, intensification and, therefore, the increase of output.[13] However, some studies indicate, perversely, that in some areas dis-intensification is occurring. Nor, for that matter, is this process restricted to Java: it is evident in Malaysia and Thailand, too.

In Preston's study of central Java, annual cultivation of some land had been replaced by cultivation in alternate years. Home gardens tended to be planted with labour-saving fruit trees rather than annual crops,[14] and dry season cropping of some riceland had stopped and the land left fallow (Preston 1989). In Rigg's study of Thailand noted above, upland formerly planted to annual crops such as cassava and kenaf (an inferior jute substitute) had been converted to labour-saving eucalyptus (Rigg 1995b). Rather more drastic is the picture revealed in Courtenay's (1988) and Kato's (1994) work on West Malaysia where there has been a widespread abandonment of rice cultivation altogether. By the end of 1986, rice land classified as 'permanently idle' in Peninsular Malaysia constituted 22 per cent of total alienated padi land (Courtenay 1988: 20) (see Box 7.2).

Preston wonders whether, in this evidence of disintensification and even adandonment of cultivation, 'we may well be witnessing an early stage in the development of . . . a major element of rural land use change' (1989: 45). In each case, the motivation for the change in cultivation methods is the need to strike a balance between the demands of agriculture and those of non-farm activities, where the 'balance' is the state which produces the greatest household benefit taking into account the totality of farm and non-farm opportunities. 'Benefit' here may not mean maximum economic return because any decision would also incorporate such factors as stability, enjoyment and gratification. In such a context, agriculture is in competition with industry for labour and the standard view that, given adequate conditions of population density and market relationships, farmers will intensify production and increase productivity (see for example Netting 1993: 27) is far from certain. The process of disintensification also coincides with, and is supported by, the growing wish among young men and women to avoid the drudgery of farming, even if it means accepting a lower wage (see Chapters 5 and 6). It is possible to say, then, that livelihoods are not so much being squeezed by inadequate farmland, but rather farming is being squeezed by alternative livelihoods.

But the sight of idle land in otherwise high pressure areas may not, necessarily, be an indication that other demands are drawing people off the land. In the Chiang Mai Valley in Northern Thailand it was estimated at the beginning of the

*Box 7.2* Out-migration and idle land in Peninsular Malaysia

Of the countries of Southeast Asia with a significant agricultural sector, Malaysia is the most advanced. The experience in Peninsular Malaysia has been of continued migration from rural areas, falling agricultural production, and increasing expanses of idle land. At the end of 1986, 90,000 hectares of paddy land was classified as permanently idle, while 432,000 hectares were cultivated. The Malaysian government clearly put this down to the 'rapid migration of rural youths to urban areas as a result of the [financial] attractiveness . . . and glamour of working in non-agricultural sectors' (quoted in Courtenay 1988: 20). Employers in the plantation sector, in agriculture more widely, in construction and in the low-skilled manufacturing sector talk of a 'labour famine' – a famine which Thais and more especially Indonesians have ameliorated but far from eradicated.[a]

There has been some debate over whether this unused paddy land should be described as 'idle' or 'abandoned'. Idle rice land, or *sawah terbiar*, is land which has been continuously uncultivated for more than three years. In the government's view this is the unintentional result of farmers' actions and the use of the term *terbiar* ('idle') 'assumes that peasants are still essentially committed to rice cultivation' (Kato 1994: 166). However, for the state of Negeri Sembilan, Kato argues that the land should really be regarded as abandoned as there is little indication that farmers intend to cultivate it in the future.

Courtenay's study of *mukim* (sub-district) Melekek north of Melaka supports Kato's conclusions. He shows that although reports of idle land in the area date from even before the 1970s, by the mid- to late-1980s, 43 per cent of paddy land was idle (?abandoned). In addition, the average age of heads of household had risen to 60 years, the population of the *mukim* had declined by a third between 1970 and 1986, and 75 per cent of incomes came from pensions or remittances. For just one household in the sample could agriculture be considered the major source of income. This process of out-migration and decline occurred in spite of concerted government support for agriculture and rice production. Furthermore, the households who were giving up farming were not willing to sell their land, notwithstanding its increasingly marginal contribution to household income. Courtenay concludes: 'A likely future scenario for *mukim* Melekek, and others like it in Melaka State, is for out-migration to continue, for all padi land to be used for alternative crops on some form of amalgamated farm or estate, or indeed for non-agricultural use, and for the villages themselves to function more as places of retirement or as dormitory settlements for residents working elsewhere in the State' (1988: 27).

*Sources*: Courtenay 1987 and 1988; Kato 1994

Notes

a Hugo estimates that somewhere between 500,000 and one million Indonesians work in Malaysia – which has a labour force of well under ten million (1993).

1990s that one third of cultivable land had been bought by land speculators and left uncultivated (Enters 1995). These plots were not even rented to land hungry farmers because of the owners' fear that this might make sale more difficult (Thomas Enters, personal communication, 1995). Around Bangkok there is evidence that land speculation and development has had serious side-effects for those families who are still committed to farming (see Ross and Anuchat Poungsomlee 1995: 138). As land is taken out of production pest infestation increases on the idle land, new housing estates and other developments may alter the drainage characteristics of an area, while factory waste and domestic sewage raise the risk of pollutants entering the agro-ecosystem. Thus speculation and land development may take on a momentum of their own as households who might wish to continue farming are encourage to sell up as land use changes undermine production.

Making the dangerous assumption that rural areas in the various countries of the region will pass through similar stages of development, the experience of Malaysia may be a mirror for the future of Thailand, Indonesia and elsewhere. Kato, basing his conclusions on work in the Malaysian state of Negeri Sembilan, identifies two key trends: 'de-agriculturalization' and 'de-*kampong*-ization' (see also Box 7.2). He writes:

> Malays of Negeri Sembilan, especially the young, are simply not interested in agriculture. Single girls prefer factory work to rice cultivation. . . . Unemployed young men are not interested in rubber tapping unless compelling circumstances . . . impel them. . . . A strange scene now exends across the countryside of Negeri Sembilan. Former *sawah* [wet rice fields] are covered with bushes and shrubs; rubber smallholdings too are sometimes full of thick undegrowth. . . . Yet one sees many large and beautiful houses.
>
> (Kato 1994: 167–8)

The riceland is untended because the young no longer wish to farm. The land is abandoned because households can afford to forgo agricultural production. The houses are large and opulent because men and women have jobs in local towns. And the average age is rising because people retire [back] to the village while the young are working elsewhere. The honourable Malay peasant toiling in the ricefields in the sun, the symbol of Malaysia which the government seems so loath to abandon as farmers have done their fields, is rapidly becoming history.

### To lose and regain the land: environmental degradation and agrarian change

The emphasis in the literature since the mid-1980s on sustainable development, demands that some attention be given to the question of the associations between agrarian change and environmental degradation. Many explanations of environmental change suggest a sequence of processes along the lines of:

> commercialization + > modern technology + < land + > population = > environmental degradation

Although such a view would fulfil received wisdom, the evidence suggests that the links between agrarian and environmental change are somewhat more varied. In a sense, this is to be expected: as rural livelihoods diversify, so it is becoming increasingly difficult to generalize about rural areas, farming systems, or people's livelihoods – and therefore their various impacts on the environment. Pretty (1995: 69–70), in a summary of the range of factors that lie behind soil and water degradation, suggests that farmers may:

- lack knowledge and skills
- lack the incentive to invest in conservation efforts if the economic costs of conservation outweigh the expected benefits
- lack the necessary labour to construct or maintain structures, both social and physical
- be hamstrung by the misguided efforts of previous conservation programmes
- be so preoccupied by and responsive to policies designed to increase productivity that they ignore the environmental costs

But the articulation between agrarian change and the environment need not be one-way. In other words, there is evidence from the field to indicate that the simple equation given above can be reversed. Allen, for example, in reviewing the literature notes that the 'evidence on sustainability of . . . upland agricultural systems is contradictory', adding that '[i]n some areas, improved access and incorporation into the market economy have apparently brought about the rehabilitation of previously degraded areas, while in other locations, the "mining" of forest and soil resources has continued unabated' (Allen 1993: 233).

The Southeast Asian region has seen a fundamental shift from land abundance to land scarcity and, in many areas, a concomitant move from extensification to intensification.[15] On richer agricultural lands this intensification has seen a sustained increase in yields founded on the application of new technologies.[16] However, in marginal areas, agricultural systems are degrading the land resource and reducing productivity. Continued population growth and settlement of marginal areas, combined with rising needs, has led farmers to cultivate their fields without employing proper land-use practices. In the 1970s and 1980s in the marginal Northeast of Thailand, for example, there was abundant evidence that farmers were 'mining' the uplands by cultivating nutrient-demanding crops like cassava with the use of virtually no inputs or land conservation measures. In the Tengger Highlands of East Java, fallowing of the land, the traditional means of replenishing soil fertility, had almost ceased by the mid-1980s. Farmers, Hefner writes, 'just cannot afford it' (1990: 105). Intensive vegetable cultivation on slopes of 40 per cent or steeper is causing severe erosion. Potatoes are triple-cropped, using deep hoeing cultivation practices and the once deep and rich volcanic soils are being stripped down to bare rock and clay (Hefner 1990:

104–12). In Blaikie's terms, this type of development is indicative of the 'desperate ecocide of the poor' (Blaikie 1985; see also the papers in Blaikie and Brookfield 1987). People living on marginal lands tend to be poor; they lack the resources to invest in their land; and are forced, from necessity, to degrade the resource on which their lives depend. But it is not only the poor who, from necessity, embrace such environmentally destructive methods. Hefner points out that on the Tengger Highlands it is the large, just as much as the small landowners who are engaged in these environmentally disasterous methods. More sustainable techniques would be feasible on their holdings, but this 'would undermine the social gains [the larger farmers] rightly feel they have earned, and would relegate them once more to the status of backward uplanders' (1990: 111).

However, there is emerging evidence that this picture of poor people being caught in an ecological trap from which they cannot escape is excessively deterministic.[17] They may be poor; there may be a degree of inertia built into the system; they may lack knowledge of new crops and methods; but they are not – necessarily – powerless. Rigg's work, for example, has shown how farmers in the Northeast of Thailand have adapted their production methods, increasing the use of organic and artificial fertilizers and sometimes reafforesting their land with indigenous and exotic species (Rigg, field research 1994). Likewise, the work of Nibbering in the uplands of Gunung Kidal spanning the provinces of Yogyakarta and Central Java, shows farmers successfully adapting their methods following rapid population growth. He writes that: 'the destitution of the population and the seemingly hopeless nature of the environmental situation some thirty years ago did not hinder but rather generated a drive for change' (1991: 130).[18] A similar process has been identified in the uplands of Cebu in the Philippines, an area of high population density (520 people per square kilometre in 1990) where severe erosion was identified as a problem as early as the mid-1930s. Farmers here have responded by adjusting their cultivation methods, embracing new crops, agro-forestry systems, contour ploughing, rock walls and hedgerows. These creative responses from farmers have, Kummer et al. argue, been largely overlooked in the analysis of the area's environmental history. 'There are reasons to believe', they write, 'that the upland environment of Cebu is actually improving' (1994: 275).

But perhaps the best example of such pressure-induced change comes not from Southeast Asia, but from the Machakos District of Kenya. Tiffen et al.'s study describes how farmers have regained the land in the face of rapid population growth and agricultural commercialization, averting – indeed reversing – an environmental crisis. Through spontaneous farmer response there has been a 'reversal of degradation . . . rising productivity and living standards, and [the] successful exploitation of lands previously deemed unfit for agricultural use' (Tiffen et al. 1993: 261). For Tiffen et al., the Akamba people were not caught in an environment–poverty impasse, but were instead 'engaged in an on the whole successful struggle with their physical and socio-political environment, [making choices] about the use of their talents and their assets' (1993: 265).

Non-farm activities need not play a determining, or even a major role in these adaptations. However, if livelihoods are increasingly met from other sources than agriculture then this may ease pressures on agricultural systems, permitting farmers to cultivate their land less intenstively. This certainly appeared to be the case in Rigg's two villages in Mahasarakham province, where alternative income generating activities had allowed farmers the opportunity to plant their uplands not with degrading annual crops, but with tree crops (Rigg, field research 1994).[19] In this instance, the move into tree crops also made economic sense given the growing shortage of labour in the area (discussed above). Farmers may not just farm less intensively; they may also invest income generated from non-farm work in land improvement. This was true, for example, of Nibbering's study mentioned above (see also Allen 1993: 233). Thus the diversification of the farm economy may induce and/or permit farmers to either farm less intensively or farm more sustainably, or both, by financing a shift to longer term, more ecologically balanced methods of cultivation. It is also important to stress that such adaptations do not depend on government assistance and 'expert' advice; in many cases farmers have embraced such changes spontaneously and on their own initiative.

But diversification is not always environmentally-friendly. There are also studies which show that when farmers engage in non-farm activities (and can rely on alternative sources of income) they tend to neglect agriculture. The 'slack' season, despite the terminology, is not a time for idleness or entertainment. It is a period when bunds and dykes are repaired, houses built, rice barns patched, farm implements made or repaired, terraces restored, and ditches dredged. There is considerable work which indicates that when these jobs are neglected, whether that be because individual households are in labour deficit or because village heads find it impossible to muster the necessary support for community projects, there occurs a long-term deterioration in the productive base and a consequent decline in output (see for example, Parnwell 1986: 110). Among the Ifugao of Banaue in Luzon, who have been carving terraces out of the mountain slopes for 2,000 years, this neglect is destroying one of the wonders of the region. Alternative occupations mean that survival no longers requires that the Ifugao challenge nature in this awe-inspiring way; the paddies and terraces are being abandoned, and without constant care they are quickly deteriorating. Somewhat ironically, it is the local Department of Tourism which is at the forefront of the effort to encourage the Ifugao to preserve their terraces, and their heritage. The motive is clear: these creations of the Ifugao have become important local tourist attractions (Rhodes 1994). Although the rice terraces of Banaue are a dramatic example of environmental degradation driven by neglect, most instances of decline are less obvious. The fear is that they will therefore tend to be overlooked, and yet the productive implications – particularly when irrigation systems and terracing are allowed to degenerate – are enormous.

Environmental deterioration induced by commercial pressures at the hands of which farmers are powerless, is taken to be self-evidently 'bad'. Yet in some places and for some people, such changes are viewed by those affected as beneficial. Lewis, working in the mountains of northern Luzon, the Philippines, for example, writes that '[e]nvironmentally, Buguias is a disaster area' (p. 79) due to the wholehearted adoption of commercial vegetable farming; yet, just a page later, is able to add that ' . . . the *entire* Buguias community fully endorses the postwar transition from subsistence cultivation to market gardening [emphasis added]' (Lewis 1992a and see Lewis 1992b). The reason for this is clear: people are infinitely better off. Under the old regime, a woman explained to Lewis, 'life was terrible – we only ate sweet potatoes' (1992a: 80).[20]

The pressures of commercialization and the demands of intensification have brought tensions to bear on environmental systems, just as they have on populations. How these tensions work themselves out through time, in environmental and human terms, is the critical issue. In terms of livelihoods it is not self-evident that environmental degradation is necessarily destructive. Equally, in terms of environmental systems it is not self-evident that degradation is self-reinforcing and irreversible. Studies have shown that farmers are adaptive and that environments are sometimes more resilient than expected. Further the role and place of non-farm activities in changing our assumptions about the 'logic' of environmental degradation reflected in the equation at the beginning of the section is growing in importance.

## INTERACTIONS IN URBAN AREAS AND INDUSTRY

> The prosperity of Bangkok is supported by rural villages in Thailand, of which one is Tonyang Village. Without this support Bangkok could not thrive.
>
> (Tomosugi 1995: 102)

The discussion so far has focused entirely on the impacts of rural–urban interactions on rural areas and people. But there is a strong argument that the economic success and industrial vitality of the market economies of Southeast Asia has been predicated on the movement of millions of country people to cities and towns. Given that the literature on rural–urban migration has tended to argue that rural areas 'suffer' from a loss of human capital as the young and more educated and motivated are lost to urban employment, the obverse would seem to be that urban areas and industries are selectively gaining from such movements (see Rigg 1988c; Tomosugi 1995: 103). Koppel and James suggest a simple distinction between those countries which have supported agriculture as a means towards achieving a successful transition to an industrial economy (the so-called East Asian model), and those that have exploited agriculture to finance industrialization. The former group includes Indonesia, Thailand and Malaysia, while the most obvious members of the latter group are Myanmar (Burma) and the Philippines which both adopted import substitution industrialization strategies (Koppel and James 1994: 286–90).

Micro-studies indicate that rural people are not absorbed into the urban economy in a random and undifferentiated manner (the proverbial 'melting pot' scenario). Many play a distinct and identifiable role, although this should, of course, not be taken to mean that the migrant–resident division in terms of employment is immutable. The problem, and this was discussed in Chapter 5 (see page 159), is that it is often difficult to decide when a 'rural migrant' makes the transition to 'urban resident'. None the less, there are some sectors where rural people make a visible, and highly significant, contribution (Table 7.2). These include casual labour, unskilled/labour-intensive factory work, the informal sector, and domestic service. A set of defining characteristics of all this work is that it is poorly-paid, rarely salaried, offers little opportunity for advancement and promotion, and conditions are often harsh and/or dangerous. Illegal labour migrants too are often absorbed into the labour force in a selective and identifiable manner (Table 7.3). It has been argued that migrants are willing to tolerate such a situation for the very reason that they are migrants sojourning in the city for just a few weeks or months (e.g. Paritta Chalermpow Koanantakool and Askew 1993: 67). Workers are often in the city with the primary intention of earning money to remit or take back to their rural households.[21] The same is true of labourers who cross international frontiers in their search for work. In Indonesia, an important distinction in factory employment is between 'employees' or *pegawati*, and 'workers' or 'labourers', known as *pekerja* or *buruh* (Kistanto 1991: 299). It is the latter role that rural people fill and, to some commentators, they are just *petani* (peasants) in the city; country bumpkins on the make.

*Table 7.2* Characteristic roles of rural people in the urban economy

**Informal sector**
- *Vendors, petty traders*: 70 per cent of street traders in Manila are migrants to the city, and about half of are temporary sojourners (Aguilar 1989:8)
- *Garbage pickers* (scavengers) and *junk buyers* in Hanoi: more than 50 per cent come from Xuan Thuy district in Nam Ha province, south of the capital (Hiebert 1993)
- *Drivers of becaks* (Indonesia), *saam lors* (Thailand), *cyclos* (Vietnam, Cambodia): all variations of tricycle taxis. Up to 40 per cent of the transportation in Bangkok is manned by seasonal migrants (Askew 1993: 28)
- *Foodstall operators*

**Casual sector**
- *Construction workers* in Bandung (Firman 1991) and Jakarta (Sjahrir 1993): the 'majority' of *tukang* (manual workers) in the construction industry in Jakarta are not natives of the capital; most come from rural areas of Central and East Java and were formerly wage labourers in wet rice agriculture (Sjahrir 1993: 245)
- *Security guards*
- *Domestic servants*

**Formal sector**
- *Workers in labour-intensive, low-skilled, export-oriented manufacturing* (garments, footwear, etc.): 55 per cent of workers in a large export-oriented garment factory on the outskirts of Bandung, Java commute daily from rural areas (Lok 1993: 161)

*Table 7.3* Characteristic roles of illegal labour migrants

| Destination | Country of origin | Work | Approximate daily wage |
|---|---|---|---|
| **Malaysia** | Indonesia, Thailand, Philippines, Bangladesh | Plantation workers Labourers in rice agriculture Construction Labour-intensive industry Service sector Timber industry (East Malaysia) | M$20 (US$9) |
| **Singapore** | Malaysia, Thailand, Philippines, Indonesia | Labour-intensive industry Construction Domestic servants (mainly Filipinos) Service sector | ≤ S$50 (US$35) |
| **Thailand** | Myanmar | Construction Sex industry | 50 baht (US$2) |
| **Brunei** | Thailand, Malaysia, Philippines | Domestic servants Construction Service sector | US$5 (minimum wage, 1992) |

This motivation has further implications. It means that such workers will keep their daily expenditures as low as possible. Those working in factories or on building sites are attracted, for example, by the opportunity of living in dormitory, shack or barrack accommodation and eating in canteens provided by the firm or from vendors permitted to operate within the factory or site precincts. In short, such workers have very little interaction with the mainstream urban economy (see for example, Firman 1991; Sjahrir 1993). They bring little money into the urban economy; but at the same time they are willing to work for low wages, creating the labour environment attractive to foreign investors and driving the process of FDI-led export-oriented industrialization. Sjahrir quotes a report from the Indonesian language magazine *Prisma* on the nature of an *asrama*, or dormitory for female workers in a textile factory in West Java:

> [it] consists of 187 rooms and each room is occupied by eight to ten female workers. Every room is headed by a senior worker. After the work is over the workers return to their *asrama*. They are not allowed to meet or to communicate with outsiders. They are practically isolated from the outside world. They are there only for work.
>
> (1993: 252; see also Wolf 1992: 155–8)[22]

*Plate 7.4* The urban dream. Bangkok, Thailand's capital, sucks in an estimated two million dry-season migrants searching for work each year from the kingdom's countryside

Not only do these dormitories or barracks represent sub-systems within the economy; they are also often culturally distinct. The workers are usually from up-country and tend to eat different foods, speak a different language or dialect, wear distinctive clothes, and in numerous other ways are identifiable from the resident population. The degree of distinctiveness is even more pronounced in the case of international labour migrants. Whether this degree of 'exclusivity' also makes these people possible recipients of the label 'excluded' is not clear. To long-term urban residents they may be almost invisible; unimportant people to be ignored. But to themselves and even more so to their fellow villagers in the countryside, they are men and women who have acquired some status by the act of moving to the city.[23] If this also generates significant income, their associated status grows still further. Thus, although *pekerja* may be poorly paid, isolated from the mainstream of urban life, non-unionized, and often regarded with something close to disdain by local urbanites, they are a critical component in the industrialization process.

Work in the informal sector embodies many of the same characteristics as that in the low-wage formal sector. But the workers are, by contrast, tightly bound into the operation of the city. They tend to interact more with other city residents at a social level; they live in slum or squatter accommodation which, though of poor quality, is an integral part of the physical fabric of the city; they are more likely to bring their families and make use of services like schooling and health facilities; and they spend (rather than remit) more of their earnings in the city. In short, while workers in many factories and in some types of casual work are semi-detached from the social and economic processes that give the city life, those in the informal sector are fully integrated into its functioning (e.g. Askew 1993: 42–3). The irony here is that the accepted designation of the former as 'formal' and the latter as 'informal', belies this division. A factory making garments for export, with labour drawn entirely from rural areas or abroad, a self-contained unit of production with few links to the city, is counted as a fully-functioning part of the urban economy. By contrast, a young woman making *krupuk* and *martabak* for sale to urban office workers for their lunch break counts, officially, for little or nothing in the urban economy.

## Betwixt and between

One of the processes which link rural and urban areas, and which fuel the forces of interaction, is human mobility: migration. Mobility, migration, commuting and circulation have been widely studied and there is little point in going over well-trodden ground again here (see Skeldon 1990). However, although the issue may have been researched and written about at great length and depth, this is not to suggest that there are no unresolved questions. Many link to the implications of mobility for sending and receiving areas, and for the people and households who are engaged in mobility. It might justifiably be argued that mobility studies, and especially those undertaken by geographers, have been

*Plate 7.5* A *bakso* cart in Java, Indonesia

overly preoccupied by the process itself. Research into such issues as frequency, patterns, distance and typologies have virtually become ends in themselves, while it is the effects of migration which are, arguably, of key importance. In others words scholars, among others, have been seduced by the process, and have paid less heed to the meaning of the process for people and places.

However, mobility studies undertaken in the countries of Southeast Asia have produced a rich seam of literature, possibly without equal in the developing world. Much of the best has taken a biographical approach to human mobility, studying migrant and non-migrant in depth, recording their life histories, investigating their motivations and aspirations, and placing the process within a detailed local context. And of all the countries of Southeast Asia, it is Indonesia where the most interesting and innovative studies of human mobility have been based (see the references in Skeldon 1995).

Mobility has been the 'hidden' process in much of the discussion in this chapter, and also in the previous one. Interactions, though they may be based on flows of ideas, money and goods, also commonly involve the movement of people. Scholars have tended to study mobility either for itself, or as a manifestation of unequal spatial and structural relations, or as a process where understanding must be rooted in the singular predicaments and characters of those who participate in it. Given that individuals with unique qualities and experiences move across an economic landscape which is partially moulded by the state, and are part of a wider social structure where questions of culture, class and community all play a role, it is sensible to assume that each approach has validity when it comes to disentangling form from process, and structure from outcome. All of these issues, it is hoped, have been evident in the discussion in this section of the book.

While human mobility represents the contact network that links people, places and economic activities, a number of scholars since the 1980s – McGee and Ginsburg particularly – have been writing about the emergence of distinct urban regions in which the divide between 'rural' and 'urban' is disappearing. This new form of urbanism incorporates a tight and vital interaction of people and activities over a wide area. The areas where these interactions occur have been termed Extended Metropolitan Regions or EMRs (Figure 7.2) (see McGee 1989 and 1991a; Ginsburg 1991; McGee and Greenberg 1992; Luxmon Wong-suphasawat 1995). The extended metropolitan regions centred on Bangkok, Jakarta (or JABOTABEK, including *Ja*karta, *Bo*gor, *Ta*ngerang and *Be*kasi) and Manila (Central Luzon) are the most striking examples, but Yangon (Rangoon), Ho Chi Minh City (Saigon) and Hanoi also share features that would suggest a similar process of development (Table 7.4). The authors postulate that advances in transportation technology, allied with the juxtaposition of zones of high density rice cultivation with the urban cores, has facilitated the emergence of EMRs. These regions have been termed *desakota*, a term pieced together from the Indonesian words for village (*desa*) and town (*kota*), and reflecting the concept's origins in studies undertaken in Java. However importantly *desakota* are not just physical entities; there is also a process of *kotadesasi* which involves:

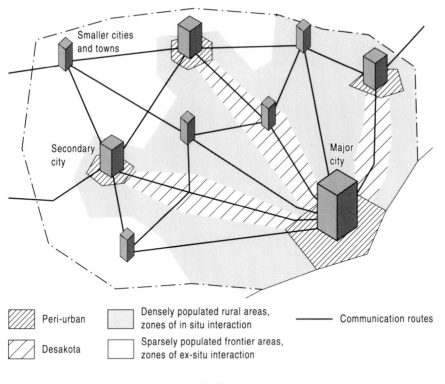

Peri-urban

Desakota

Densely populated rural areas,
zones of in situ interaction

Sparsely populated frontier areas,
zones of ex-situ interaction

—————— Communication routes

*Figure 7.2* Transcending the urban world: EMRs
*Source*: adapted from McGee 1991a: 6.

*Table 7.4* Features of Extended Metropolitan Regions (EMRs)

- Large and dense population engaged in wet rice cultivation
- Good transport networks
- Highly mobile population
- An increase in non-farm (non-agricultural) activities
- A mosaic of interlocking land uses
- Increased female participation in the labour force
- Lack of planning controls

*Source*: adapted from McGee 1991a: 16-17

the growth of distinct regions of agricultural and non-agricultural activity
characterized by intense interaction of commodities and people. . . . The
*kotadesasi* regions are generally characterized by extreme fluidity and
mobility of the population.

(McGee quoted in Ginsburg 1991: 38)

What is notable about Table 7.4 in the context of the discussion in this chapter
is that the features highlighted as characteristic of EMRs are those self-same

characteristics that have been highlighted and discussed here at a national level. In other words, it could be said that in functional terms there is no 'edge' to the EMR, and certainly it has been suggested that the whole of Java is rapidly becoming an EMR (Jamieson 1991: 277). McGee argues that the 'central processes that shape these [EMR] regions are the dynamic linkages between agriculture and nonagriculture' (1991a: 17). Thus the formerly distinctive attributes of 'rural' and 'urban' are becoming increasingly blurred. In terms of physical fabric this can be seen in the expansion of Southeast Asia's metropolitan regions; at a functional level it can be seen in the spatial intrusion of industry into rural areas; and in human and household terms in the diversification of livelihoods and the increasing movement of people between regions and jobs. It is in interpreting these changes that the real challenge lies: what is driving the process and what are its implications, particularly for the people who are caught up in the maelstrom of change. There is, as Koppel and Hawkins argue, 'a tension weaving through all the major components', between different scales and the 'imperatives of structure and culture, the domains of discipline and area, and the possibilities of universalism and particularism' (1994: 21). Although geographers may prefer to identify EMRs with trangible, physical developments – the presence of transport networks and rural factories for instance – many individuals in the region are part and parcel of the process whether they live close to the capital city or distant from it. The flows of goods, people, money, ideas, desires and aspirations will transcend any arbitrary boundary delimiting an EMR (Figure 7.2).

### Rural–urban interactions in perspective

A major theme in this section of the book has been to highlight the interactions – the linkages and their effects – between rural and urban areas. The emphasis in much of the geographical literature has been on describing the physical presence of places fulfilling the designated concept of 'city' or 'urban area' and places which dovetail with notions of rurality. Just as the Nuaula of Seram in Maluku, Indonesia construct houses for the building, not for the product, so scholars have been seduced by the existence of rural and urban areas without considering the processes that create, and sustain, them.[24] Bruce Koppel suggests that the whole rural–urban dichotomy debate is 'ersatz' because '[t]here are good reasons to argue . . . for the reality of rural–urban linkages not as derivations or reflections, but as representative and indicative of independent social facts' (Koppel 1991: 48). Macro perspectives have tended to dominate the literature, viewing development largely in terms of broad-ranging structural changes in the economy. Yet, rural work (and urban work for that matter) is 'a very complex and profoundly contextual social phenomenon' which requires micro-level understanding, not macro-scale aggregation (Koppel and James 1994: 295). The studies that represent the foundations on which this section of the book has been based, if they show one thing, it is the degree to which scholars must build flexible and mutable visions of change, sensitive to local and national particularities and informed by the need to reflect the dynamism that is part of both people's lives and national (and international) economies.

For Evans, there is a potential 'virtuous circle' of rural–urban relations where rising agricultural incomes create a demand for more consumer goods and services, which spurs the development of non-farm activities such as motorcycle and watch repair shops, which in turn absorb surplus farm labour, which further boosts demand for agricultural output while contributing cash for investment in agriculture, thus both stimulating and permitting yet further increases in agricultural production (1992: 641; see also Evans and Ngau 1991, Nipon Poapongsakorn 1994; Effendi and Manning 1994: 230; and Koppel and James 1994: 290) (see Figure 7.3a). Richard Grabowski takes this a step further in arguing that the development of a vibrant non-farm sector in rural areas is critical to the development of the industrial sector in urban areas. In his view, only when farmers have access to alternative income generating activities outside agriculture will they be willing to risk innovating in agriculture. And only when they innovate in agriculture, will rural incomes rise sufficiently to

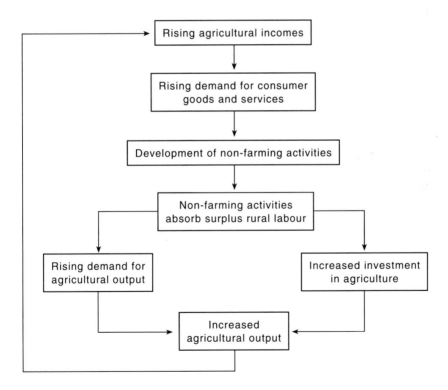

*Figure 7.3a* The virtuous cycle of rural–urban relations I

*Sources*: based on Evans 1992; Evans and Ngau 1991

create a domestic market large enough to stimulate the emergence of a modern, urban industrial sector (Figure 7.3b) (1995: 49):

> The spread of nonagricultural activities in rural areas allows farm families to diversify their sources of income and reduce risk, and this predisposes farmers to choose more innovative activities [in agriculture]. The resulting income growth stimulates further growth in cottage manufacturing and nonagricultural activities, which in turn stimulates further agricultural growth, eventually resulting in the establishment of modern industry as domestic markets expand dramatically.
>
> (Grabowski 1995: 50)

Grabowki's developmental sequence could be criticized for being too spatially restricted (with its focus on local cottage industries) and also excessively dualistic in envisaging two industrial sectors – a rural-based, 'cottage' industrial sector, and an urban-based, modern one.[25] None the less, the work does emphasize the point that farm and non-farm activities can play a critical role in stimulating modern industrial growth. In Southeast Asia, as this chapter has tried to show, the vital spatial articulation of 'rural' and 'urban' and the degree to which modern activities are to be found in rural areas and – though to a lesser extent – cottage industries in urban areas would seem to call for a rather more subtle model of how different activities complement, cross-subsidize and interrelate.

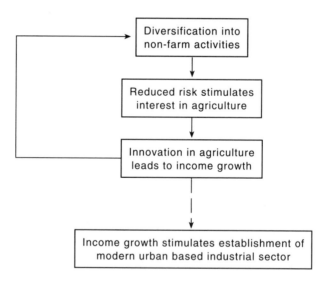

*Figure 7.3b* The virtuous cycle of rural–urban relations II
*Source*: based on Grabowski 1995.

## Notes

1 Most remittance flows are from urban to rural areas. However, during the initial period of residence in the city, rural migrants often need to be supported and this leads to a short-term flow of funds and resources in the reverse direction. It is also true that among the small number of (usually) wealthy rural migrants who might, for example, be undertaking further education in the city, there is a longer-term reverse flow.

2 Leinbach and Bowen write: 'given the prevalence of employment diversification in peasant survival strategies, the full implications [of such diversification] upon agricultural productivity must be examined', adding that it is of critical importance 'given the widespread incidence of marginality among peasant populations and the changing role of peasant forms of production within the global economy' (1992: 350).

3 This view does gloss over the existence of a very pronounced city structure in Southeast Asia long before the colonial period. In fact, cities in the region like Ayutthaya in Siam (Thailand), Pegu and Pagan in Burma (Myanmar), Melaka on the Malay Peninsular, Demak, Gresik, and Banten on Java, Makassar (Ujung Pandang) on Sulawesi, and Hué and Thang Long (Hanoi) in Tonkin (Vietnam) were impressive and cosmopolitan places compared with those of Europe at the same time. However, the historical evidence seems to indicate that the colonial period undermined these indigenous urban centres, and the hierarchy that evolved during the colonial period did not build on the traditional form, but undermined and then supplanted it. See Reid (1993).

4 The continued pertinence of such theoretical perspectives is discussed in greater detail in the last chapter.

5 De Koninck notes a rise in 'housewife-ization' in his study of two villages in West Malaysia. The 'retreat' to the house is seen, in the local context, often as an 'advance' – a measure of success. A woman who does not work is sometimes referred to as a *seri rumah*, or a 'princess of the house' and – although still rare – is linked to the attainment of affluence (De Koninck 1992: 178). Kato in Negeri Sembilan reports a similar trend with women identifying themselves as *seri rumah tangga*, or 'fulltime housewives' (1994: 163). (*Seri* is a royal honorific.) However, Wong, working in the same area as De Koninck, reveals that '[c]ontrol over the labour of unmarried daughters is rigidly exercised but this labour is applied almost exclusively to the domestic sphere and not made available for padi cultivation. The availability of unmarried daughters frees the mother for work on the family farm' (1987: 119).

6 In Trankell's study, she notes that few mechanical innovations, with the exception of the rice mill, have liberated women from the drudgery of manual farm work (Figure 7.1).

7 A shortcoming of many studies of mechanization in rice agriculture – and particularly with reference to the deleterious social and economic effects of the Green Revolution – is that 'mechanization' is taken to be undifferentiated in its effects on rich and poor farmers, large and small landholders, and men and women. This chapter does not rehearse the question of 'who benefits from the new technology'. The literature on the subject is massive. For general surveys see: Barker *et al.* 1985; Chambers 1984; Freebairn 1995; Hayami 1981; Lipton 1987; Lipton and Longhurst 1989; Pearse 1980; Rigg 1989b.

8 Women ploughing has also been noted in Laos' Xieng Khouang province which suffered selective male de-population during the war in Indochina (Schenk-Sandbergen and Outhaki Choulamany-Khamphoui 1995: 17). The authors also note that women are involved in the clearing and burning of trees and undergrowth, another task which has tended to be viewed as a male preserve (ibid.: 18).

9 Transplanters would have as severe an impact, although it seems that most farmers tend to save labour by turning to broadcasting rather than through adopting mechanized transplanting.

10 Farms have undergone a continual process of capitalization as wage labour has replaced exchange and family labour, and the use of cash inputs like chemical fertilizers and pesticides have increased.

11 This is taken from Preston's (1989) paper: 'Too busy to farm: under-utilisation of farm land in central Java'.

12 Given the degree to which rural-based households are being supported on Java by income generated from non-farm sources (see Chapter 5) it is increasingly questionable whether we can talk of Java supporting 'agricultural' population densities of up to 2,000 per square kilometre. It is probably more accurate to talk of 'rural' population densities, thereby allowing livelihoods to incorporate both farm and non-farm activities.

13 Indonesia was forced to import rice again during the mid-1990s, partly because the very success of the Green Revolution had encouraged the government to promote diversification into non-rice crops.

14 Schweizer notes in his study of a village in Klaten, Central Java, that even farm labourers tend to buy their fruit and vegetables from *warung* (market stalls) rather than grow them on their home gardens such is the availability of income earning opportunities and its conflicts with agricultural work (1987: 45–7).

15 Although note the above discussion on land abandonment and land use changes linked to labour shortages.

16 At least, so far. Environmentalists would argue that the application of the technology of the Green Revolution is creating the conditions for massive ecological failure in the future – notwithstanding the production increases that have been achieved to date. They would point, as evidence of the ecological dangers of high-input/high output monocrop production, to the effects of the brown plant hopper (BPH – *Nilaparvata lugens Stal.*) in Indonesia (see Fox 1991).

17 It comes back to the point made at the beginning of the chapter concerning whether poor people should be regarded as 'victims' of events beyond their control.

18 Hefner also notes that mid-slope farmers in the Tengger Highlands of East Java have recently taken to planting more environmentally sustainable tree crops (1990: 108).

19 Though the use – largely – of eucalyptus, an exotic species, has been extensively criticized for its deleterious environmental effects. Eucalyptus is said to 'mine' the soil of nutrients and create impoverished forest environments lacking the non-timber forest products which communities traditionally collected. Large-scale eucalyptus plantations have also been established on forest reserve land which, if not already under cultivation and regarded as 'owned' by the farmers working it, is used as a community resource (see Hirsch and Lohmann 1989; Apichai Puntasen *et al.* 1992).

20 'My time in northern Luzon demolished my most cherished beliefs. Not only had I previously espoused all the tenets of radical environmentalism, but I had been fully inculcated into the school of cultural geography that regards indigenous peoples as lacking commercial sensibilities and as always respectful of their environments' (Lewis 1992a: 80).

21 As was emphasized in Chapter 6 though, money may not be the only factor or, indeed, the determining factor. This is particularly true of younger, unmarried migrants.

22 Elmhirst, in her account of young female factory workers in Tangerang, Java paints a picture of women who are comparatively free – certainly far more so that in their home villages (see Box 6.2).

23 Silverman reports that so many men leave the island of Bawean in Indonesia to work in Malaysia and Singapore that it has become known as the Island of Women. Men are expected to work abroad – it has effectively become a rite of passage to adulthood – and some women have reportedly justified leaving their husbands on the grounds that they have failed to take such work (Silverman 1996: 61).

24 Waterson writes: 'Houses are continually in process of construction, but are rarely actually finished, the process of building itself, with its attendant rituals, being more important' (1990: 45).

25 He begins the paper by raising doubts about the usefulness of dualistic models where the economy is divided into industrial and agricultural sectors, and then proceeds to draw up a threefold model where there are two industrial sectors and an agricultural one. This subdividing of the industrial sector would not seem to help much in producing a more sensitive and nuanced understanding of the shaded interrelationships between farm and non-farm, urban and rural, and agriculture and industry.

# Part IV

# CHASING THE WIND

Modernization and development in Southeast
Asia

Development has irreparably and irreversibly changed Southeast Asia's human
landscape, and at a rate which sometimes defies belief. Even the 'forgotten' countries of the region, Laos, Myanmar and Cambodia, are opening up to rapid
change. There can be little doubt that modernization has undermined traditional
culture, and some scholars and commentators have identified a cultural crisis
looming as local identities are lost. There is also little doubt that development
has brought cumulative material gains to the great majority of the population
who have been touched by it. How the balance between, for instance, cultural
erosion, improving material standards of living, widening inequalities, better
health, environmental degradation, rising educational standards, and growing
dependency is drawn is the key issue which divides those concerned with the
region's modernization and development. The view presented in this final
chapter is that even taking account of the very real tensions and problems created
by the development process, development in Southeast Asia can be counted a
success.

# 8

# CHASING THE WIND

## Introduction

Many of the terms referred to in this book – terms like post-development, new social theory, new social movements, development alternatives, alternatives to development – resonate with the sense that something novel is unfolding; that a new landscape of development is appearing or that the existing landscape is being remodelled by strong erosive powers. And yet there remains the nagging sense that these are old ideas dressed up in the clothes of post-modernity. In addition, at the same time as the discourse of development is under sustained scrutiny, so the forces of modernization are bringing new powers to bear on people's lives. Thus while modernization is reshaping the human landscape, scholarly advances are reshaping the academic landscape. It would be reassuring to write that these two avenues of change inform one another – that there is a creative energy which links the landscape of the mind with the landscape of experience.

At the meta-scale, economic progress and social change in Southeast Asia can be seen as a process where continuities are more important than discontinuities. In other words, that there is an identifiable narrative of development in the region over the last 100 years that can be identified and analysed. This sense of continuity is reflected in the tendency to talk of economic 'trends' and 'trajectories'. Yet, at the meso- and micro-scales the processes of change require a rather more nuanced interpretation that respects not only important differences between countries, cultures, places, households and individuals, but also the particularities that can be associated with recent experience. Partly this distinction can be seen as one simply of scale: that from a distance colours merge. It is also, though, one of time: that recent events, by dint of their very freshness, embody greater detail. Only with the passage of time do common threads merge and are discontinuities shed. Even so, there are some developments this book has described, such as those associated with rural–urban interactions, which represent, arguably, a break with established trends and which therefore require fresh analysis and assessment.

The later chapters of this book, from Chapters 3 to 7, took this meso/micro-scale perspective. The intention was to analyse, through the use of local studies, not just how modernization is variously affecting the countries of the region, but also – and perhaps more importantly – to describe how people are constructing varied responses to the challenge of modernization. This challenge tends to be interpreted as one where people either respond to a widening of choice brought about by modernization, or struggle against a narrowing of choice as modernization closes off traditional livelihood strategies. Often both trends can be discerned in single areas – even in single households. Chapters 5, 6 and 7 focused on this issue with reference to evolving interactions between rural and urban areas. What was not addressed in any detail however was how this affects our interpretation of 'winners' and 'losers' in the development game and what implications it has for development theory.

## Identifying winners and losers in the rural–urban interaction game

Much of the debate over agrarian change and urban transformation has centred on the twin questions of 'who wins?' and 'who loses?'. Has development brought not only unequal benefits, but also real losses to some groups and some areas? Or, to put it in terms of the urban bias thesis, is the city 'parasitic' or 'generative'?[1] It should have become clear from the discussion in Chapters 5, 6 and 7, that as the very participants in the game change places, it is difficult to associate particular people or households with particular outcomes. The eldest son of a rice farmer who quits his home to work in a factory leaves his family short of an agricultural worker. He provides the urban-industrial sector with cheap labour that can be paid at less than the cost of its reproduction leading to charges of super-exploitation. Yet, at the same time, the income he remits to his parents allows them to save labour by renting a rotavator while sending a younger daughter to secondary school. The money may also be used to intensify production by investing in yield-enhancing technologies. This young man may, functionally, remain a critical part of the rural household, yet spatially and administratively he may be counted an urbanite. In such fluid and spatially dispersed situations there are numerous questions to be asked about how to characterize people and households. Should this young man's 'sacrifice' (if, indeed, we can view it as a individual sacrifice in the context of collective household gain) be viewed in household terms? Should he be characterized as a countryman in the city; as a sojourning farmer; as a temporarily detached rural dweller; as an urbanite with tenuous family links with the countryside; or as embodying all (or none) of these things simultaneously?[2] The difficulty is that conceptual structures and semantic divisons force scholars to prise people into categories which are no longer – or never were – valid.

Not only do the very participants in this game refuse to stay where they are seated, but there is confusion over whether there are distinct seats at all. Numerous studies have pointed out that there are rarely homogeneous and easily identifiable rural and urban economic sectors (see for example, Moore 1984; and *JDS* 1984 and 1993). As the discussion of Southeast Asia's Extended

Metropolitan Regions (EMRs) in Chapter 7 illustrated, in some key areas there is a mess of interlocking activities. Even beyond such EMRs, many analysts would prefer to talk of a core–periphery continuum, rather than a sharp divide; and still others a rather more nuanced subdivision of rural areas taking into account their varied intergration into the regional, national and international economies. Lipton may have claimed that there is a clear-cut city/country divide in his influential work *Why poor people stay poor: urban bias in world development* (1977), but numerous critics (e.g. Moore 1984) have voiced concern over such a sweeping generalization based largely on work from a single country – India. Certainly, in the case of Southeast Asia, it is hard to sustain such a view at a general level. In his own defence, Lipton states that '"[r]ural–urban" is not a categorisation of space alone. . . . The overlap "rural, agricultural, labour-using, dispersed" *vis-à-vis* the overlap "urban, non-farm, capital-using, concentrated", while imperfect (and complex in operation), does define a central class conflict' (Lipton 1984: 155).[3] The contention here, with reference to Southeast Asia, is that not only is it imperfect – whether the division is formulated in terms of space, interest group (class) or sector – but that the imperfections are deepening and multiplying.[4] McGee concurs with such a view, asking, in the context of the debate over EMRs, 'what happens if the rural–urban dichotomy ceases to exist . . . ? Then the whole policy debate on urban and rural allocation of resources becomes fuzzy and meaningless' (1991a: 20). We are landed with a twofold division of space – rural and urban – into which people must be slotted, yet at the same time knowing full well that increasingly people should be placed in a third category, one of mixed membership.

It seems that the sources of rural poverty, and certainly the antidotes to rural poverty, are not to be found in agriculture, and perhaps not even in the countryside. The usual gamut of rural development policies – land improvement, irrigation, credit, extension, land reform – are akin, to coin a phrase, to shifting chairs on the deck of the *Titanic*. On those occasions when non-farm measures are brought into the rural development equation, there has been a tendency to conceive of them as being subsidiary and supplementary to farming, placed under opaque headings such as 'community development', and organized as income generating activities to be undertaken by women or youth groups, and rarely regarded as mainstream, let alone alternative. In this way, rural labour processes have been viewed as 'offshoots of fundamentally agrarian processes' and theories of agrarian change have been preoccupied with viewing differentiation within a context where small-scale agricultural production provides the basis for life and livelihood (Koppel and Hawkins 1994: 4 and 23).

Nor is it only a case of old divisions breaking down (or, arguably, never having been as clear-cut as the ideal of town versus country would lead one to believe), but also of new developments rendering such divisions obsolete. The broad pattern of economic change in the countryside, and more particularly technological change in agriculture, has made the rural sector less marginal, and therefore more powerful, *vis-à-vis* the urban sector (Varshney 1993: 5). An important factor in this relative shift in the balance of power has been the advent of modern communications. Rural people in Southeast Asia are not, generally, isolated and remote. The spatial marginalization thesis that holds that because farmers are remote in spatial terms they are also politically and economically marginal is hard to sustain when farmers in the poor Northeast of Thailand cut gems for sale in Bangkok, and women in rural Java take on piece-work for foreign-owned garment and footwear companies based in the Jakarta extended metropolitan region. These people may still be exploited, but probably no more so than other gem-cutters or pieceworkers in Bangkok and Jakarta. Rural people have increasing access to information, and burgeoning and cheapening transport links allow them to tap into this information. Thus a root cause of urban bias, sheer isolation, has been substantially eroded.

Many theories of rural neglect, urban bias among them, not only suffer from spatial over-generalization (i.e. they do not account for significant differences between countries), but also tend to lack, or have misconstrued, the temporal dimension of change. Urban bias, core–periphery and other similar models may have had some validity a decade or two ago, and may still do so in some specific areas, but recent economic change has made them increasingly irrelevant. As was described in some detail in Chapter 5, the rural world in Southeast Asia is a *new* world. As Bates argues we '*must* understand the politics of the profoundly paradoxical transformation of agriculture's location in the political economy of nations, as these nations develop: its transformation from an embattled majority that is taxed into a minority powerful enough to be subsidised [emphasis in original]' (1993: 221).

At a policy level there is still a tendency to conceptualize and implement development policies in terms of 'urban development policies' and 'rural development policies'. The above discussion implies that such a distinction – at both the conceptual and empirical levels – is obsolete. Ginsburg talks, with reference to EMRs, of the 'urbanization of the countryside' and postulates that if it is occurring 'then it will be necessary to think through again many of the generalizations and assumptions that underlie the formulation of urban policies in most Asian countries' (1991: 38). He might have added agricultural policies, too. He questions the logic of trying to control big city growth both on the grounds that such controls have manifestly failed and because if they did succeed they would undermine economic growth (1991: 42–3). However, there are also implications for policies aimed at promoting regional development and rural industrialization, the latter discussed in Chapter 5 (see page 191).

## POST-DEVELOPMENTAL SLEIGHT OF HAND

The opening chapters of this book dealt at some length with the ideas of the post-developmentalists. Arturo Escobar's assertion that the so-styled discourse and strategy of development has produced 'massive underdevelopment and improverishment, untold exploitation and oppression', was highlighted as illustrative of the developmental pessimism that informs much of the recent literature (1995a: 4). Escobar's is not a lone voice. Sachs, for example, writes apocalyptically:

> The idea of development stands like a ruin in the intellectual landscape. Delusion and disappointment, failures and crimes have been the steady companions of development and they tell a common story: it did not work.
>
> (Sachs 1992: 1)

There is no doubt that recent post-developmental writing has been a necessary and welcome antidote to the more laudatory assessments of development over the post-war period. The way in which scholars have challenged notions of progress, have reinterpreted poverty, have analysed development as a hegemonic discourse, and have examined the historical roots of modernization, for instance, are all valuable new perspectives. It is in making the leap from these new ways of 'looking', to the declaration that development has failed (as reflected in the quotations above), that post-developmentalism, it is suggested here, has oversold itself. But before investigating this more fully, it is necessary to address the important question whether Southeast Asia is enjoying a unique experience of development. Is it, in other words, the exception that proves the rule?

Every country and every region is clearly exceptional and in that sense it is *always* possible to reject contrary experiences as not applicable. Yet this rejection of anything that does not conform – for the very reason that it does not conform – prevents any incisive critique of the assumptions on which a thesis may be built. It is a way of fencing-off a paradigm from critical debate. It also seems that though post-developmental texts may be fairly narrowly based in spatial terms – i.e. they draw their evidence from a handful of countries and regions – they do attempt to build a generic argument. As such it would seem valid to confront that argument in similarly narrow geographical terms. This is not to say that Southeast Asia is a small region. It constitutes ten countries with a combined population of over 450 million – equivalent to all the countries of Central and South America.

## The landscape of modernization and impoverishment in Southeast Asia

In the case of the growth economies of Southeast Asia, it is hard to think of one indicator of human well-being that has not improved during the course of modernization. It is notable that those countries which have experienced stagnation or decline in such indicators are those that have experienced an *absence* of development as modernization whether that be due to prolonged war (Vietnam [1945–1975]), failed attempts at state socialist development (Laos [1975–1986], Vietnam [1954–1979], Cambodia [1975–1979]), economic mismanagement (Myanmar [1962–]), or some combination of these. In other words, the evidence at the national level at least is quite the reverse from that proposed by some post-developmentalists: development has led to real, substantial and, in some cases, sustained improvements in human well-being (Table 8.1). Nor can this be rejected as a case of the benefits accruing to just a small segment of the population, leaving the majority mired in poverty. Research has repeatedly shown that improvements in livelihood have been broadly based, even if they have not been equally distributed (see Chapter 3). Booth, on Indonesia, contends that:

> The road that was sealed for the first time in living memory, the new school, the new health clinic, the improved irrigation system, all these were convincing evidence that *pembangunan* (economic development) meant improved access to public amenities, which in turn could lead to increased earning opportunities, and higher family incomes and living standards. By the latter part of the 1970s, it was impossible to doubt that incomes and living standards were improving, especially in Java, where the great majority of the rural population were concentrated. . . . These were remarkable achievements for a country as backward as Indonesia was in the mid-1960s.
>
> (Booth A. 1995b: 109)

A valid post-developmental retort might be that these indicators of human well-being are themselves part of the problem: they are intimately associated with the pervading discourse of development and tell us more about what 'development as modernization' defines as important than what local people consider to be the essence of a good life. The difficulty with this perspective is that the modernization ethic, broadly defined, has been internalized by most people in Southeast Asia. Although scholars like Hettne (1995), as well as some politicians and practioners, may suggest that there are alternative avenues to development than that mapped out by modernization it is difficult to see these as much more than paper alternatives. It is also hard, any longer, to argue that modernization is an alien ethic in the countries of Southeast Asia. In large part it has become part of the local developmental landscape. This is evident, for example, in participatory wealth rankings of the type described in Chapter 3 which endeavour to bring to light the values and priorities of local people (see page 113). For although they illuminate a slightly different vision of what is important, they do not fundamentally challenge the modernization ethos. Local people are concerned about remoteness from markets, with stability of income, and with an inability to save, for instance. People's desires and aspirations, for themselves and their children, are framed in terms of modernization. Importantly, failure is expressed in terms of 'missing out' on development. Hettne, anticipating just such a response, argues that the failure of his alternative – or 'another' – development (small-scale, need-oriented, community-based, endogenous, appropriate, environmentally sustainable and benign . . . the familiar recipe) to take root in many countries of the South is due to power relations, corruption and lack of representativeness of governments there (1990: 155 and 1995:

Table 8.1 Development, modernization and well-being

| | Real GDP/ cap ($PPP)^a (1992) | Life expectancy at birth (years)^b 1960 | 1994 | Maternal mortality^c 1980–92 | Infant mortality^d 1960 | 1994 | Under-5 mortality^e 1992 | Underweight children (%, of under 5s)^f 1975 | 1989–95 | Percentage of age group in secondary education 1992 | Adult literacy (%)^g 1970 | 1995 |
|---|---|---|---|---|---|---|---|---|---|---|---|---|
| Modernizers | | | | | | | | | | | | |
| Brunei | 20,589 | 62.3 | 74 | n.a. | 63 | 8 | n.a. | n.a. | n.a. | n.a. | n.a. | n.a. |
| Indonesia | 2,950 | 41.2 | 63 | 450 | 139 | 53 | 111 | 51 | 39 | 38 | 54 | 84 |
| Malaysia | 7,790 | 53.9 | 71 | 59 | 73 | 12 | 13 | 31 | 23 | 58 | 60 | 83 |
| Philippines | 2,550 | 52.8 | 65 | 100 | 80 | 40 | 59 | 39 | 30 | 74 | 83 | 95 |
| Singapore | 18,330 | 64.5 | 75 | 10 | 36 | 5 | 6 | n.a. | n.a. | n.a. | n.a. | 91 |
| Thailand | 5,950 | 52.3 | 69 | 50 | 103 | 36 | 33 | 36 | 13 | 33 | 79 | 94 |
| Average by country | | 54.5 | 70 | 134 | 82 | 26 | 44 | 39 | 26 | 51 | 69 | 89 |
| Reformers | | | | | | | | | | | | |
| Cambodia | 1,250 | 42.4 | 52 | 120 | 155 | 97 | 181 | 43 | 38 | n.a. | n.a. | 38 |
| Laos | 1,760 | 40.4 | 52 | 300 | 146 | 92 | 141 | 41 | 40 | 22 | n.a. | 57 |
| Myanmar | 751 | 43.8 | 58 | 460 | 158 | 80 | 111 | 41 | 31 | n.a. | 71 | 83 |
| Vietnam | 1,010 | 44.2 | 68 | 500 | 147 | 42 | 48 | 55 | 45 | 33 | n.a. | 94 |
| Average by country | | 42.7 | 58 | 345 | 152 | 78 | 120 | 45 | 39 | 28 | 71 | 68 |

Sources: Human Development Report and World Development Report

Notes

a An international comparable scale of real GDPs using Purchasing Power Parities (PPPs) rather than exchange rates as conversion factors. The purchasing power of 'currencies' – as it is more accurately described – is calculated according to the real cost of a fixed basket of goods and services between countries.

b 1994 data, except for Cambodia and Brunei, which are 1992 figures

c The annual number of deaths from pregnancy-related causes per 100,000 live births

d The annual number of deaths before the age of 1 per 1,000 live births. Figures are 1994, except for Cambodia which is 1992.

e The annual number of deaths of children aged 5 and under, averaged over the previous five years

f A measure of moderate and severe child malnutrition based on median weight-for-age. Figures are 1989–95, except for Cambodia which is a 1992 figure. The Myanmar figure refers to children aged 3 years and under.

g 1995 data, except for Cambodia which is a 1993 figure

161–63). However, to discard modernization as a drug of the élite rather than as the opium of the masses seems, certainly in the context of Southeast Asia, to be an epic misreading of the direction of change and the prevalence of the modernization ethic. There also seems to be tendency to retreat into the cosy realm of normative science where the world can be reordered to fit a predetermined idea of what it *should* look like (see Hettne 1995: 160–206).[5] Clearly there is no self-evident 'should' that can be used as a template to build this new world of development; people hold many visions. 'God forbid', as the philosopher Francis Bacon warned 'that we should give out a dream of our own imagination for a pattern of the world'.

Escobar would seem to see part of the problem lying in the tendency to frame the answer to the development dilemma in terms of 'development alternatives' rather than 'alternatives to development' (Escobar 1995b: 217). But here again there would seem to be little real desire – whether among national élites or local groups – to embrace such alternatives to development, although many would subscribe to development alternatives. Even within the grass-roots movement in the region the emphasis is less on alternatives to development than with making existing development fairer, more effective, and less environmentally destructive, for example. In other words, to work within the existing system rather than to subvert it; to search for new and better ways to achieve what is fundamentally the same end as that mapped out by modernization. Thus the growing public agitation in Indonesia during the early 1990s over working conditions, wages, widespread corruption and the distribution of the benefits of rapid growth should not be read as a rejection of development, but rather as a rejection of the way in which political and economic élites have preferentially 'captured' development for themselves.

Two examples illustrate the practical and moral difficulties of applying post-developmental concepts to the realities of people's lives. In 1996 Stan Sesser, a journalist and writer, travelled to the Lao village of Takho, close to the border with Cambodia. This area, near the Phaphaeng Falls on the Mekong, has been designated as the site for a massive hotel–casino–golf course development. Located on the main road (just a track at present) to Cambodia, Ban Takho is on the front line of this development and the 130 households that live in the village were not consulted in any way. 'A Lao official simply came to town one day last year [1995] and told them at some point the entire village would have to be relocated', Sesser reports. But when questioned about their impending relocation, Sesser was shocked to find that not only were the villagers unconcerned at this dictatorial approach, but they welcomed the investment. 'Now we grow rice, and at the end of the year we've barely made a living' one villager observed. 'We want a better life than we have. It would be much better for my children to work in a hotel. Everyone here wants development' (Sesser 1996). The second example comes from the Philippines. Nelson Tula, a Bagobos *datu* (chief) lives on the slopes of Mount Apo in the southern Philippine island of Mindanao. The area near his village had been selected by the Philippine

National Oil Corporation as the site for a geo-thermal power plant. Already a new road had brought his village within an hour's travel of medical attention, rather than a day, and he looked forward to be able to cook his rice in an electric rice cooker rather than spend hours looking for fuelwood. When asked the innocuous question whether he will miss the stars at night when electricity arrives he replied, 'I can always switch off the lights if I want to see the heavens' (Knipp 1996).

It would be easy enough to dismiss the comments of Nelson Tula and the Ban Takho villager as the uninformed ramblings of two men who knew no better. Certainly, in the case of the Lao PDR, there are numerous studies which show that the trajectory of development being embraced there, and Laos' position of relative powerlessness *vis-à-vis* its more powerful neighbours, and especially Thailand, puts it in an unenviable position regarding the developmental choices available to the government (see for example, Hirsch 1995; Usher 1996; Rigg and Jerndal 1996; Rigg 1995c). But putting, for a moment, the wider question of the regional political economy of development to one side, what these men do fully appreciate are the conditions in which they currently lead their lives. And across the region, people are embracing the modernization ethic because they are dissatisfied with the present (see page 165). So while villagers may be unaware of the future that modernization holds for them, and may have been dazzled by promises of wealth and supposed well-being, foreign researchers are ill-placed to comment – at the experiential level – on the present. The key difference, of course, is that while development or a lack of it has a direct and profound impact on Nelson Tula and the Ban Takho villager, for the researcher, post-developmental or otherwise, it is often merely academic.

## Modernization and the creation of poverty and affluence

It has also been repeatedly argued that modernization has created poverty. Anthropologists, for decades, have observed that poverty is a social condition. Thus development – paradoxically – creates poverty by creating the conditions in which the absence of the rudiments of modernity is characterized as poverty. Again, though, this academic sleight of hand would seem to carry little weight in the context of a village where life expectancy is less than 50, the majority of children suffer from some degree of malnutrition, and infant mortality may be well over 100 per 1,000 births. Laos is probably the country in Southeast Asia where villages come closest to the peasant ideal. Modernization has scarcely touched some areas. Yet in the village of Ban Bung San in Khammouane province, of 150 children born, 47 died before the age of 5 (31 per cent), mostly of disease and fever. In Ban Nong Jama, a Khamu village in Xieng Khouang province, 48 out of 136 died (35 per cent) (Schenk-Sandbergen and Outhaki Choulamany-Khamphoui 1995). As Popkin remarks in his critique of the 'myth of the village': 'A way of life that may have existed only for lack of alternatives is extolled as a virtue. Peasants who had little or nothing to eat are assumed to have had a rich spiritual life. . . . Somehow, what might only have been the necessities or oppressions of one era come to be interpreted as traditional values during the next' (Popkin 1979: 3). Few would disagree that such a catalogue of 'tradition' should be targeted for development and yet the evidence for Southeast Asia demonstrably shows that the best way to achieve this end is through pursuing (prudent) modernization.

Nor, for that matter, does the historical evidence from the region support the view that before modernization there were no notions of poverty and affluence, no divisions between classes, and no marked commercialization of production. Indeed, quite the reverse:

> Within rural society, the pattern of access to land was highly differentiated . . . and a class of dependent landless households existed long before the full-scale colonial exploitation of Java.
>
> (Hart 1986: 20)

> [I]t is clear that the structure of Javanese peasant society, at least until the first decades of the nineteenth century, gave important advantages to the group of "landowning" peasants who could prosper independently by drawing on the labour services of a resident work-force of dependents and landless peasants.
>
> (Carey 1986: 86)

> [T]his examination [of nineteenth century northern Siam] . . . reveals a society with a complex division of labor, serious class stratification, dire poverty, a wide-ranging trade network, and an unanticipated dynamism.
>
> (Bowie 1992: 815, 819)

> [I]f there is any constant element in the history of Java, it is its well-attested function as a large-scale rice exporter perhaps as early as [AD] 900 . . . [peasants] were producing rice far in excess of their own requirements for a considerable time.
>
> (Boomgaard 1991: 16)

As Vandergeest cautions in a paper on Thailand, 'we should be suspicious of theories based on contrasts between the past and the present, in which the past is constructed as that which the present is not' (1991: 423). Anthony Reid's volumes (1988 and 1993) on Southeast Asia in the 'age of commerce' (1450–1680) represent perhaps the most convincing argument that during this period the region shared the experience of Europe in terms of its integration into gobal trade, the commercialization of production and consumption, economic specialization, urbanization, the monetization of taxation, and political absolutism (1993: 268). He concludes:

> Because it has resumed that path [of commerce] (again for better or worse), the first age of commerce is now more immediately relevant. As

Southeast Asians dramatically reshape their present, they need not be inhibited by their immediate past, with its memories of political diffusion, social stratification, and resignation to others of the high economic ground. An earlier period offers abundant evidence of a variety of creative responses to rapid economic change, a variety of social forms, a variety of political and intellectual possibilities.

(Reid 1993: 329–30)

## Alternatives to development, development alternatives

The discussion above about alternatives to development and development alternatives raises the question of whether they are alternative in the sense of being local and endogenous. It is sometimes assumed that anything local must, virtually by it being local, be better and more appropriate than that which is exogenous or Eurocentric. Cowen and Shenton question whether Hettne's call for another development which will 'transcend the European model' has not ascribed a false identity on 'alternatives' which themselves are largely products of Western thinking. Even those alternatives which appear to be explicitly generated in the South – Islamic and Buddhist economics, for example – appear, on examination, to be guilty of conceptual cross-dressing.

In a sense there is neither a purely 'indigenous' nor a solely 'exogenous'. The tendency to highlight what is local and what is imported sets up a false dichotomy which is then used to argue a case for development discourse imperialism. This appears to be Sachs' approach when he suggests that the development discourse is made up of a web of key concepts like poverty, production and equality which arose in the West and have been projected onto the rest of the world, thus crystallizing 'a set of tacit assumptions which reinforce the Occidental world view'. The idea of development has been so pervasive and so powerful, he seems to argue, that it has created a dominant and domineering reality that has snuffed out alternative (i.e. non-Occidental) world views. The sequence of logic appears irrefutable. Yet is based on the assumption that such concepts have purely Western roots and that they have colonized the rest of the world in their original form. It can be argued that not only do notions of 'poverty' and 'equality' have indigenous roots, but the discourse of development has not been imported or imposed wholesale and without modification.

It is also true that while some post-developmentalists would seem to wish to reject the language of modernization, arguing that the discourse of development has been a false prophet, the basis by which modernization is shown to have failed is, itself, often modernist in complexion. Thus Peet and Watts feel able to write that:

[f]or good reason . . . many intellectuals and activists from South America [have] come to see development discourse as a cruel hoax . . . It is precisely the ground swell of *antidevelopment thinking*, oppositional discourses that

have as their starting point the rejection of development, of rationality, and the Western modernist project.

(1993: 238 [emphasis in original])

And yet, earlier in the same paragraph, they find support for this view in the very modernist reasons that they would seem to wish to reject. They write, for example, of the 'appalling spectacle detailed blandly every year in the World Bank *Development Report* of deepening global polarization', continuing that '[i]n the fin de siècle world economy, 82.7 percent of global income is accounted for by the wealthiest 20 percent, while the poorest 20 percent account for 1.4 percent' (1993: 238). This identification of failure in terms of the distribution of global income is surely much the same as measuring success in terms of economic growth.

## CHASING THE WIND

On the basis of the evidence presented here at least, the suggestion that development 'did not work' does not stand up to scrutiny in the Southeast Asian context. It looks good; it has polemic appeal; it is founded on some exciting and worthwhile new perspectives; but ultimately, and in totality rather than in its constituent parts, it does not wash. Such a view will be unfashionable in many quarters of the academic community. The tendency is to see development (as modernization) itself as the problem, rather than as a solution to the problem of under-development. This claim, that the problem lies in the 'discourse of development', does a disservice because it detracts attention away from the very real failures and inconsistencies of development in Southeast Asia. These have been addressed in the earlier chapters of this book: widening human and regional inequalities, severe environmental problems, the exclusion from development of particular groups in society, and cronyism and corruption, for example. At root the difference between the post-developmentalists and the developmentalists is that while the latter would see the solution to these problems lying within a growth/modernization context, many post-developmentalists would wish to contruct 'another development' or, more radically, 'alternatives to development'.

Yet, and here the views become informed by personal experience, the twin claims that first, lack of development is not a problem and second, that modernization has not improved human well-being, are hard to reconcile with the process and nature of change in Southeast Asia. In summary, I cannot bring myself, both on the basis of the 'evidence' and for reasons of human decency, to reject the struggles of glutinous-rice farmers in Salavan in southern Laos, of cyclo drivers in Ho Chi Minh City, or of garment factory workers outside Jakarta as misguided and ultimately flawed endeavours motivated by the chimera of 'development'.

## Notes

1 Two good collections of papers reviewing the progress of thought on the urban bias thesis are to be found in issues of the *Journal of Development Studies: JDS* (1984) 'Development and the rural–urban divide' (special issue), *Journal of Development Studies* 20(3); and *JDS* (1993) 'Beyond urban bias' (special issue), *Journal of Development Studies* 29(4).

2 This links back to Chapman's observation that 'the essence and meaning of people's mobility become far more comprehensible when conceived as an active dialog between different places, some urban, some rural, some both and some neither' (1995: 257) (see page 161).

3 In a later paper he admits that he had been 'too prone to see the rural rich as crypto-townspeople, and the urban poor as temporarily sojourning countryfolk' (Lipton 1993: 253). Lipton goes on to highlight a number of shortcomings in his original 1977 book. He admits, for example, that he underestimated the political, administrative and fiscal unsustainability of urban bias, tended to gloss over important differences in the experience of urban bias between places and over time, and paid insufficient attention to establishing exceptions to the urban bias thesis (Lipton 1993: 254–6).

4 Bates argues that at a general, global level a clear division is absent, 'calling into question one of the basic assumptions underlying the literature on urban bias: the clear separation between town and country (1993: 222).

5 This links back to the discussion in Chapter 1 regarding the 'impasse' in Marxist development sociology and Booth D.'s (1985) critique of normative approaches (see page 30).

# BIBLIOGRAPHY

ActionAid Vietnam (1995) 'Nine profile villages: Mai Son district, Son La province', unpublished Action Aid report, London.

Aguilar, Filomeno V., Jr. (1989) 'Curbside capitalism: the social relations of street trading in Metro Manila', *Philippine Sociological Review* 37(3–4): 6–25.

Akin Rabibhadana (1993) *Social inequity: a source of conflict in the future?*, Bangkok: Thai Development Research Institute.

Alatas, Syed Farid (1995) 'The theme of "relevance" in third world human sciences', *Singapore Journal of Tropical Geography* 16(2): 123–40.

Alexander, Paul, Boomgaard, Peter and White, Benjamin (1991) (eds) *In the shadow of agriculture: non-farm activities in the Javanese economy, past and present*, Amsterdam: Royal Tropical Institute.

Allen, Bryant J. (1993) 'The problems of upland land management', in: Harold Brookfield and Yvonne Byron (eds) *South-East Asia's environmental future: the search for sustainability*, Tokyo: United Nations University Press, pp. 225–37.

Alpha (1994) *Thailand in figures 1994*, Bangkok: Alpha Research.

—— (1996) *Thailand in figures 1996*, Bangkok: Alpha Research.

Amara Pongsapich, Chantana Bansirichote, Phinit Lapthananon and Suriya Veeravongse (1993) *Socio-cultural change and political development in Central Thailand, 1950–1990*, Bangkok: Thai Development Research Institute.

Amsden, Alice H. (1994) 'Why isn't the whole world experimenting with the East Asian model to develop?: Review of *The East Asian miracle*', *World Development* 22(4): 627–33.

Anderson D. and Leiserson, M.W. (1980) 'Rural non-farm employment in developing countries', *Economic Development and Cultural Change* 28(2): 227–48.

Angeles-Reyes, Edna (1994) 'Nonfarm work in the Philippine rural economy: an omen of change or a change of omens?' in: Bruce Koppel, John Hawkins and William James (eds) *Development or deterioration: work in rural Asia*, Boulder, CO and London: Lynne Rienner, pp. 133–65.

Antlöv, Hans and Svensson, Thommy (1991) 'From rural home weavers to factory labour: the industrialization of textile manufacturing in Majalaya', in: Paul Alexander, Peter Boomgaard and Ben White (eds) *In the shadow of agriculture: non-farm activities in the Javanese economy, past and present*, Amsterdam: Royal Tropical Institute, pp. 113–26.

Aphicat Chamratrithirong *et al.* (1995) *National migration survey of Thailand*, Bangkok: Institute for Population and Social Research, Mahidol University.

Apichai Puntasen, Somboon Siriprachai and Chaiyuth Punyasavatsut (1992) 'Political economy of eucalyptus: business, bureaucracy and the Thai government', *Journal of Contemporary Asia* 22(2): 187–206.

288

Arghiros, Daniel and Wathana Wongsekiarttirat (1996) 'Development in Thailand's extended metropolitan region: the socio-economic and political implications of rapid change in an Ayutthaya district, central Thailand', in: Michael J.G. Parnwell (ed.) *Uneven development in Thailand*, Aldershot: Avebury, pp. 125–45.

Arndt, H.W. (1987) *Economic development: the history of an idea*, Chicago: Chicago University Press.

Askew, Marc (1993) *The making of modern Bangkok: state, market and people in the shaping of the Thai metropolis*, Bangkok: Thai Development Research Institute.

Aung-Thwin, Michael (1995) 'The "Classical" in Southeast Asia: the present in the past', *Journal of Southeast Asian Studies* 26(1): 75–91.

Awanohara, Susumu (1993) 'Hard labour: dispute over workers' rights sours US–Indonesia ties', *Far Eastern Economic Review*, 13 May, p. 13.

Balisacan, Arsenio M. (1992) 'Rural poverty in the Philippines: incidence, determinants and policies', *Asian Development Review* 10(1): 125–63.

—— (1994) 'Urban poverty in the Philippines: nature, causes and policy measures', *Asian Development Review* 12(1): 117–52.

Bardacke, Ted (1995) 'The people's new stake in policy-making', *Financial Times* (Thailand survey), 14 December, p. 2.

Barker, Randolph, Herdt, Robert W. and Rose, Beth (1985) *The rice economy of Asia*, Washington DC: Resources for the Future.

Bates, Robert H. (1993) 'Urban bias: a fresh look', *Journal of Development Studies* 29(4): 219–28.

Baulch, Bob (1996) 'The new poverty agenda: a disputed consensus', *IDS Bulletin* 27(1): 1–10.

Bebbington, Anthony (1993) 'Modernization from below: an alternative indigenous development', *Economic Geography* 69(3): 274–92.

Bell, Peter F. (1992) 'Gender and economic development in Thailand', in: Penny van Esterik and John van Esterik (eds) *Gender and development in Southeast Asia*, Toronto: Canadian Council for Southeast Asian Studies, pp. 61–81.

—— (1996) 'Development or maldevelopment? The contradictions of Thailand's economic growth', in: Michael J.G. Parnwell (ed.) *Uneven development in Thailand*, Aldershot: Avebury, pp. 49–62.

Bencha Yoddumnern-Attig (1992) 'Thai family structure and organization: changing roles and duties in historical perspective', in: Bencha Yoddumnern-Attig *et al.* (eds) *Changing roles and statuses of women in Thailand: a documentary assessment*, Bangkok: Institute for Population and Social Research, Mahidol University, pp. 8–24.

Beresford, Melanie (1993) 'The Vietnamese economy 1979–1993: reforming or revolutionising Asian socialism?', *Asian Studies Review* 17(2): 33–46.

Beresford, Melanie and McFarlane, Bruce (1995) 'Regional inequality and regionalism in Vietnam and China', *Journal of Contemporary Asia* 25(1):50–72.

Bessell, Sharon (1996) 'Children at work', *Inside Indonesia* 46 (March): 18–21.

Black, Richard (1993) 'Geography and refugees: current issues', in: Richard Black and Vaughan Robinson (eds) *Geography and refugees: patterns and processes of change*, London: Belhaven, pp. 3–13.

Blaikie, Piers (1985) *The political economy of soil erosion in developing countries*, London: Longman.

Blaikie, Piers and Brookfield, Harold (1987) (edits) *Land degradation and society*, London: Methuen.

Blussé, Leonard (1991) 'The role of Indonesian Chinese in shaping modern Indonesian life', *Indonesia*: 1–11.

Boomgaard, Peter (1989) *Children of the colonial state: population growth and economic development in Java, 1795–1880*, Amsterdam: CASA Monographs 1, Free University Press.

—— (1991) 'The non-agricultural side of an agricultural economy, Java 1500–1900', in: Paul Alexander, Peter Boomgaard and Ben White (eds) *In the shadow of agriculture: non-farm activities in the Javanese economy, past and present*, Amsterdam: Royal Tropical Institute, pp. 14–40.

Booth, Anne (1985) 'Accommodating a growing population in Javanese agriculture', *Bulletin of Indonesian Economic Studies* 21(2): 115–45.

—— (1988) *Agricultural development in Indonesia*, Sydney: Allen and Unwin.

—— (1992a) 'The World Bank and Indonesian poverty', *Journal of International Development* 4(6): 633–42.

—— (1992b) 'Income distribution and poverty', in: Anne Booth (ed.) *The oil boom and after: Indonesian economic policy and performance in the Soeharto era*, Singapore: Oxford University Press, pp. 323–62.

—— (1993) 'Counting the poor in Indonesia', *Bulletin of Indonesian Economic Studies* 29(1): 53–83.

—— (1995a) 'Southeast Asian economic growth: can the momentum be maintained?', *Southeast Asian Affairs 1995*, Singapore: Institute of Southeast Asian Studies, pp. 28–47.

—— (1995b) 'Regional disparities and inter-governmental fiscal relations in Indonesia', in: Ian G. Cook, Marcus A. Doel and Rex Li (eds) *Fragmented Asia: regional integration and national disintegration in Pacific Asia*, Aldershot: Avebury, pp. 102–36.

—— (forthcoming) 'Rapid economic growth and poverty decline: a comparison of Indonesia and Thailand', in Chris Dixon and David Drakakis-Smith (eds) *Uneven development in South East Asia*, Aldershot: Avebury.

Booth, David (1985) 'Marxism and development sociology: interpreting the impasse', *World Development* 13(7): 761–87.

—— (1993) 'Development research: from impasse to new agenda', in: Frans J. Schuurman (ed.) *Beyond the impasse: new directions in development theory*, London: Zed Books, pp. 49–76.

Bowie, Katherine A. (1992) 'Unraveling the myth of the subsistence economy: textile production in nineteenth century Northern Thailand', *Journal of Asian Studies* 51(4): 797–823.

BPS (1993) *Statistik Indonesia 1992*, Jakarta: Biro Pusat Statistik.

Breman, Jan (1980) *The village on Java and the early-colonial state*, Comparative Asian Studies Programme (CASP), Rotterdam: Erasmus University.

Brennan, M. (1985) 'Class, politics and race in modern Malaysia', in: R. Higgott and R. Robison (eds) *Southeast Asia: essays in the political economy of structural change*, London: Routledge and Kegan Paul, pp. 93–127.

Bresnan, John (1993) *Managing Indonesia: the modern political economy*, New York: Columbia University Press.

Brillantes, Alex Bello, Jr. (1991) 'National politics viewed from Smokey Mountain', in: Benedict J. Kerkvliet and Resil B. Mojares (eds) *From Marcos to Aquino: local perspectives on political transition in the Philippines*, Honolulu: University of Hawaii Press, pp. 187–311.

Brohman, John (1995) 'Universalism, Eurocentrism, and ideological bias in development studies: from modernisation to neoliberalism', *Third World Quarterly* 16(1): 121–40.

Brookfield, Harold, Hadi, Abdul Samad and Mahmud, Zaharah (1991) *The city in the village: the in-situ urbanization of villages, villagers and their land around Kuala Lumpur, Malaysia*, Singapore: Oxford University Press.

Brookfield, Harold, Potter, Lesley and Byron, Yvonne (1995) *In the place of the forest: environmental and socio-economic transformation in Borneo and the eastern Malay Peninsula*, Tokyo: United Nations University Press.

Brown, Tim and Werasit Sittitrai (1995a) 'The impact of HIV on children in Thailand', Programme on AIDS, Thai Red Cross Society, Bangkok.

290

—— (1995b) 'The HIV/AIDS epidemic in Thailand: addressing the impact on children', *Asia-Pacific Population and Policy*, 35 (July–August).

Brown, Tim and Xenos, Peter (1994) 'AIDS in Asia: the gathering storm', *Asia Pacific Issues* 16 (August).

Bryant, Raymond L. and Parnwell, Michael J.G. (1996) 'Politics, sustainable development and environmental change in South-East Asia', in: Michael Parnwell and Raymond Bryant (eds) *Environmental change in South East Asia: people, politics and sustainable development*, London: Routledge, pp. 1–20.

Bunnag, Tej (1977) *The provincial administration of Siam 1892–1915, the Ministry of the Interior under Prince Damrong Rajanubhab*, Kuala Lumpur: Oxford University Press.

Buruma, Ian (1995) 'The Singapore way', *The New York Review*, 19 October, pp. 66–71.

Caldwell, J.A. (1974) *American economic aid to Thailand*, Lexington, MA: Lexington Books.

Carey, Peter (1986) 'Waiting for the "Just King": the agrarian world of south-central Java from Giyanti (1755) to the Java War (1825–1830)', *Modern Asian Studies* 20(1): 59–137.

Carter, S.E., Chidiamassamba, A., Jeranyama, P., Mafukidze, B., Malakela, G.P., Mvena, M., Nabane N., Van Oosterhout-Campbell, S.A.M., Price, L. and Sithole, N. (1993) 'Some observations on wealth ranking after an RRA looking at soil fertility management in Northeastern Zimbabwe', *RRA Notes* 18 (June): 47–52.

Castles, Lance (1991) 'Jakarta: the growing centre', in: Hal Hill (ed.) *Unity and diversity: regional economic development in Indonesia since 1970*, Singapore: Oxford University Press, pp. 232–53.

Cederroth, Sven (1995) *Survival and profit in rural Java: the case of an East Javanese village*, Richmond, Surrey: Curzon Press.

Cederroth, Sven and Gerdin, Ingela (1986) 'Cultivating poverty: the case of the Green Revolution in Lombok', in: I. Norlund, S. Cederroth and I. Gerdin (eds) *Rice societies: Asian problems and prospects*, Scandanavian Institute of Asian Studies, London: Curzon Press, pp. 124–50.

Chamberlain, James R. (ed.) (1991) *The Ram Khamhaeng controversy: collected papers*, Bangkok: The Siam Society.

Chambers, Robert (1984) 'Beyond the Green Revolution: a selective essay', in: T.P. Bayliss-Smith and S. Wanmali (eds) *Understanding Green Revolutions: agrarian change and development planning in South Asia, essays in honour of B.H. Farmer*, Cambridge: Cambridge University Press, pp. 362–79.

—— (1995) 'Poverty and livelihoods: whose reality counts?', *Environment and Urbanization* 7(1): 173–204.

Chanda, Nayan (1995) 'War and peace', *Far Eastern Economic Review*, 4 May, pp. 20–4.

Chant, Sylvia and McIlwaine, Cathy (1995a) *Women of a lesser cost: female labour, foreign exchange and Philippine development*, London: Pluto Press.

—— (1995b) 'Gender and export manufacturing in the Philippines: continuity or change in female employment? The case of the Mactan Export Processing Zone', *Gender, Place and Culture* 2(2): 147–76.

Chantana Banpasirichote (1993) *Community integration into regional industrial development: a case study of Klong Ban Pho, Chachoengsao*, Bangkok: Thai Development Research Institute.

Chapman, Murray (1995) 'Island autobiographies of movement: alternative ways of knowing?', in: Paul Claval and Singaravelou (eds) *Ethnogéographies*, Paris: L'Harmattan, pp. 247–59.

Chatsumarn Kabilsingh (1987) 'How Buddhism can help protect nature', in: S. Davies (ed.) *Tree of life: Buddhism and protection of nature*, Hong Kong: Buddhist Perception of Nature Project, pp. 7–16.

291

Chatthip Nartsupha (1986) 'The village economy in pre-capitalist Thailand', in: Seri Phongphit (ed.) *Back to the roots: village and self-reliance in a Thai context*, Bangkok: Rural Development Documentation Centre, pp. 155–65.

—— (1991) 'The "community culture" school of thought', in: Manas Chitakasem and Andrew Turton (eds) *Thai constructions of knowledge*, London: School of Oriental and African Studies, pp. 118–41.

Chaumeau, Christine (1996) 'Road to Thailand: one of hope, despair', *Phnom Penh Post*, 14–27 June, p. 5.

Chin, James (1996) 'The 1991 Sarawak election: continuity of ethnic politics', *South East Asia Research* 4(1): 23–40.

Choice, Winita (1995) 'Uprooting of the rice farming community: from poor farmers to worse-off labourers', *Thai Development Newsletter* no. 27–28: 39–48.

Clad, James (1989) *Behind the myth: business, money and power in South East Asia*, London: Unwin Hyman.

Clover, Helen (1995) 'Background to migration in Thailand', draft PhD chapter, School of Oriental and African Studies, London.

Cohen, Margot (1994) 'Sharing the wealth', *Far Eastern Economic Review*, 28 April, p. 58.

—— (1995) 'Seed money: poverty scheme fuels both enterprise and red tape', *Far Eastern Economic Review*, 9 February, pp. 24–6.

—— (1996) 'Twisting arms for alms', *Far Eastern Economic Review*, 2 May, pp. 25–9.

Cohen, Paul T. (1984) 'The sovereignty of *dhamma* and economic development: Buddhist social ethics in rural Thailand', *Journal of the Siam Society* 72(1–2): 197–211.

Conway, Susan (1992) *Thai textiles*, British Museum Press: London.

Cook, Paul (1994) 'Policy, reform, privatization, and private sector development in Myanmar', *South East Asia Research* 2(2): 117–40.

Cook, Paul and Minogue, Martin (1993) 'Economic reform and political change in Myanmar (Burma)', *World Development* 21(7): 1151–61.

Corbridge, Stuart (1986) *Capitalist world development: a critique of radical development geography*, Basingstoke: Macmillan.

—— (1990) 'Post-Marxism and development studies: beyond the impasse', *World Development* 18(5): 623–39.

Courtenay, P.P. (1987) 'Out-migration and idle land – a survey of Mukim Melekek, Malacca State, Malaysia', unpublished research report.

—— (1988) 'Farm size, out-migration and abandoned padi land in Mukim Melekek, Melaka (Peninsular Malaysia)', *Malaysian Journal of Tropical Geography* 17: 18–28.

Cowen, Michael and Shenton, Robert (1995) 'The invention of development', in: Jonathan Crush (ed.) *Power of development*, London: Routledge, pp. 27–43.

Crush, Jonathan (1995a) 'Introduction: imagining development', in: Jonathan Crush (ed.) *Power of development*, London: Routledge, pp. 1–23.

—— (ed.) (1995b) *Power of development*, London: Routledge.

Dang Phong (1995) 'Aspects of agricultural economy and rural life in 1993', in: Benedict J. Tria Kerkvliet and Doug J. Porter (eds) *Vietnam's rural transformation*, Boulder, CO: Westview and Singapore: Institute of Southeast Asian Studies, pp. 165–84.

Dao The Tuan (1995) 'The peasant household economy and social change', in: Benedict J. Tria Kerkvliet and Doug J. Porter (eds) *Vietnam's rural transformation*, Boulder, CO: Westview and Singapore: Institute of Southeast Asian Studies, pp. 139–163.

Davidson, Gillian M. and Drakakis-Smith, David (1995) 'The price of success: disadvantaged groups in Singapore', paper presented at the first EUROSEAS Conference, Leiden, The Netherlands, 29 June – 1 July.

De Koninck, Rodolphe (1992) *Malay peasants coping with the world: breaking the community circle?*, Singapore: Institute of Southeast Asian Studies.

Dearden, Philip (1995) 'Development, the environment and social differentiation in Northern Thailand', in: Jonathan Rigg (ed.) *Counting the costs: economic growth and environmental change in Thailand*, Singapore: Institute of Southeast Asian Studies, pp. 111–30.

Demaine, Harvey (1986) '*Kanpatthana*: Thai views of development', in: M. Hobart and R.H. Taylor (eds) *Context, meaning and power in Southeast Asia*, Ithaca, New York: Southeast Asia Program, Cornell University, pp. 93–114.

Desbarats, J. (1987) 'Population redistribution in the Socialist Republic of Vietnam', *Population and Development Review* 13(1): 43–76.

Dick, Howard and Forbes, Dean (1992) 'Transport and communications: a quiet revolution', in: Anne Booth (ed.) *The oil boom and after: Indonesian economic policy and performance in the Soeharto era*, Singapore: Oxford University Press, pp. 258–82.

Dixon, Chris (1978) 'Development, regional disparity and planning: the experience of Northeast Thailand', *Journal of Southeast Asian Studies* 8(2): 210–23.

Dixon, Chris and Drakakis-Smith, David (eds) (forthcoming) *Uneven development in South East Asia*, Aldershot: Avebury.

Dommen, Arthur J (1994) 'Laos: consolidating the economy', *Southeast Asian Affairs 1994*, Singapore: Institute of Southeast Asian Studies, pp. 167–76.

Echols, John M. and Shadily, Hassan (1987) *An Indonesian–English dictionary*, Ithaca, New York: Cornell University Press.

*Economist* (1993a) 'The price of Thailand's prosperity', *The Economist*, 15 May, pp. 81–2.

—— (1993b) 'A survey of Asia', *The Economist*, 30 October.

—— (1993c) 'How cheap can you get?', *The Economist*, 21 August, pp. 49–50.

—— (1995a) 'Smoky Mountain blues', *The Economist*, 9 September, pp. 75–6.

—— (1995b) 'The miracle of the sausage-makers', *The Economist*, 9 December, pp. 71–2.

Eder, James F. (1993) 'Family farming and household enterprise in a Philippine community, 1971–1988: persistence or proletarianization?' *Journal of Asian Studies* 53(3): 647–71.

Edmundson, Wade C. (1994) 'Do the rich get richer, do the poor get poorer? East Java, two decades, three villages, 46 people', *Bulletin of Indonesian Economic Studies* 30(2): 133–48.

Edwards, Michael (1989) 'The irrelevance of development studies', *Third World Quarterly* 11(1): 116–35.

Effendi, Tadjuddin Noer (1993) 'Diversification of the rural economy: non-farm employment and incomes in Jatinom, Central Java', in: Chris Manning and Joan Hardjono (eds) *Indonesia assessment 1993 – Labour: sharing in the benefits of growth?*, Political and Social Change Monograph no. 20, Research School of Pacific Studies, Canberra: Australian National University, pp. 290–302.

Effendi, Tadjuddin Noer and Manning, Chris (1994) 'Rural development and nonfarm employment in Java', in: Bruce Koppel, John Hawkins and William James (eds) *Development or deterioration: work in rural Asia*, Boulder, CO and London: Lynne Rienner, pp. 211–47.

Eliot, Joshua (1995a) *Indonesia, Malaysia and Singapore handbook*, Bath: Trade and Travel.

—— (1995b) *Vietnam, Laos and Cambodia handbook*, Bath: Trade and Travel.

Elmhirst, Becky (1995a) 'Gender, environment and transmigration: comparing migrant and *pribumi* household strategies in Lampung, Indonesia', paper presented to the Third WIVS conference on Indonesian Women in the Household and Beyond, Royal Institute of Linguistics and Anthropology, Leiden, 25–29 September.

——— (1995b) '*Anak mas, anak tiri*: difference and convergence in the livelihood strategies of transmigrants and indigenous people in Indonesia', an outline of provisional findings presented to ICRAF, Bogor, Indonesia, 18 April.

——— (1996) 'Transmigration and local communities in North Lampung: exploring identity politics and resource control in Indonesia', paper presented at the Association of South East Asian Studies' (ASEASUK) Conference, School of Oriental and African Studies, London, 25–27 April.

Engineer, Asghar Ali (1992) 'Islamic economics: a progressive perspective', in: K.S. Jomo (ed.) *Islamic economic alternatives: critical perspectives and new directions*, Basingstoke: Macmillan, pp. 117–24.

Enters, Thomas (1995) 'The economics of land degradation and resource conservation in Northern Thailand: challenging the assumptions', in: Jonathan Rigg (ed.) *Counting the costs: economic growth and environmental change in Thailand*, Singapore: Institute of Southeast Asian Studies, pp. 90–110.

*Environment and Urbanization* (1995) 'The under-estimation and misrepresentation of urban poverty', *Environment and Urbanization* 7(1): 3–10.

Escobar, Arturo (1995a) *Encountering development: the making and unmaking of the Third World*, Princeton, NJ: Princeton University Press.

——— (1995b) 'Imagining a post-development era', in: Jonathan Crush (ed.) *Power of development*, London: Routledge, pp. 211–27.

Esteva, Gustavo (1992) 'Development', in: Wolfgang Sachs (ed.) *The development dictionary: a guide to knowledge as power*, London: Zed Books, pp. 6–25.

Evans, Grant (1994) 'Deadly debris', *Far Eastern Economic Review*, 22 September: 58–9.

Evans, Hugh Emrys (1992) 'A virtuous circle model of rural–urban development: evidence from a Kenyan small town and its hinterland', *Journal of Development Studies* 28(4): 640–67.

Evans, Hugh Emrys and Ngau, Peter (1991) 'Rural–urban relations, household income diversification, and agricultural productivity', *Development and Change* 22: 519–45.

Evers, Hans-Dieter (1991) 'Trade as off-farm employment in Central Java', *Sojourn* 6(1): 1–21.

Fairclough, Gordon (1994) 'Industrial revolution', *Far Eastern Economic Review*, 14 April, p. 31.

——— (1995a) 'Doing the dirty work: Asia's brothels thrive on migrant labour', *Far Eastern Economic Review*, 14 December, pp. 27–8.

——— (1995b) 'Expensive and difficult: voting rules are tough on migrant workers', *Far Eastern Economic Review*, 6 July, p. 17.

——— (1996) 'Dangerous addiction', *Far Eastern Economic Review*, 23 May, p. 67.

Fairclough, Gordon and Tasker, Rodney (1994) 'Separate and unequal', *Far Eastern Economic Review*, 14 April, pp. 22–3.

Fane, George (1994) 'Survey of recent developments', *Bulletin of Indonesian Economic Studies* 30(1): 3–38.

Farrington, John, Lewis, David J. with Satish, S. and Miclat-Teves, Aurea (1993) (eds) *Non-governmental organizations and the state in Asia: rethinking roles in sustainable agricultural development*, London: Routledge.

*FEER* (1992) 'The pen and the saw', *Far Eastern Economic Review*, 27 August, pp. 8–9.

——— (1994) 'Stuck at the bottom', *Far Eastern Economic Review*, 13 January, pp. 70–1.

——— (1995) 'Vietnam revisited', *Far Eastern Economic Review*, 4 May, pp. 28–9.

——— (1996) 'Child labour: it isn't black and white', *Far Eastern Economic Review*, 7 March, pp. 54–7.

Fforde, Adam (1994) 'The institutions of transition from central planning: the case of Vietnam', Economic Division Working Papers 94/4, Research School of Pacific Studies, Australian National University, Canberra.

Fforde, Adam and Sénèque, Steve (1995) 'The economy and the countryside: the relevance of rural development policies', in: Benedict J. Tria Kerkvliet and Doug J. Porter (eds) *Vietnam's rural transformation*, Boulder, CO: Westview and Singapore: Institute of Southeast Asian Studies, pp. 97–138.

Firdausy, Carunia (1994) 'Urban poverty in Indonesia: trends, issues and policies', *Asian Development Review* 12(1): 68–89.

Firdausy, Carunia and Tisdell, Clem (1992) 'Rural poverty and its measurement: a comparative study of villages in Nusa Penida, Bali', *Bulletin of Indonesian Economic Studies* 28(2): 75–93.

Firman, Tommy (1991) 'Rural households, labor flows and the housing construction industry in Bandung, Indonesia', *Tijdschrift voor Econ. en Soc. Geografie* 82(2): 94–105.

—— (1994) 'Labour allocation, mobility, and remittances in rural households: a case from Central Java, Indonesia', *Sojourn* 9(1): 81–101.

Fitzgerald, Tricia (1995) 'Child labour to get special scrutiny', *Phnom Penh Post*, 3–16 November, p. 12.

Forbes, Dean (1993) 'What's in it for us? Images of Pacific Asian development', in: Chris Dixon and David Drakakis-Smith (eds) *Economic and social development in Pacific Asia*, London: Routledge.

—— (1995) 'The urban network and economic reform in Vietnam', *Environment and Planning* A, 27(5): 793–808.

—— (1996) 'Urbanization, migration, and Vietnam's spatial structure', *Sojourn* 11(1): 24–51.

Fox, James J. (1991) 'Managing the ecology of rice production in Indonesia', in: Joan Hardjono (ed.) *Indonesia: resources, ecology and environment*, Singapore: Oxford University Press, pp. 61–84.

Freebairn, D.K. (1995) 'Did the green revolution concentrate incomes? A quantative study of research reports', *World Development* 23(2): 265–79.

Friedmann, John (1992) *Empowerment: the politics of alternative development*, Oxford: Basil Blackwell.

Gates, Carolyn L. (1995) 'Foreign direct investment, institutional change, and Vietnam's gradualist approach to reform', *Southeast Asian Affairs 1995*, Singapore: Institute of Southeast Asian Studies, pp. 382–400.

Gilbert, Alan (1994) 'Third World cities: poverty, employment, gender roles and the environment during a time of restructuring', *Urban Studies* 31 (4–5): 605–33.

Ginsburg, Norton (1991) 'Extended metropolitan regions in Asia: a new spatial paradigm', in: Norton Ginsburg, Bruce Koppel and T.G. McGee (eds) *The extended metropolis: settlement transition in Asia*, Honolulu: University of Hawaii Press, pp. 27–46.

Glewwe, Paul and Gaag, Jacques van der (1990) 'Identifying the poor in developing countries: do different definitions matter?', *World Development* 18(6): 803–14.

Goodkind, Daniel (1995) 'Rising gender inequality in Vietnam since reunification', *Pacific Affairs* 68(3): 342–59.

Gosling, Betty (1991) *Sukhothai: its history, culture and art*, Singapore: Oxford University Press.

Grabowski, Richard (1995) 'Commercialization, nonagricultural production, agricultural innovation, and economic development', *The Journal of Developing Areas* 30: 41–62.

Grabowskiy, Volker (1993) 'Forced resettlement campaigns in Northern Thailand during the early Bangkok period', paper presented at the 5th International Conference on Thai Studies, School of Oriental and African Studies, London, July.

Grandstaff, Terry (1988) 'Environment and economic diversity in Northeast Thailand', in: Terd Charoenwatana and A. Terry Rambo (eds) *Sustainable rural development in Asia*, Khon Kaen, Thailand: Khon Kaen University, pp. 11–22.

Griffin, Keith (1989) *Alternative strategies for economic development*, Basingstoke: Macmillan.

Grijns, Mies and van Velzen, Anita (1993) 'Working women: differentiation and marginalisation', in: Chris Manning and Joan Hardjono (eds) *Indonesia assessment 1993 – Labour: sharing in the benefits of growth?*, Political and Social Change Monograph no. 20, Research School of Pacific Studies, Canberra: Australian National University, pp. 214–228.

Gujit, Irene (1992) 'The elusive poor: a wealth of ways to find them', *RRA Notes*, no. 15 (May): 7–13.

Hadiz, Vedi R. (1993) 'Workers and working class politics in the 1990s', in: Chris Manning and Joan Hardjono (eds) *Indonesia assessment 1993 – Labour: sharing in the benefits of growth?*, Political and Social Change Monograph no. 20, Research School of Pacific Studies, Canberra: Australian National University, pp. 186–200.

Håkangård, Agneta (1992), *Road 13: A Socio-economic Study of Villagers, Transport and Use of Road 13 S, Lao P.D.R.*, Stockholm: Development Studies Unit, Department of Social Anthropology, Stockholm University.

Halliday, Fred (1996) *Islam and the myth of confrontation: religion and politics in the Middle East*, London: I.B. Tauris.

Handley, Paul (1992) 'Rich Thais, poor Thais', *Far Eastern Economic Review*, 20 August, p. 48.

Hardjono, Joan (1993) 'From farm to factory: transition in rural employment in Majalaya sub-district, West Java', in: Chris Manning and Joan Hardjono (eds) *Indonesia assessment 1993 – Labour: sharing in the benefits of growth?*, Political and Social Change Monograph no. 20, Research School of Pacific Studies, Canberra: Australian National University, pp. 273–89.

Harrison, Rachel (1995) 'The writer, the horseshoe crab, his "golden blossom" and her clients: tales of prostitution in contemporary Thai short stories', *South East Asia Research* 3(2): 125–52.

Hart, Gillian (1986) *Power, labor and livelihood: processes of change in rural Java*, Berkeley: University of California Press.

—— (1994) 'The dynamics of diversification in an Asian rice region', in: Bruce Koppel, John Hawkins and William James (eds) *Development or deterioration: work in rural Asia*, Boulder, CO and London: Lynne Rienner, pp. 47–71.

Hayami, Yujiro (1981) 'Induced innovation, Green Revolution, and income distribution: comment', *Economic Development and Cultural Change* 30(1): 169–76.

Hefner, Robert W. (1990) *The political economy of mountain Java: an interpretive history*, Berkeley: University of California Press.

Henderson, Jeffrey (1993) 'Against the economic orthodoxy: on the making of the East Asian miracle', *Economy and Society* 22(2): 200–17.

Heng Hiang Khng (1995) 'Vietnam: taking stock of reforms and dogma', *Southeast Asian Affairs 1995*, Singapore: Institute of Southeast Asian Studies, pp. 365–81.

Hermalin, Albert I (1995) 'Aging in Asia: setting the research foundation', *Asia Pacific Population Research Reports* 4 (April).

Hettne, Björn (1990) *Development theory and the three worlds*, Harlow: Longman.

—— (1995) *Development theory and the three worlds: towards an international political economy of development*, 2nd edition, Harlow: Longman.

Hewison, Kevin (1993) 'Nongovernmental organizations and the cultural development perspective in Thailand: a comment on Rigg (1991)', *World Development* 21(10): 1699–708.

Hewison, Kevin and Brown, Andrew (1994) 'Labour and unions in an industrialising Thailand', *Journal of Contemporary Asia* 24(4): 483–514.

Hewison, Kevin, Robison, Richard and Rodan, Garry (1993) (eds) *Southeast Asia in the 1990s: authoritarianism, democracy and capitalism*, Sydney: Allen and Unwin.

Hiebert, Murray (1993) 'A fortune in waste: scarcity forces Vietnam to reuse its resources', *Far Eastern Economic Review*, 23 December, p. 36.

Hill, Hal (1991) 'The emperor's clothes can now be made in Indonesia', *Bulletin of Indonesian Economic Studies* 27(3): 89–127.

—— (1993) 'Southeast Asian economic development: an analytical survey', Economics Division Working Papers no. 93/4, Research School of Pacific Studies, Australian National University, Canberra.

Hill, Hal and Weidemann, Anna (1991) 'Regional development in Indonesia: patterns and issues', in: Hal Hill (ed.) *Unity and diversity: regional economic development in Indonesia since 1970*, Singapore: Oxford University Press, pp. 3–54.

Hirsch, Philip (1989) 'The state in the village: interpreting rural development in Thailand', *Development and Change* 20(1): 35–56.

—— (1990) *Development dilemmas in rural Thailand*, Singapore: Oxford University Press.

—— (1995) 'Thailand and the new geopolitics of Southeast Asia: resource and environmental issues', in: Jonathan Rigg (ed.) *Counting the costs: economic growth and environmental change in Thailand*, Singapore: Institute of Southeast Asian Studies, pp. 235–59.

Hirsch, Philip and Lohmann, Larry (1989) 'Contemporary politics of environment in Thailand', *Asian Survey* 29(4): 439–51.

Hoang Thi Thanh Nhan (1995) 'Poverty and social polarization in Vietnam: reality and solution', *Vietnam Economic Review* 2(28): 18–23.

Hobart, Mark (1993) 'Introduction: the growth of ignorance', in: Mark Hobart (ed.) *An anthropological critique of development: the growth of ignorance*, London: Routledge, pp. 1–30.

Hong Lysa (1991) '*Warasan Setthasat Kan'muang*: critical scholarship in post-1976 Thailand', in: Andrew Turton and Manas Chitakasem (eds) *Thai constructions of knowledge*, London: School of Oriental and African Studies, pp. 99–117.

Hugo, Graeme (1993) 'Indonesian labour migration to Malaysia: trends and policy implications', *Southeast Asian Journal of Social Science* 21(1): 36–70.

Huxley, Tim (1996) 'Southeast Aisa in the study of international relations: the rise and decline of a region', *Pacific Review* 9(2): 199–228.

Ilich, Ivan (1992) 'Needs', in: Wolfgang Sachs (ed.) *The development dictionary: a guide to knowledge as power*, London: Zed Books, pp. 88–101.

*Inside Indonesia* (1995) 'Freeport under siege', *Inside Indonesia* 45 (December), pp. 16–27.

Irvin, George (1995) 'Vietnam: assessing the achievements of *doi moi*', *Journal of Development Studies* 31(5): 725–50.

James, Mark (1993) 'Tribes and tourists', *Inside Indonesia* 37 (December): 23–4.

Jamieson, Neil (1991) 'The dispersed metropolis in Asia: attitudes and trends in Java', in: Norton Ginsburg, Bruce Koppel and T.G. McGee (eds) *The extended metropolis: settlement transition in Asia*, Honolulu: University of Hawaii Press, pp. 275–97.

Jay, Sian (1996) 'Indonesian river systems', in: Jonathan Rigg (ed.) *The human environment*, Singapore: Editions Didier Millet, pp. 90–1.

Jayasankaran, S. (1995) 'Balancing act', *Far Eastern Economic Review*, 21 December, pp. 24–6.

*JCA* (1978) 'Editorial: Thailand special issue', *Journal of Contemporary Asia* 8(1): 3–4.

*JDS* (1984) 'Development and the rural–urban divide' (special issue), *Journal of Development Studies* 20(3).

—— (1993) 'Beyond urban bias' (special issue), *Journal of Development Studies* 29(4).

Jenkins, Rhys (1994) 'Capitalist development in the NICs', in: Leslie Sklair (ed.) *Capitalism and development*, London: Routledge, pp.72–86.

Jerndal, Randi and Rigg, Jonathan (forthcoming) 'From buffer state to crossroads state:

spaces of human activity and integration in the Lao PDR', in: Grant Evans (ed.) *Laos: culture and society*, Bangkok: Silkworm Books.

Jomo, K.S. (1984) 'Malaysia's New Economic Policy: a class perspective', *Pacific Viewpoint* 25(2): 153–72.

—— (1992) 'Islam and capitalist development: a critique of Rodinson and Weber', in: K.S. Jomo (ed.) *Islamic economic alternatives: critical perspectives and new directions*, Basingstoke: Macmillan, pp. 125–38.

Jones, Gavin W. and Manning, Chris (1992) 'Labour force and employment during the 1980s', in: Anne Booth (ed.) *The oil boom and after: Indonesian economic policy and performance during the Soeharto era*, Singapore: Oxford University Press, pp. 363–410.

*JSEAS* (1995) '25th anniversary special issue: perspectives on Southeast Asian studies', *Journal of Southeast Asian Studies* 26(1).

Kamphol Adulavidhaya and Tongroj Onchan (1985) 'Migration and agricultural development in Thailand: past and future', in: P.M. Hauser, D.B. Suits and N. Ogawa (eds) *Urbanization and migration in Asean development*, Tokyo: National Institute for Research Advancement, pp. 427–52.

Kanok Rerkasem and Benjavan Rerkasem (1994) *Shifting cultivation in Thailand: its current situation and dynamics in the context of highland development*, London: International Institute for Environment and Development.

Kato, Tsuyoshi (1994) 'The emergence of abandoned paddy fields in Negeri Sembilan, Malaysia', *Southeast Asian Studies (Tonan Ajia Kenky)* 32(2): 145–72.

Kaufman, Howard K. (1977) *Bangkhuad: a community study in Thailand*, Rutland, Vermont: Charles Tuttle.

Kemp, Jeremy (1988) *Seductive mirage: the search for the village community in Southeast Asia*, Dordrecht: Foris.

—— (1991) 'The dialectics of village and state in modern Thailand', *Journal of Southeast Asian Studies* 22(2): 312–26.

Kemp, Melody (1993) 'The unknown industrial prisoner: women, modernisation and industrial health', *Inside Indonesia* 37 (December): 20–2.

—— (1994) 'Factory life', unpublished paper.

Kerkvliet, Benedict J. Tria (1995a) 'Rural society and state relations', in: Benedict J. Tria Kerkvliet and Doug J. Porter (eds) *Vietnam's rural transformation*, Boulder, CO: Westview and Singapore: Institute of Southeast Asian Studies, pp. 65–96.

—— (1995b) 'Village–state relations in Vietnam: the effects of everyday politics of decollectivization', *Journal of Asian Studies* 54(2): 396–418.

Kerkvliet, Benedict J. Tria and Porter, Doug J. (1995) 'Rural Vietnam in rural Asia', in: Benedict J. Tria Kerkvliet and Doug J. Porter (eds) *Vietnam's rural transformation*, Boulder, CO: Westview and Singapore: Institute of Southeast Asian Studies, pp. 1–37.

Keyes, Charles F. (1989) *Thailand: Buddhist kingdom as modern nation state*, Boulder, CO: Westview Press.

Kiely, Ray (1995) *Sociology and development: the impasse and beyond*, London: UCL Press.

Kim, Kyong-Dong (1994) 'Confucianism and capitalist development in East Asia', in: Leslie Sklair (ed.) *Capitalism and development*, London: Routledge, pp.87–106.

King, Victor T. (1993a) *The peoples of Borneo*, Oxford: Blackwell.

—— (1993b) '*Politik pembangunan*: the political economy of rainforest exploitation and development in Sarawak, East Malaysia', *Global Ecology and Biogeography Letters* 3 (4–6): 235–44.

King, Victor R. and Jawan, Jayum A. (1996) 'The Ibans of Sarawak, Malaysia: ethnicity, marginalisation and development', in: Denis Dwyer and David Drakakis-Smith (eds) *Ethnicity and development*, London: Wiley, pp. 195–214.

Kistanto, Nurdien H. (1991) 'Peasants, civil servants, and industrial workers in Java: a preliminary note', *Sojourn* 6(2): 290–306.

Knipp, Steven (1996) 'Power generator', *Far Eastern Economic Review*, 2 May, p. 86.

Koizumi, Junko (1992) 'The commutation of Suai from Northeast Siam in the middle of the nineteenth century', *Journal of Southeast Asian Studies* 23(2): 276–307.

Kolko, Gabriel (1995) 'Vietnam since 1975: winning a war and losing the peace', *Journal of Contemporary Asia* 25(1): 3–49.

Koppel, Bruce (1991) 'The rural–urban dichotomy reexamined: beyond the ersatz debate?', in: Norton Ginsburg, Bruce Koppel and T.G. McGee (eds) *The extended metropolis: settlement transition in Asia*, Honolulu: University of Hawaii Press, pp. 47–70.

Koppel, Bruce and Hawkins, John (1994) 'Rural transformation and the future of work in rural Asia', in: Bruce Koppel, John Hawkins and William James (eds) *Development or deterioration: work in rural Asia*, Boulder, CO and London: Lynne Rienner, pp. 1–46.

Koppel, Bruce and James, William (1994) 'Development or deterioration? Understanding employment diversification in rural Asia', in: Bruce Koppel, John Hawkins and William James (eds) *Development or deterioration: work in rural Asia*, Boulder, CO and London: Lynne Rienner, pp. 275–301.

Krannich, C.R. and Krannich, R.L. (1980) 'Family planning policy and community-based innovations in Thailand', *Asian Survey* 20(10): 1023–37.

Krugman, Paul (1994) 'The myth of Asia's miracle', *Foreign Affairs* 73(6): 62–78.

Kummer, David, Concepcion, Roger and Canizares, Bernado (1994) 'Environmental degradation in the uplands of Cebu', *Geographical Review* 84(3): 266–76.

Kunio, Yoshihara (1988) *The rise of ersatz capitalism in South-East Asia*, Singapore: Oxford University Press.

Kuran, Timur (1992) 'The economic system in contemporary Islamic thought', in: K.S. Jomo (ed.) *Islamic economic alternatives: critical perspectives and new directions*, Basingstoke: Macmillan, pp. 9–47.

Kwon, Jene (1994) 'The East Asia challenge to neoclassical orthodoxy', *World Development* 22(4): 635–44.

Lall, Sanjaya (1994) '*The East Asian miracle*: does the bell toll for industrial strategy?', *World Development* 22(4): 645–54.

Lao PDR (1989) *Report on the economic and social situation, development strategy, and assistance needs of the Lao PDR*, volume 1, Geneva: Lao PDR.

Laslett, Peter (1984) 'The family as a knot of individual interests', in: Robert McC. Netting, Richard R. Wilk and Eric J. Arnould (eds) *Households: comparative and historical studies of the domestic group*, Berkeley: University of California Press, pp. 353–79.

Le Dang Doanh (1992) 'Economic reform and development in Vietnam', Economic Division Working Papers 92/1, Research School of Pacific Studies, Australian National University, Canberra.

Leach, Martin (1994) 'Targeting aid to the poorest in urban Ethiopia – is it possible? Rapid urban appraisal', *RRA Notes* 21 (November): 48–54.

Lee, Raymond L.M. (1986) 'Social networks and ethnic interaction in urban Malaysia: an exploratory survey', *Sojourn* 1(1): 109–24.

Lee-Wright, P. (1990) *Child slaves*, London: Earthscan.

Lefferts, Leedom (1975) 'Change and population in a Northeastern Thai village', in: John F. Kantner and Lee McCaffrey (eds) *Population and development in Southeast Asia*, Lexington, MA: D.C. Heath and Co., pp. 173–8.

Leinbach, Thomas R. and Bowen, John T. (1992) 'Diversity in peasant economic behavior: transmigrant households in South Sumatra, Indonesia', *Geographical Analysis* 24(4): 335–51.

Leinbach, Thomas R., Watkins, John F. and Bowen, J. (1992) 'Employment behavior and the family in Indonesian transmigration', *Annals of the Association of American Geographers* 82(1): 23–47.

Lewis, Martin W. (1992a) *Green delusions: an environmentalist critique of radical environmentalism*, Durham, NC: Duke University Press.

—— (1992b) *Wagering the land: ritual, capital, and environmental degradation in the Cordillera of northern Luzon 1900–1986*, Berkeley: University of California Press.

Liew Kim Siong (1994) 'Welfarism and an affluent Singapore' (first published in the *Business Times*, 25 June 1992) in: Derek da Cunha (ed.) *Debating Singapore: reflective essays*, Singapore: Institute of Southeast Asian Studies, pp. 51–4.

Lindberg, Clas, Loiske, Vesa-Matti, Östberg, Wilhelm and Mung'ong'o, Claude (1995) 'Handle with care! Rapid studies and the poor', *PLA Notes* 22 (February): 11–16.

Lipton, Michael (1977) *Why poor people stay poor: urban bias in world development*, London: Temple Smith.

—— (1984) 'Urban bias revisited', *Journal of Development Studies* 20(3): 139–66.

—— (1987) 'Development studies: findings, frontiers and fights', *World Development* 15(4): 517–25.

—— (1993) 'Urban bias: of consequences, causes and causality', *Journal of Development Studies* 29(4): 229–58.

Lipton, Michael with Longhurst, Richard (1989) *New seeds and poor people*, London: Unwin Hyman.

Livingston, Carol and Ker Munthit (1993) 'Fighting spawns new wave of refugees', *Phnom Penh Post* 2(19) (10–23 Septmber): 2.

Lohmann, Larry (1991) 'Peasants, plantations and pulp: the politics of eucalyptus in Thailand', *Bulletin of Concerned Asian Scholars* 24(4): 3–17.

—— (1992) 'Land, power and forest colonisation in Thailand', in: *Agrarian reform and environment in the Philippines and Southeast Asia*, London: Catholic Institute for International Relations, pp. 85–99.

—— (1995) 'No rules of engagement: interest groups, centralization and the creative politics of "environment" in Thailand', in: Jonathan Rigg (ed.) *Counting the costs: economic growth and environmental change in Thailand*, Singapore: Institute of Southeast Asian Studies, pp. 211–34.

Lok, Helen (1993) 'Labour in the garment industry: an employer's perspective', in: Chris Manning and Joan Hardjono (eds) *Indonesia assessment 1993 – Labour: sharing in the benefits of growth?*, Political and Social Change Monograph no. 20, Research School of Pacific Studies, Canberra: Australian National University, pp. 155–72.

Luxmon Wongsuphasawat (1995) 'The extended metropolitan region and uneven industrial development in Thailand', paper presented at the first EUROSEAS Conference, Leiden, 29 June – 1 July.

McAndrew, John P. (1989) 'Urbanization and social differentiation in a Philippine village', *Philippine Sociological Review* 37(1–2): 26–37.

McBeth, John (1995) 'Open wound: labour activist's murder haunts military', *Far Eastern Economic Review*, 22 June, p. 32.

—— (1996a) 'Coming together: new groupings point to widespread political unease', *Far Eastern Economic Review*, 15 February, p. 24.

—— (1996b) 'What's yours is mine', *Far Eastern Economic Review*, 28 March, pp. 60–1.

McGee, T.G. (1989) '*Urbanisasi* or *kotadesasi*? Evolving patterns of urbanization in Asia', in: Frank J. Costa, Ashok K. Dutt, Lawrence J.C. Ma and Allen G. Noble (eds) *Urbanization in Asia: spatial dimensions and policy issues*, Honolulu: University of Hawaii Press, pp. 93–108.

—— (1991a) 'The emergence of *desakota* regions in Asia: expanding a hypothesis', in: Norton Ginsburg, Bruce Koppel and T.G. McGee (eds) *The extended metropolis: settlement transition in Asia*, Honolulu: University of Hawaii Press, pp. 3–25.

—— (1991b) 'Presidential address: Eurocentrism in geography – the case of Asian urbanization', *Canadian Geographer* 35(4): 332–44.

McGee, T.G. and Greenberg, Charles (1992) 'The emergence of extended metropolitan regions in ASEAN', *ASEAN Economic Bulletin* 9(1): 22–44.

McIlwaine, Cathy (1995) 'Fringes or frontiers? export-oriented development and the implications for gender: the case of the Philippines', paper presented at the first EUROSEAS Conference, Leiden, the Netherlands, 29 June – 1 July.

Mackie, Jamie (1991) 'Towkays and tycoons: the Chinese in Indonesian economic life in the 1920 and 1980s', *Indonesia* 83–96.

McVey, Ruth (1992) 'The materialization of the Southeast Asian entrepreneur', in: Ruth McVey (ed.) *Southeast Asian capitalists*, Studies on Southeast Asia, Ithaca, NY: Cornell University Press, pp. 7–33.

—— (1995) 'Change and continuity in Southeast Asian Studies', *Journal of Southeast Asian Studies* 26(1): 1–9.

Mahbubani, Kishore (1995) 'The Pacific way', *Foreign Affairs* 74(1): 100–11.

Manning, Chris (1987) 'Rural economic change and labour mobility: a case study from West Java', *Bulletin of Indonesian Economic Studies* 23(3): 52–79.

—— (1993a) 'Introduction', in: Chris Manning and Joan Hardjono (eds) *Indonesia assessment 1993 – Labour: sharing in the benefits of growth?*, Political and Social Change Monograph no. 20, Research School of Pacific Studies, Canberra: Australian National University, pp. 1–9.

—— (1993b) 'Examining both sides of the ledger: employment and wages during the New Order', in: Chris Manning and Joan Hardjono (eds) *Indonesia assessment 1993 – Labour: sharing in the benefits of growth?*, Political and Social Change Monograph no. 20, Research School of Pacific Studies, Canberra: Australian National University, pp. 61–87.

—— (1993c) 'Structural change and industrial relations during the Soeharto period: an approaching crisis?', *Bulletin of Indonesian Economic Studies* 29(2): 59–95.

Mason, Andrew D. (1996) 'Targeting the poor in rural Java', *IDS Bulletin* 27(1): 67–82.

Mason, Karen (1995) 'Is the situation of women in Asia improving or deteriorating?' (with the assistance of Amy Cardamone, Jill Holdren and Leah Retherford), *Asia Pacific Population Research Reports no. 6*, Honolulu: East–West Center.

Mason, Richard (1995) 'Parti Bansa Dayak Sarawak and the Sarawak state elections of 1987 and 1991', *Kajian Malaysia* 13(1): 26–58.

Mather, Celia E. (1983) 'Industrialization in the Tangerang Regency of West Java: women workers and the Islamic patriarchy', *Bulletin of Concerned Asian Scholars* 15(2): 2–17.

Maurer, Jean-Luc (1991) 'Beyond the sawah: economic diversification in four Bantul villages, 1972–1987', in: Paul Alexander, Peter Boomgaard, and Ben White (eds) *In the shadow of agriculture: non-farm activities in the Javanese economy, past and present*, Amsterdam: Royal Tropical Institute, pp. 92–112.

Medhi Krongkaew (1993) 'Poverty and income distribution', in: Peter G. Warr (ed.) *The Thai economy in transition*, Cambridge: Cambridge University Press, pp. 401–37.

—— (1995) 'Growth hides rising poverty', *Bangkok Post*, 29 December, pp. 63–5.

Medhi Krongkaew, Pranee Tinakorn and Suphat Suphachalasai (1992) 'Rural poverty in Thailand: policy issues and responses', *Asian Development Review* 10(1): 199–225.

Mehmet, Ozay (1995) *Westernizing the Third World: the Eurocentricity of economic development theories*, London: Routledge.

Mendis, Patrick (1994) 'Buddhist economics and community development strategies', *Community Development Journal* 29(3): 195–202.

Menembu, Angel (1995) 'The first people', *Inside Indonesia* 45 (December), pp. 23–4.

Miranda, Mariano, Jr. (1988) 'The economics of poverty and the poverty of economics: the Philippine experience', in: Mamerto Canlas, Mariano Miranda Jr. and James Putzel (eds) *Land, poverty and politics in the Philippines*, London: Catholic Institute for International Relations, pp. 11–46.

Moerman, Michael (1968) *Agricultural change and peasant choice in a Thai village*, Berkeley: University of California Press.

Monzel, Kristen L. (1993) '"Only the women know": powerlessness and marginality in three Hmong women's lives', in: Richard Black and Vaughan Robinson (eds) *Geography and refugees: patterns and processes of change*, London: Belhaven, pp. 118–33.

Moore, Mick (1984) 'Categorising space: urban–rural or core–periphery in Sri Lanka', *Journal of Development Studies* 20(3): 102–22.

Morrison, Philip S. (1993) 'Transitions in rural Sarawak: off-farm employment in the Kamena Basin', *Pacific Viewpoint* 34(1): 45–68.

Muijzenberg, Otto van den (1991) 'Tenant emancipation, diversification and social differentiation in Central Luzon', in: Jan Breman and Sudipto Mundle (eds) *Rural transformation in Asia*, Delhi: Oxford University Press, pp. 313–37.

Mulder, Niels (1989) *Individual and society in Java: a cultural analysis*, Yogyakarta, Indonesia: Gadjah Mada University Press.

Murray, Alison J. (1991) *No money, no honey: a study of street traders and prostitutes in Jakarta*, Singapore: Oxford University Press.

Muscat, Robert J. (1994) *The fifth tiger: a study of Thai development policy*, Tokyo: United Nations University Press.

Myers, W.E. (1991) 'Introduction', in: W.E. Myers (ed.) *Protecting working children*, London: Zed Books, pp. 3–10.

Myrdal, Gunnar (1968) *Asian drama: an inquiry into the poverty of nations* (3 volumes), Harmondsworth: Penguin.

Naqvi, S.H.N., Beg, H.U., Ahmed, R., and Nazeer, M.N. (1992) 'Principles of Islamic econmic reform', in: K.S. Jomo (ed.) *Islamic economic alternatives: critical perspectives and new directions*, Basingstoke: Macmillan, pp. 153–87.

Naylor, Rosamond (1992) 'Labour-saving technologies in the Javanese rice economy: recent developments and a look into the 1990s', *Bulletin of Indonesian Economic Studies* 28(3): 71–89.

Nederveen Pieterse, Jan (1991) 'Dilemmas of development discourse: the crisis of developmentalism and the comparative method', *Development and Change* 22: 5–29.

Netting, Robert McC. (1993) *Smallholders, householders: farm families and the ecology of intensive, sustainable agriculture*, Stanford, CA: Stanford University Press.

Netting, Robert McC., Wilks, Richard R. and Arnould, Eric J. (1984) 'Introduction', in: Robert McC. Netting, Richard R. Wilk and Eric J. Arnould (eds) *Households: comparative and historical studies of the domestic group*, Berkeley: University of California Press, pp. xiii–xxxviii.

Nibbering, Jan Willem (1991) 'Crisis and resilience in upland land use in Java', in: Joan Hardjono (ed.) *Indonesia: resources, ecology, and environment*, Singapore: Oxford University Press, pp. 104–32.

Nielsen, Preben (1994) 'Transportation network: current status and future plans', in: Chi Do Pham (ed.) *Economic development in Lao PDR*, Vientiane: Horizon 2000, pp. 183–91.

Nipon Poapongsakorn (1994) 'Transformations in the Thai rural labor market', in: Bruce Koppel, John Hawkins and William James (eds) *Development or deterioration: work in rural Asia*, Boulder, CO and London: Lynne Rienner, pp. 167–210.

NSC (1995) *Expenditure and consumption survey and social indicator survey (1992–1993)*, Vientiane, Laos: Committee for Planning and Cooperation, National Statistical Centre.

NSO (1993) *Statistical handbook of Thailand 1993*, Bangkok: National Statistical Office.

O'Brien, Leslie (1983) 'Four paces behind: women's work in Peninsular Malaysia', in Lenore Manderson (ed.) *Women's work and women's roles: economics and everyday life in Indonesia, Malaysia and Singapore*, Development Studies Centre Monograph no. 32, Canberra: Australian National University, pp. 193–215.

O'Connor, Richard A. (1993) 'Interpreting Thai religious change: temples, Sangha reform and social change', *Journal of Southeast Asian Studies* 24(2): 330–39.

Ong, Aihwa (1987) *Spirits of resistance and capitalist discipline: factory women in Malaysia*, New York: SUNY Press.

—— (1990) 'Japanese factories, Malay workers: class and sexual metaphors in West Malaysia', in: Jane Monnig Atkinson and Shelly Errington (eds) *Power and difference: gender in Island Southeast Asia*, Stanford, CA: Stanford University Press, pp. 395–422.

Otsuka, Keijiro, Gascon, Fe and Asano, Seki (1994) 'Green revolution and labour demand in rice farming: the case of central Luzon, 1966–1990', *Journal of Development Studies* 31(1): 82–109.

Page, John M. (1994) 'The East Asian miracle: an introduction', *World Development* 22(4): 615–25.

Paritta Chalermpow Koanantakool and Askew, Marc (1993) *Urban life and urban people in transition*, The 1993 Year End Conference on Who Gets What and How? – Challenges for the Future, Bangkok: Thai Development Research Institute.

Parnwell, Michael J.G. (1986) 'Migration and the development of agriculture: a case study of Northeast Thailand', in: Michael J.G. Parnwell (ed.) *Rural development in North-East Thailand: case studies of migration, irrigation and rural credit*, Occasional paper no. 12, Centre for South-East Asian Studies, University of Hull, pp. 93–140.

—— (1988) 'Rural poverty, development and the environment: the case of Northeast Thailand', *Journal of Biogeography* 15: 199–208.

—— (1990) *Rural industrialisation in Thailand*, Hull Paper in Developing Area Studies no. 1, Centre of Developing Area Studies, University of Hull.

—— (1992) 'Confronting uneven development in Thailand: the potential role of rural industries', *Malaysian Journal of Tropical Geography* 22(1): 51–62.

—— (1993) 'Tourism, handicrafts and development in North-East Thailand', paper presented at the 5th International Thai Studies Conference, SOAS, London, July.

—— (1994) 'Rural industrialisation and sustainable development in Thailand', *Thai Environment Institute Quarterly Environment Journal* 1(2): 24–39.

—— (ed.) (1996) *Thailand: uneven development*, Aldershot: Avebury.

Parnwell, Michael J.G. and Arghiros, Daniel A. (1996) 'Uneven development in Thailand', in: Michael J.G. Parnwell (ed.) *Uneven development in Thailand*, Aldershot: Avebury, pp.1–27.

Parnwell, Michael J.G. and Rigg, Jonathan (1996) 'The people of Isan, Northeast Thailand: missing out on the economic boom?', in: Denis Dwyer and David Drakakis-Smith (eds) *Ethnicity and development*, London: Wiley, pp. 215–48.

Pasuk Phongpaichit (1984) 'The Bangkok masseuses: origins, status and prospects', in: G.W. Jones (ed.) *Women in the urban and industrial workforce: Southeast and East Asia*, Canberra: Australian National University.

Pawadee Tonguthai (1987) 'Implicit policies affecting urbanization in Thailand', in: Roland J. Fuchs, Gavin W. Jones and Ernesto M. Pernia (eds) *Urbanization and urban policies in Pacific Asia*, Boulder, CO: Westview Press, pp. 183–92.

Pearse, Andrew (1980) *Seeds of plenty, seeds of want: social and economic implications of the Green Revolution*, Oxford: Clarendon Press.

Pedersen, Poul (1992) 'The study of perception of nature – towards a sociology of knowledge about nature', in: Ole Bruun and Arne Kalland (eds) *Asian perceptions of nature*, Nordic Proceedings in Asian Studies no. 3, Copenhagen: Nordic Institute of Asian Studies, pp. 148–58.

Peet, Richard and Watts, Michael (1993) 'Introduction: development theory and environment in an age of market triumphalism', *Economic Geography* 69(3): 227–53.

Perkins, Dwight (1994) 'There are at least three models of East Asian development', *World Development* 22(4): 655–61.

Pettus, Ashley S. (1995) 'Vietnam's learning curve', *Far Eastern Economic Review*, 18 August, pp. 36–7.

Picard, Michel (1993) 'Cultural tourism in Bali: national integration and regional differentiation', in: Michael Hitchcock, Victor T. King and Michael J.G. Parnwell (eds) *Tourism in South-East Asia*, London: Routledge, pp. 71–88.

Popkin, Samuel L. (1979) *The rational peasant: the political economy of rural society in Vietnam*, Berkeley: University of California Press.

Porter, Doug J. (1995a) 'Economic liberalization, marginality and the local state', in: Benedict J. Tria Kerkvliet and Doug J. Porter (eds) *Vietnam's rural transformation*, Boulder, CO: Westview and Singapore: Institute of Southeast Asian Studies, pp. 215–46.

—— (1995b) 'Scenes from childhood: the homesickness of development discourses', in: Jonathan Crush (ed.) *Power of development*, London: Routledge, pp. 63–86.

Porter, Gina (1995) 'Mobility and inequality in rural Nigeria: the case of off-road communities'. Paper presented at the Institute of British Geographers Conference, January, University of Northumbria at Newcastle, UK.

Pottier, J. (ed.) (1992) *Practising development*, London: Routledge.

Pravit Rojanaphruk (1992) 'Villagers vow to hold their ground', *The Nation*, 4 December.

Prawase Wasi (1991) 'Tourism and child prostitution', in: Koson Srisang (ed.) *Caught in modern slavery: tourism and child prostitution in Asia*, Bangkok: Ecumenical Coalition on Third World Tourism, pp. 26–8.

Prayudh Payutto (1994) *Buddhist economics: a middle way for the market place*, Bangkok: Buddhadhamma Foundation.

—— (1995) *Buddhadhamma: natural laws and values for life* (translated by Grant A. Olson), New York: State University of New York Press.

Preston, David A. (1989) 'Too busy to farm: under-utilisation of farm land in Central Java', *Journal of Development Studies* 26(1): 43–57.

Pretty, Jules N. (1995) *Regenerating agriculture: policies and practice for sustainability and self-reliance*, London: Earthscan.

Price, Susanna (1983) 'Rich woman, poor woman: occupation differences in a textile producing village in Central Java', in: Lenore Manderson (ed.) *Women's work and women's roles: economics and everyday life in Indonesia, Malaysia and Singapore*, Development Studies Centre Monograph no. 32, Canberra: Australian National University, pp. 97–110.

Quibria, M.G. (1991) 'Understanding poverty: an introduction to conceptual and measurement issues', *Asian Development Review* 9(2): 90–112.

Reid, Anthony (1988) *Southeast Asia in the age of commerce 1450–1680, the lands below the winds* (volume 1), New Haven, CT: Yale University Press.

—— (1993) *Southeast Asia in the age of commerce 1450–1680: expansion and crisis* (volume 2), New Haven, CT: Yale University Press.

Reynolds, Craig J. (1987) *Thai radical discourse: the real face of Thai feudalism today*, Ithaca, NY: Cornell University Southeast Asia Program.

—— (1995) 'A new look at old Southeast Asia', *Journal of Asian Studies* 52(2): 419–46.

Rhodes, Belinda (1994) 'Step by step: the Philippines attempts to save Banaue's crumbling rice paddies', *Far Eastern Economic Review*, 15 September, pp. 50–1.

Richter, Kerry and Bencha Yoddumnern-Attig (1992) 'Framing a study of Thai women's changing roles and statuses', in: Bencha Yoddumnern-Attig *et al.* (eds) *Changing roles and statuses of women in Thailand: a documentary assessment*, Bangkok: Institute for Population and Social Research, Mahidol University, pp. 1–7.

Ricklefs, M.C. (1981) *A history of modern Indonesia*, Basingstoke: Macmillan.

Rigg, Jonathan (1986) 'Innovation and intensification in Northeastern Thailand: Brookfield applied', *Pacific Viewpoint* 27: 29–45.

—— (1988a) "Land ownership and land tenure as measures of wealth and marginalisation: evidence from Northeast Thailand', *Area* 20(4): 339–45.

—— (1988b) 'Singapore and the recession of 1985', *Asian Survey* 28(3): 340–52.

—— (1988c) 'Perspectives on migrant labouring and the village economy in developing countries: the Asian experience in a world context', *Progress in Human Geography* 12(1): 66–86.

—— (1989a) *International contract labor migration and the village economy: the case of Tambon Don Han, Northeastern Thailand*, papers of East–West Population Institute no. 112, East-West Center: Honolulu.

—— (1989b) 'The new rice technology and agrarian change: guilt by association?', *Progress in Human Geography* 13(3):374–99.

—— (1991a) 'Grass-roots development in rural Thailand: a lost cause?', *World Development* 19(2/3): 199–211.

—— (1991b) *Southeast Asia: a region in transition*, London: Unwin Hyman.

—— (1993) 'A reply to Kevin Hewison', *World Development* 21(10): 1709–13.

—— (1994a) 'Redefining the village and rural life: lessons from South East Asia', *Geographical Journal* 160(2): 123–35.

—— (1994b) 'Alternative development strategies, NGOs and the environment in Thailand: a critique', *TEI Quarterly Environment Journal* 2(2): 16–26.

—— (1995a) 'Counting the costs: economic growth and environmental change in Thailand', in: Jonathan Rigg (ed.) *Counting the costs: economic growth and environmental change in Thailand*, Singapore: Institute of Southeast Asian Studies, pp. 3–24.

—— (1995b) 'Errors in the making: rice, knowledge, technological change and "applied" research in Northeastern Thailand', *Malaysian Journal of Tropical Geography* 26(1): 19–33.

—— (1995c) 'Managing dependency in a reforming economy: the Lao PDR', *Contemporary Southeast Asia* 17(2): 147–72.

—— (forthcoming) 'Uneven development and the (re-)engagement of Laos', in: Chris Dixon and David Drakakis-Smith (eds) *Uneven development in South East Asia*, Aldershot: Avebury.

Rigg, Jonathan and Jerndal, Randi (1996) 'Plenty in the context of scarcity: forest management in Laos', in: Michael J.G. Parnwell and Raymond Bryant (eds) *Environmental change in South East Asia: people, politics and sustainable development*, London: Routledge, pp. 145–62.

Rigg, Jonathan and Stott, Philip (forthcoming) 'Forest tales: politics, environmental policies and their implementation in Thailand', in: Uday Desai (ed.) *Ecological policy and politics in developing countries: economic growth, democracy and environmental protection*, New York: State University of New York (SUNY) Press.

Ritchie, Mark A. (1993) 'The "village" in context: arenas of social action and historical change in Northern Thai peasant classes', paper presented at the 5th International Thai Studies Conference, SOAS, London, July.

Rodan, Garry (1989) *The political economy of Singapore's industrialization: national state and international capital*, Basingstoke: Macmillan.

Ross, Helen and Anuchat Poungsomlee (1995) 'Environmental and social impact of urbanization in Bangkok', in: Jonathan Rigg (ed.) *Counting the costs: economic growth and environmental change in Thailand*, Singapore: Institute of Southeast Asian Studies, pp. 131–51.

Rostow, W.W. (1995) 'Letters to the editor', *Foreign Affairs* 74(1): 183–4.

Rotgé, Vincent L. (1992) 'Rural employment shift in the context of growing rural–urban linkages: trends and prosects for DIY [Daerah Istimewa Yogyakarta]', paper presented at the International Conference on Geography in the ASEAN Region, Gadjah Mada University, Yogyakarta, 31 August–3 September.

Sachs, Wolfgang (1992) 'Introduction', in: Wolfgang Sachs (ed.) *The development dictionary: a guide to knowledge as power*, London: Zed Books, pp. 1–5.

Said, Edward (1979) *Orientalism*, New York: Vintage Books.

Sairin, Sjafri (1996) 'The appeal of plantation labour: economic imperatives and cultural considerations among Javanese workers in North Sumatra', *Sojourn* 11(1): 1–23.

Sanitsuda Ekachai (1988) 'Mental casualties among the Roi-Et migrants', *Bangkok Post*, 29 February, p. 36.

—— (1990) *Behind the smile: voices of Thailand*, Bangkok: Post Publishing.

Schenk-Sandbergen, Loes and Outhaki Choulamany-Khamphoui (1995) *Women in rice fields and offices: irrigation in Laos*, Heiloo, The Netherlands: Empowerment.

Schmidt, Johannes D. (1996) 'Paternalism and planning in Thailand: facilitating growth without social benefits', in: Michael J.G. Parnwell (ed.) *Uneven development in Thailand*, Aldershot: Avebury, pp. 63–81.

—— (forthcoming) 'The challenge from South East Asia: between equity and growth', in: Chris Dixon and David Drakakis-Smith (eds) *Uneven development in South East Asia*, Aldershot: Avebury.

Schmidt-Vogt, Dietrich (1995) 'Swidden farming and secondary vegetation: two case studies from Northern Thailand', in: Jonathan Rigg (ed.) *Counting the costs: economic growth and environmental change in Thailand*, Singapore: Institute of Southeast Asian Studies, pp. 47–64.

Schober, Juliane (1995) 'The Theravada Buddhist engagement with modernity in Southeast Asia: whither the social paradigm of the galactic polity?', *Journal of Southeast Asian Studies* 26(2): 307–25.

Schumacher, E.F. (1973) *Small is beautiful: a study of economics as if people mattered*, London: Blond and Briggs.

Schuurman, Frans J. (1993) 'Introduction: development theory in the 1990s', in: Frans J. Schuurman (ed.) *Beyond the impasse: new directions in development theory*, London: Zed Books, pp. 1–48.

Schwarz, Adam (1995) 'Listen up: UN urges Vietnam to rethink poverty spending', *Far Eastern Economic Review*, 16 November, p. 99.

—— (1996) 'Proletarian blues', *Far Eastern Economic Review*, 25 January, pp. 21–2.

Schweizer, Thomas (1987) 'Agrarian transformation? rice production in a Javanese village', *Bulletin of Indonesian Economic Studies* 23(2): 38–70.

Scott, James C. (1976) *The moral economy of the peasant: rebellion and subsistence in Southeast Asia*, New Haven, CT: Yale University Press.

—— (1985) *Weapons of the weak: everyday forms of peasant resistance*, New Haven, CT: Yale University Press.

Sen, Amartya (1992) *Inequality reexamined*, Oxford: Clarendon Press.

Seri Phongphit (1988) *Religion in a changing society: Buddhism, reform and the role of monks in community development in Thailand*, Hong Kong: Arena Press.

Sesser, Stan (1996) 'Khong Island, Laos', Mekong Diary Dispatches, Internet.

Shamsul, A.B. (1989) *Village: the imposed social construct in Malaysia's developmental initiatives*, Working paper no. 115, Sociology of Development Research Centre, Bielefeld: University of Bielefeld.

Shari, I. and Jomo, K.S. (1984) 'The New Economic Policy and "national unity": development and inequality 25 years after independence', in: S. Husin Ali (ed.) *Ethnicity, class and development: Malaysia*, Kuala Lumpur: Persatuan Sains Sosial, pp. 329–55.

Shari, M.H. (1988) *Culture and environment in Thailand: dynamics of a complex relationship*, Bangkok: The Siam Society.

Silverman, Gary (1995) 'Honour thy father', *Far Eastern Economic Review*, 2 March, pp. 50–2.

—— (1996) 'Vital and vulnerable', *Far Eastern Economic Review* 23 May, pp. 60–6.

Singarimbun, Masri (1993) 'The opening of the village labour market: changes in employment and welfare in Sriharjo', in: Chris Manning and Joan Hardjono (eds) *Indonesia assessment 1993 – Labour: sharing in the benefits of growth?*, Political and Social Change Monograph no. 20, Research School of Pacific Studies, Canberra: Australian National University, pp. 261–72.

Sinit Sitthirak (1995) 'Prostitution in Thailand: a North–South dialogue on neo-colonialism, militarism, and consumerism', *Thai Development Newsletter*, nos 27–28: 62–8.

Sjahrir, Kartini (1993) 'The informality of employment in construction: the case of Jakarta', in: Chris Manning and Joan Hardjono (eds) *Indonesia assessment 1993 – Labour: sharing in the benefits of growth?*, Political and Social Change Monograph no. 20, Research School of Pacific Studies, Canberra: Australian National University, pp. 240–58.

Skeldon, Ronald (1990) *Population mobility in developing countries: a reinterpretation*, London: Belhaven.

—— (1995) 'The challenge facing migration research: a case for greater awareness', *Progress in Human Geography* 19(1): 91–6.

Smalley, William A. (1994) *Linguistic diversity and national unity: language ecology in Thailand*, Chicago: University of Chicago Press.

Smart, J.E. (1986) 'Worker circulation between Asia and the Middle East', *Pacific Viewpoint* 27: 1–28.

Smith, William (1995) 'Impementing the 1993 Land Law: the impact of land allocation on rural households in Son La and Ha Tinh provinces', unpublished ActionAid report, London (June).

Smith, William and Tran Thanh Binh (1994) 'The impact of the 1993 Land Law on rural households in the Mai Son District of Son La province', unpublished ActionAid report, London (May).

Soesastro, M. Hadi (1991) 'East Timor: questions of economic viability', in: Hal Hill (ed.) *Unity and diversity: regional economic development in Indonesia since 1970*, Singapore: Oxford University Press, pp. 207–29.

Somchai Ratanakomut, Charuma Ashakul and Thienchay Kirananda (1994) 'Urban poverty in Thailand: critical issues and policy measures', *Asian Development Review* 12(1): 204–24.

Sondakh, Luck (1994) 'Employment patterns and the role of off-farm labour in rural north Sulawesi', in: Helmut Buchholt and Ulrich Mai (eds) *Continuity, change and aspirations: social and cultural life in Minahasa, Indonesia*, Singapore: Institute of Southeast Asian Studies, pp. 167–75.

Sponsel, Leslie E. and Poranee Natadecha-Sponsel (1994) 'The potential contribution of Buddhism in developing an environmental ethic for the conservation of bio-diversity', in: Lawrence S. Hamilton (ed.) *Ethics, religion and biodiversity: relations between conservation and cultural values*, Knapwell, Cambridge: White Horse Press, pp. 75–97.

—— (1995) 'The role of Buddhism in creating a more sustainable society in Thailand', in: Jonathan Rigg (ed.) *Counting the costs: economic growth and environmental change in Thailand*, Singapore: Institute of Southeast Asian Studies, pp. 27–46.

Steinberg, David I. (1982) *Burma: a socialist nation of Southeast Asia*, Boulder, CO: Westview Press.

Steinberg, D.J. *et al.* (1985) *In search of Southeast Asia: a modern history*, Sydney: Allen and Unwin.

Stott, Philip (1991) '*Mu'ang* and *pa*: élite views of nature in a changing Thailand', in: Manas Chitakasem and Andrew Turton (eds) *Thai constructions of knowledge*, London: School of Oriental and African Studies, pp. 142–54.

Suganya Hutaserani (1990) 'The trends of income inequality and poverty and profile of the urban poor in Thailand', *TDRI Quarterly Review* 5(4): 14–19.

Sulak Sivaraksa (1990) *Siam in crisis, a collection of articles by S. Sivaraksa*, Bangkok: Thai Inter-Religious Commission for Development.

—— (1996) 'Buddhism in crisis', *Far Eastern Economic Review*, 9 May, p. 31.

Svensson, Thommy (1996) 'The impact of the rise of Pacific Asia: the Asia–Europe summit in perspective', *Nordic Newsletter of Asian Studies (NIAS nytt)*, no. 2 (July): 5–9.

Tambunan, Tulus (1995) 'Forces behind the growth of rural industries in developing countries: a survey of literature and a case study from Indonesia', *Journal of Rural Studies* 11(2): 203–15.

Tan, Mély G. (1991) 'The social and cultural dimensions of the role of ethnic Chinese in Indonesian society', *Indonesia*: 113–25.

Tanabe, Shigeharu (1994) *Ecology and practical technology: peasant farming systems in Thailand*, Bangkok: White Lotus.

Tapp, Nicholas (1988) 'Squatters or refugees: development and the Hmong', paper presented at the Symposium on the Peripheral Areas and Minority Groups of South-East Asia, Centre for South-East Asian Studies, University of Hull, 20–22 April.

—— (1989) *Sovereignty and rebellion: the White Hmong of Northern Thailand*, Singapore: Oxford University Press.

Tasker, Rodney (1994a) 'Home town jobs', *Far Eastern Economic Review*, 14 April, pp. 24–8.

—— (1994b) 'Trees and jobs', *Far Eastern Economic Review*, 3 March, p. 19.

Tasker, Rodney and Handley, Paul (1993) 'Economic hit list', *Far Eastern Economic Review*, 5 August, pp. 38–44.

Taylor, Jim (1991) 'Living on the rim: ecology and forest monks in Northeast Thailand', *Sojourn* 6(1): 106–25.

—— (1993) 'Social action and resistance on the Thai frontier: the case of Phra Prajak Khuttajitto', *Bulletin of Concerned Asian Scholars* 25(2): 3–16.

Taylor, Robert H. (1987) *The state in Burma*, London: Hurst.

TDN (1993) 'Ethnic minorities in changing Thai society' (special issue), *Thai Development Newsletter*, 23: 16–51.

—— (1995a) 'Industrialisation rears its ugly head in Songkhla', *Thai Development Newsletter* 29: 70–1.

—— (1995b) "After three decades of development: it's time to RETHINK', *Thai Development Newsletter* no. 27–28.

TDRI (1987) *Thailand: natural resources profile*, Bangkok: Thailand Development Research Institute.

TDSC (1992) 'No democracy, no land rights for the landless', *TDSC Information Sheets* no. 5, August, Bangkok: Thai Development Support Committee.

Terwiel, B.J. (1989) *Through travellers' eyes: an approach to early nineteenth-century Thai history*, Bangkok: Duang Kamol.

Thamrin, Juni (1993) 'Labour in small-scale manufacturing: the footwear industry in West Java', in: Chris Manning and Joan Hardjono (eds) *Indonesia assessment 1993 – Labour: sharing in the benefits of growth?*, Political and Social Change Monograph no. 20, Research School of Pacific Studies, Canberra: Australian National University, pp. 139–54.

Thayer, Carlyle A. (1995) 'Mono-organizational socialism and the state', in: Benedict J. Tria Kerkvliet and Doug J. Porter (eds) *Vietnam's rural transformation*, Boulder, CO: Westview and Singapore: Institute of Southeast Asian Studies, pp. 39–64.

Thee Kian Wie (1995) 'Foreign direct investment in Indonesia since Independence', paper presented at the first EUROSEAS Conference, Leiden, The Netherlands, June–July.

Thomas, David (1988) *Village land use in Northeast Thailand: predicting the effects of development policy on village use of wildlands*, Ann Arbor, MI: University Microfilms International.

Thongchai Winichakul (1994) *Siam mapped: a history of the geo-body of a nation*, Honolulu: University of Hawaii Press.

—— (1995) 'The changing landscape of the past: new histories in Thailand since 1973', *Journal of Southeast Asian Studies* 26(1): 99–120.

Thrift, Nigel and Forbes, Dean (1986) *The price of war: urbanization in Vietnam, 1854–1985*, London: Allen and Unwin.

Tiffen, M., Mortimore, M., and Gichuki, F. (1993) *More people, less erosion: environmental recovery in Kenya*, London: Wiley.

Timberg, Thomas (1995) 'The poor versus the disenfranchised: welfare versus empowerment', *Economic Development and Cultural Change* 43(3): 651–62.

Tisdell, Clem and Firdausy, Carunia (unpublished) 'Regional development in Indonesia'.

Tjondronegoro, Sediono M.P., Soejono, Irlan and Hardjono, Joan (1992) 'Rural poverty in Indonesia: trends, issues and policies', *Asian Development Review* 10(1): 67–90.

Toh Mun Heng and Low, Linda (1988) 'Economic planning and policy-making in Singapore', *Economic Bulletin for Asia and the Pacific* 39(1): 22–32.

Tomosugi, Takashi (1995) *Changing features of a rice-growing village in Central Thailand: a fixed-point study from 1967 to 1993*, Tokyo: Centre for East Asian Cultural Studies.

Tongroj Onchan (1985) 'Migrants and characteristics of migrants in selected areas of Thailand', in: P.M. Hauser, D.B. Suits and N. Ogawa (eds) *Urbanization and migration in Asean development*, Tokyo: National Institute for Research Advancement, pp. 455–73.

Toye, John (1987) *Dilemmas of development: reflections on the counter-revolution in development theory and policy*, Oxford: Basil Blackwell.

Toyota, Mika (1996) 'The effects of tourism development on an Akha community: a Chiang Rai village case study', in: Michael J.G. Parnwell (ed.) *Uneven development in Thailand*, Aldershot: Avebury, pp. 226–40.

Tran Khanh (1993) *The ethnic Chinese and economic development in Vietnam*, Singapore: Institute of Southeast Asian Studies.

Trankell, Ing-Britt (1993) *On the road in Laos: an anthropological study of road construction and rural communities*, Uppsala Research Reports in Cultural Anthropology No. 12, Uppsala: Uppsala University.

Tremewan, Christopher (1994) *The political economy of social control in Singapore*, Basingstoke: St Martin's Press.

Truong, Thanh-Dam (1990) *Sex, money and morality: prostitution and tourism in Southeast Asia*, London: Zed Books.

Turk, Carrie (1995) 'Identifying and tackling poverty: ActionAid's experience in Vietnam', *PLA Notes* 23 (June): 37–41.

Turton, Andrew and Tanabe, Shigeharu (eds) (1984) *History and peasant consciousness in South East Asia*, Senri Ethnological Studies no. 13, Osaka: National Museum of Ethnology.

Ulack, Richard (1983) 'Migration and intra-urban mobility: characteristics of squatters and urban dwellers', *Crossroads* 1(1): 49–59.

UNDP (1994) *Human development report 1994*, Oxford University Press: New York.

Ungphakorn, Ji (1995) 'Time for a rethink on the fight against poverty in Thailand', *Thai Development Newsletter* no. 29: 54–5.

US Department of Labor (1995) 'By the sweat and toil of children', US Department of Labor report released in 1994 and re-printed in *Inside Indonesia*, 43: 25–7.

Usher, Ann Danaiya (1996) 'The race for power in Laos: the Nordic connections', in: Michael J.G. Parnwell and Raymond Bryant (eds) *Environmental change in South East Asia: people, politics and sustainable development*, London: Routledge, pp. 123–44.

van de Walle, Dominique (1996) *Infrastructure and poverty in Viet Nam*, LSMS Working Paper no. 121, Washington DC: World Bank.

Van Schendel, Willem (1991) *Three deltas: accumulation and poverty in rural Burma, Bengal and South India*, New Delhi: Sage.

Vandergeest, Peter (1991) 'Gifts and rights: cautionary notes on community self-help in Thailand', *Development and change* 22: 421–43.

Vandergeest, Peter and Buttel, Frederick H. (1988) 'Marx, Weber and development sociology: beyond the impasse', *World Development* 16(6): 683–95.

Varshney, Ashutosh (1993) 'Urban bias in perspective', *Journal of Development Studies* 29(4): 3–22.

Vatikiotis, Michael (1996) 'Sino chic', *Far Eastern Economic Review*, 11 January, pp. 22–4.

Vickers, Adrian (1989) *Bali: a paradise created*, Berkeley, CA: Periplus.

Vitit Muntarbhorn (1986) 'Child prostitution and Thailand', in: Patricia Hyndman (ed.) *The meeting of experts on the exploitation of the child and the conference on child labour and prostitution*, Sydney: Law Association for Asia and the Pacific, pp. 405–21.

Vittachi, Imran (1996) 'Cambodian dilemma: labor rights or investors', *Phnom Penh Post*, 14–27 June, p. 15.

Vu Tuan Anh (1995) 'Economic policy reforms: an introductory overview', in: Irene Nørlund, Carolyn L. Gates and Vu Cao Dam (eds) *Vietnam in a changing world*, Richmond, Surrey: Curzon Press, pp. 17–30.

Vylder, Stefan de (1995) 'State and market in Vietnam: some issues for an economy in transition', in: Irene Nørlund, Carolyn L. Gates and Vu Cao Dam (eds) *Vietnam in a changing world*, Richmond, Surrey: Curzon Press, pp. 31–70.

Walker, Anthony R. (1995) 'From the mountains and the interiors: a quarter of a century of research among fourth world peoples in Southeast Asia (with special reference to Northern Thailand and Peninsular Malaysia)', *Journal of Southeast Asian Studies* 26(2): 326–65.

Warr, Peter G. (1993) *Thailand's economic miracle*, Thailand Information Papers, National Studies Centre, Australian National University, Canberra.

Waters, Benjamin (1993) 'The tragedy of Marsinah: industrialisation and worker's rights', *Inside Indonesia* 36: 12–13.

Waterson, Roxana (1990) *The living house: an anthropology of architecture in South-East Asia*, Singapore: Oxford University Press.

Watts, Michael (1995) '"A new deal in emotions": theory and practice and the crisis of development', in: Jonathan Crush (ed.) *Power of development*, London: Routledge, pp. 44–62.

Weber, Helmut (1994) 'The Indonesian concept of development and its impact on the process of social transformation', in: Helmut Buchholt and Ulrich Mai (eds) *Continuity, change and aspirations: social and cultural life in Minahasa, Indonesia*, Singapore: Institute of Southeast Asian Studies, pp. 194–210.

White, Benjamin (1976) 'Population, employment and involution in a Javanese village', *Development and Change* 7: 267–90.

—— (1979) 'Political aspects of poverty, income distribution and their measurement: some examples from rural Java', *Development and Change* 10: 91–114.

—— (1991) 'Economic diversification and agrarian change in rural Java, 1900–1990', in: Paul Alexander, Peter Boomgaard, and Ben White (eds) *In the shadow of agriculture: non-farm activities in the Javanese economy, past and present*, Amsterdam: Royal Tropical Institute, pp. 41–69.

—— (1993) 'Industrial workers on West Java's urban fringe', in: Chris Manning and Joan Hardjono (eds) *Indonesia assessment 1993 – Labour: sharing in the benefits of growth?*, Political and Social Change Monograph no. 20, Research School of Pacific Studies, Canberra: Australian National University, pp. 127–138.

White, Benjamin and Wiradi, Gunawan (1989) 'Agrarian and nonagrarian bases of inequality in nine Javanese villages', in: Gillian Hart, Andrew Turton and Benjamin White (eds) *Agrarian transformations: local processes and the state in Southeast Asia*, Berkeley: University of California Press, pp. 266–302.

Wilks, Richard R. and Netting, Robert McC. (1984) 'Households: changing forms and functions', in: Robert McC. Netting, Richard R. Wilk and Eric J. Arnould (eds) *Households: comparative and historical studies of the domestic group*, Berkeley: University of California Press, pp. 1–28.

Williams, Michael C. (1992) *Vietnam at the crossroads*, London: Pinter.

Wolf, Diane Lauren (1990) 'Daughters, decisions and domination: an empirical and conceptual critique of household strategies', *Development and Change* 21: 43–74.

—— (1992) *Factory daughters: gender, household dynamics, and rural industrialization in Java*, Berkeley: University of California Press.

Wolf, Eric R. (1967) 'Closed corporate peasant communities in Mesoamerica and Central Java', in: Jack M. Potter, May N. Diaz and George M. Foster (eds) *Peasant society: a reader*, Boston: Little, Brown and Company, pp. 230–46 (first published in *South Western Journal of Anthropology* 13(1): 1–18, 1957).

Wolpe, Harold (1972) 'Capitalism and cheap labour-power in South Africa: from segregation to apartheid', *Economy and Society* 1: 425–56.

Wong, Diana (1987) *Peasants in the making: Malaysia's green revolution*, Singapore: Institute of Southeast Asian Studies.

World Bank (1978) *Thailand: towards a development strategy of full participation*, report no. 2059-TH, Washington DC: World Bank.

—— (1993a) *The East Asian Miracle: economic growth and public policy*, Oxford: Oxford University Press.

—— (1993b) *World development report 1993*, New York: Oxford University Press.

—— (1995) *World development report 1995: workers in an integrating world*, New York: Oxford University Press.

—— (1996) *World development report 1996: from plan to market*, New York: Oxford University Press.

Wratten, Ellen (1995) 'Conceptualizing urban poverty', *Environment and Urbanization* 7(1): 11–36.

Wyatt, David (1982) *Thailand: a short history*, New Haven, CT: Yale University Press.

Yamazawa, Ippei (1992) 'On Pacific economic integration', *Economic Journal* 102 (November): 1519–29.

Yanagihara, Toru (1994) 'Anything new in the *Miracle* report? Yes and no', *World Development* 22(4): 663–70.

Yenchai Laohavanich (1989) 'A Thai Buddhist view of nature', in Siam Society (ed.) *Culture and environment in Thailand*, Bangkok: Siam Society, pp. 259–63.

Yong Mun Cheong (1992) 'The political structures of the independent states', in: Nicholas Tarling (ed.) *The Cambridge history of Southeast Asia: the nineteenth and twentieth centuries*, volume 2, Cambridge: Cambridge University Press, pp. 387–465.

Zakaria, Fareed (1994) 'Culture is destiny: a conversation with Lee Kuan Yew', *Foreign Affairs* 73(2): 109–26.

Zasloff, Joseph J. and Brown, MacAlister (1991) 'Laos 1990: socialism postponed but leadership intact', *Southeast Asian Affairs 1991*, Singapore: Institute of Southeast Asian Studies, pp. 141–58.

# INDEX

Marshall, D. 67
Marsinah 229–30
*martabak* 262
Marxism and post-Marxism 30, 55, 64, 101
*masalah cina* 123
masculinization 244–5, 247
Mason, A. 105, 106, 172
Mason, K. 222–3
Mason, R. 120
maternal mortality 281
Mather, C. 204, 214, 217, 237
matriarchal 221
matrifocal 221
Maurer, J–L. 189, 200, 206
Mechai Viravaidya 57
mechanization in agriculture 147,
    153–4, 157, 174, 178, 181, 183,
    206, 241–51
Medhi Krongkaew 71, 73, 74, 80, 81, 82,
    88, 92, 106, 107
media 59, 124, 226
Mehmet, O. 40, 215
Mekong 101, 103, 169, 242, 282
Melaka 253
Melanau 120
Melanesia 161
Mendis, P. 52
*merantau* 184
middle classes 59, 146
Middle East 45, 222
Middle Way 52, 56
Mien 118
migrants and migration 83, 92, 96, 101,
    126–30, 157–8, 161, 178, 181–2,
    184–7, 192, 194–7, 208, 214, 239, 241,
    243, 253, 258–66, 276
military 58–9, 89, 94–5
Miller, J. 177
Minahasa 182
Minangkabau 184, 221
Mindanao (Philippines) 87, 214, 250,
    282–3
Mindon, King 45, 66
minimum wage 205, 208, 211, 215–16,
    223–5, 234, 244, 260 (*see also* wages)
minimum working age 224
mining 94–5
Ministry of Labour and Social Welfare 233
Minogue, M. 17, 22
minorities 21, 95, 115–26, 131, 139, 172
miracle economies 3, 5, 9, 25, 26, 40, 62,
    111, 202
Miranda, M. 84
missionaries 121
Mlabri 118
Mobile Development Units (MDUs) 89

mobility (*see* migrants and migration)
modern varieties (MVs) 250–1 (*see also*
    Green Revolution)
modernity 210–11
modernization 3–4, 23, 31–3, 37, 40, 41,
    45, 50, 53, 56, 61, 63, 66, 69, 95,
    109–12, 115, 119, 121, 133, 134, 139,
    142, 146, 147, 154, 157–8, 165, 189,
    197, 235, 239, 273–87, 279–86
modernization theory 31
Moerman, M. 157–8, 198
Mongkut, King 35, 45, 66
monks and monkhood 50, 53–4, 55, 57,
    58–9, 66
Monzel, K. 150
Moore, M. 277
morality 51, 52, 55–9
morbidity (*see* health)
Morrison, P. 163, 184, 185, 189
mortality 113
*muban* (*see ban*)
Muda Irrigation Scheme 181, 185, 189,
    250 (*see also* Kedah)
Muijzenberg, O. 197
Mulder, N. 29
multinational companies 24, 32, 216–17
murder rate 62
Murray, A. 37, 131, 132, 210–11
Muscat, R. 83, 89, 90
Muslim (see Islam)
Myanmar (Burma) 10–11, 15–18, 19, 20,
    22, 45, 46, 47, 48–9, 50, 54–5, 64, 66,
    72, 80, 98, 115, 117, 120, 123, 128,
    131, 136, 155, 170, 183, 231, 258, 260,
    273, 279, 281
Myers, W. 225
Myrdal, G. 5, 37

*nai naa* 212
Nakhon Ratchasima 50, 59
Nan 248
Naqvi, S. 52
narcotics 119
*nat* 66
Natadecha-Sponsel, P. 56
National Development Planning Board
    (Indonesia) 116
National Economic and Social
    Development Board (Thailand) 33, 48,
    73, 74, 87, 90–1
National Economic Development Board
    (Thailand) 33, 48
National Housing Authority (Philippines)
    136
National Ideology (Singapore) 60, 62
national parks 47–8